J. E. Marsden
M. McCracken

The Hopf Bifurcation and Its Applications

with contributions by

P. Chernoff, G. Childs, S. Chow, J. R. Dorroh, J. Guckenheimer, L. Howard, N. Kopell, O. Lanford, J. Mallet-Paret, G. Oster, O. Ruiz, S. Schecter, D. Schmidt, and S. Smale

Springer-Verlag New York
1976

J. E. Marsden
Department of Mathematics
University of California
at Berkeley

M. McCracken
Department of Mathematics
University of California
at Santa Cruz

AMS Classifications: 34C15, 58F10, 35G25, 76E30

Library of Congress Cataloging in Publication Data

Marsden, Jerrold E.
The Hopf bifurcation and its applications.

(Applied mathematical sciences; v. 19)
Bibliography
Includes index.
1. Differential equations. 2. Differential equations, Partial. 3. Differentiable dynamical systems. 4. Stability. I. McCracken, Marjorie, 1949- joint author. II. Title. III. Series.
QA1.A647 vol. 19 [QA372] 510'.8s [515'.35]
76-21727

Printed in the United States of America

ISBN 0-387-90200-7 Springer-Verlag New York · Heidelberg · Berlin
ISBN 3-540-90200-7 Springer-Verlag Berlin · Heidelberg · New York

To the courage of

G. Oyarzún

PREFACE

The goal of these notes is to give a reasonably complete, although not exhaustive, discussion of what is commonly referred to as the Hopf bifurcation with applications to specific problems, including stability calculations. Historically, the subject had its origins in the works of Poincaré [1] around 1892 and was extensively discussed by Andronov and Witt [1] and their co-workers starting around 1930. Hopf's basic paper [1] appeared in 1942. Although the term "Poincaré-Andronov-Hopf bifurcation" is more accurate (sometimes Friedrichs is also included), the name "Hopf Bifurcation" seems more common, so we have used it. Hopf's crucial contribution was the extension from two dimensions to higher dimensions.

The principal technique employed in the body of the text is that of invariant manifolds. The method of Ruelle-Takens [1] is followed, with details, examples and proofs added. Several parts of the exposition in the main text come from papers of P. Chernoff, J. Dorroh, O. Lanford and F. Weissler to whom we are grateful.

The general method of invariant manifolds is common in dynamical systems and in ordinary differential equations; see for example, Hale [1,2] and Hartman [1]. Of course, other methods are also available. In an attempt to keep the picture balanced, we have included samples of alternative approaches. Specifically, we have included a translation (by L. Howard and N. Kopell) of Hopf's original (and generally unavailable) paper. These original methods, using power series and scaling are used in fluid mechanics by, amongst many others, Joseph and Sattinger [1]; two sections on these ideas from papers of Iooss [1-6] and

Kirchgässner and Kielhoffer [1] (contributed by G. Childs and O. Ruiz) are given.

The contributions of S. Smale, J. Guckenheimer and G. Oster indicate applications to the biological sciences and that of D. Schmidt to Hamiltonian systems. For other applications and related topics, we refer to the monographs of Andronov and Chaiken [1], Minorsky [1] and Thom [1].

The Hopf bifurcation refers to the development of periodic orbits ("self-oscillations") from a stable fixed point, as a parameter crosses a critical value. In Hopf's original approach, the determination of the stability of the resulting periodic orbits is, in concrete problems, an unpleasant calculation. We have given explicit algorithms for this calculation which are easy to apply in examples. (See Section 4, and Section 5A for comparison with Hopf's formulae). The method of averaging, exposed here by S. Chow and J. Mallet-Paret in Section 4C gives another method of determining this stability, and seems to be especially useful for the next bifurcation to invariant tori where the only recourse may be to numerical methods, since the periodic orbit is not normally known explicitly.

In applications to partial differential equations, the key assumption is that the semi-flow defined by the equations be smooth in all variables for $t > 0$. This enables the invariant manifold machinery, and hence the bifurcation theorems to go through (Marsden [2]). To aid in determining smoothness in examples we have presented parts of the results of Dorroh-Marsden [1]. Similar ideas for utilizing smoothness have been introduced independently by other authors, such as D. Henry [1].

Some further directions of research and generalization are given in papers of Jost and Zehnder [1], Takens [1, 2], Crandall-Rabinowitz [1, 2], Arnold [2], and Kopell-Howard [1-6] to mention just a few that are noted but are not discussed in any detail here. We have selected results of Chafee [1] and Ruelle [3] (the latter is exposed here by S. Schecter) to indicate some generalizations that are possible.

The subject is by no means closed. Applications to instabilities in biology (see, e.g. Zeeman [2], Gurel [1-12] and Section 10, 11); engineering (for example, spontaneous "flutter" or oscillations in structural, electrical, nuclear or other engineering systems; cf. Aronson [1], Ziegler [1] and Knops and Wilkes [1]), and oscillations in the atmosphere and the earth's magnetic field (cf. Durand [1]) are appearing at a rapid rate. Also, the qualitative theory proposed by Ruelle-Takens [1] to describe turbulence is not yet well understood (see Section 9). In this direction, the papers of Newhouse and Peixoto [1] and Alexander and Yorke [1] seem to be important. Stable oscillations in nonlinear waves may be another fruitful area for application; cf.Whitham [1]. We hope these notes provide some guidance to the field and will be useful to those who wish to study or apply these fascinating methods.

After we completed our stability calculations we were happy to learn that others had found similar difficulty in applying Hopf's result as it had existed in the literature to concrete examples in dimension ≥ 3. They have developed similar formulae to deal with the problem; cf. Hsü and Kazarinoff [1, 2] and Poore [1].

The other main new result here is our proof of the validity of the Hopf bifurcation theory for nonlinear partial differential equations of parabolic type. The new proof, relying on invariant manifold theory, is considerably simpler than existing proofs and should be useful in a variety of situations involving bifurcation theory for evolution equations.

These notes originated in a seminar given at Berkeley in 1973-4. We wish to thank those who contributed to this volume and wish to apologize in advance for the many important contributions to the field which are not discussed here; those we are aware of are listed in the bibliography which is, admittedly, not exhaustive. Many other references are contained in the lengthy bibliography in Cesari [1]. We also thank those who have taken an interest in the notes and have contributed valuable comments. These include R. Abraham, D. Aronson, A. Chorin, M. Crandall, R. Cushman, C. Desoer, A. Fischer, L. Glass, J. M. Greenberg, O. Gurel, J. Hale, B. Hassard, S. Hastings, M. Hirsch, E. Hopf, N. D. Kazarinoff, J. P. LaSalle, A. Mees, C. Pugh, D. Ruelle, F. Takens, Y. Wan and A. Weinstein. Special thanks go to J. A. Yorke for informing us of the material in Section 3C and to both he and D. Ruelle for pointing out the example of the Lorentz equations (See Example 4B.8). Finally, we thank Barbara Komatsu and Jody Anderson for the beautiful job they did in typing the manuscript.

Jerrold Marsden
Marjorie McCracken

TABLE OF CONTENTS

SECTION 1

INTRODUCTION TO STABILITY AND BIFURCATION IN DYNAMICAL SYSTEMS AND FLUID MECHANICS

Suppose we are studying a physical system whose state x is governed by an evolution equation $\frac{dx}{dt} = X(x)$ which has unique integral curves. Let x_0 be a fixed point of the flow of X; i.e., $X(x_0) = 0$. Imagine that we perform an experiment upon the system at time $t = 0$ and conclude that it is then in state x_0. Are we justified in predicting that the system will remain at x_0 for all future time? The mathematical answer to this question is obviously yes, but unfortunately it is probably not the question we really wished to ask. Experiments in real life seldom yield exact answers to our idealized models, so in most cases we will have to ask whether the system will remain near x_0 if it started near x_0. The answer to the revised question is not always yes, but even so, by examining the evolution equation at hand more minutely, one can sometimes make predictions about the future behavior of a system starting near x_0. A trivial example will illustrate some of the problems involved. Consider the following two differential equations on the real line:

$$x'(t) = -x(t) \tag{1.1}$$

and

$$x'(t) = x(t). \tag{1.2}$$

The solutions are respectively:

$$x(x_0,t) = x_0e^{-t} \tag{1.1'}$$

and

$$x(x_0,t) = x_0e^{+t}. \tag{1.2'}$$

Note that 0 is a fixed point of both flows. In the first case, for all $x_0 \in \mathbb{R}$, $\lim_{t\to\infty} x(x_0,t) = 0$. The whole real line moves toward the origin, and the prediction that if x_0 is near 0, then $x(x_0,t)$ is near 0 is obviously justified. On the other hand, suppose we are observing a system whose state x is governed by (1.2). An experiment telling us that at time $t = 0$, $x(0)$ is approximately zero will certainly not permit us to conclude that $x(t)$ stays near the origin for all time, since all points except 0 move rapidly away from 0. Furthermore, our experiment is unlikely to allow us to make an accurate prediction about $x(t)$ because if $x(0) < 0$, $x(t)$ moves rapidly away from the origin toward $-\infty$ but if $x(0) > 0$, $x(t)$ moves toward $+\infty$. Thus, an observer watching such a system would expect sometimes to observe $x(t) \xrightarrow[t\to\infty]{} -\infty$ and sometimes $x(t) \xrightarrow[t\to\infty]{} +\infty$. The solution $x(t) = 0$ for all t would probably never be observed to occur because a slight perturbation of the system would destroy this solution. This sort of behavior is frequently observed in nature. It is not due to any nonuniqueness in the solution to the differential equation involved, but to the <u>instability</u> of that solution under small perturbations in initial data.

Indeed, it is only stable mathematical models, or features of models that can be relevant in "describing" nature.[+]

Consider the following example.[*] A rigid hoop hangs from the ceiling and a small ball rests in the bottom of the hoop. The hoop rotates with frequency ω about a vertical axis through its center (Figure 1.1a).

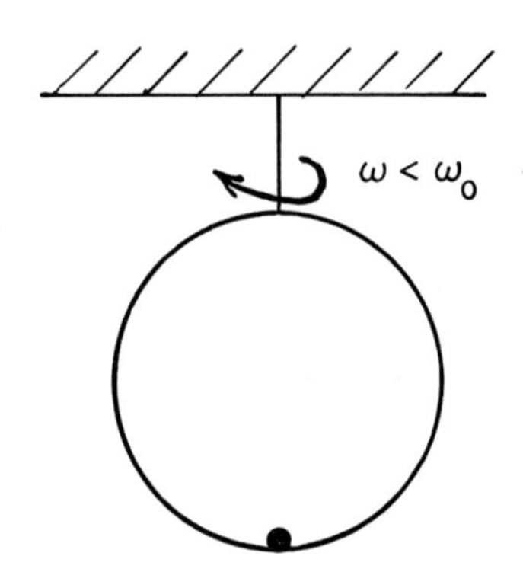

Figure 1.1a

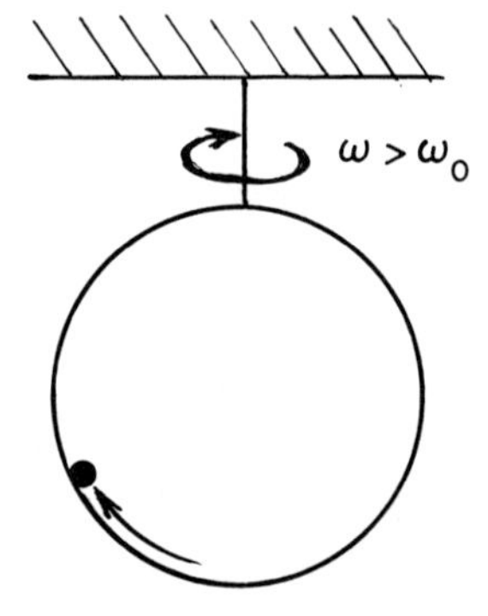

Figure 1.1b

For small values of ω, the ball stays at the bottom of the hoop and that position is stable. However, when ω reaches some critical value ω_0, the ball rolls up the side of the hoop to a new position $x(\omega)$, which is stable. The ball may roll to the left or to the right, depending to which side of the vertical axis it was initially leaning (Figure 1.1b). The position at the bottom of the hoop is still a fixed point, but it has become unstable, and, in practice, is never observed to occur. The solutions to the differential equations governing the ball's motion are unique for all values of ω,

---

[+]For further discussion, see the conclusion of Abraham-Marsden [1].

[*]This example was first pointed out to us by E. Calabi.

but for $\omega > \omega_0$, this uniqueness is irrelevant to us, for we cannot predict which way the ball will roll. Mathematically, we say that the original stable fixed point has become unstable and has split into two stable fixed points. See Figure 1.2 and Exercise 1.16 below.

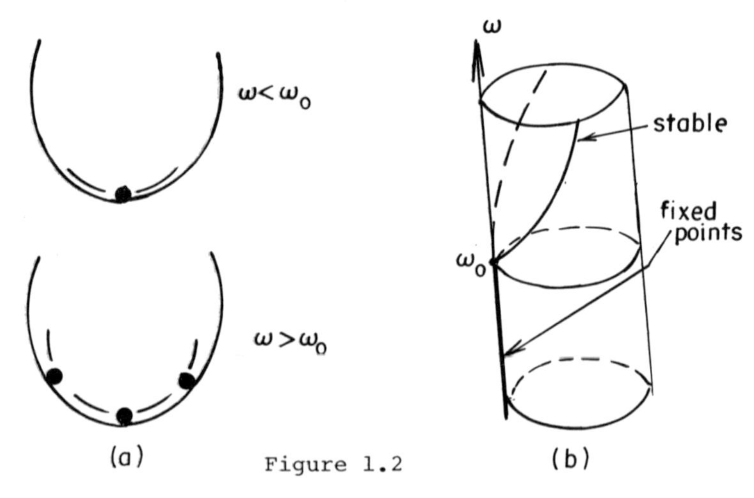

Figure 1.2

Since questions of stability are of overwhelming practical importance, we will want to define the concept of stability precisely and develop criteria for determining it.

(1.1) Definition. Let F_t be a C^0 flow (or semiflow)* on a topological space M and let A be an invariant set; i.e., $F_t(A) \subset A$ for all t. We say A is stable (resp. asymptotically stable or an attractor) if for any neighborhood U of A there is a neighborhood V of A such

---

*i.e., $F_t: M \to M$, F_0 = identity, and $F_{t+s} = F_s \circ F_t$ for all $t, s \in \mathbb{R}$. C^0 means $F_t(x)$ is continuous in (t,x). A semiflow is one defined only for $t \geq 0$. Consult, e.g., Lang [1], Hartman [1], or Abraham-Marsden [1] for a discussion of flows of vector fields. See Section 8A, or Chernoff-Marsden [1] for the infinite dimensional case.

that the flow lines (integral curves) $x(x_0,t) \equiv F_t(x_0)$ belong to U if $x_0 \in V$ (resp. $\bigcap_{t \geq 0} F_t(V) = A$).

Thus A is stable (resp. attracting) when an initial condition slightly perturbed from A remains near A (resp. tends towards A). (See Figure 1.3).

If A is not stable it is called unstable.

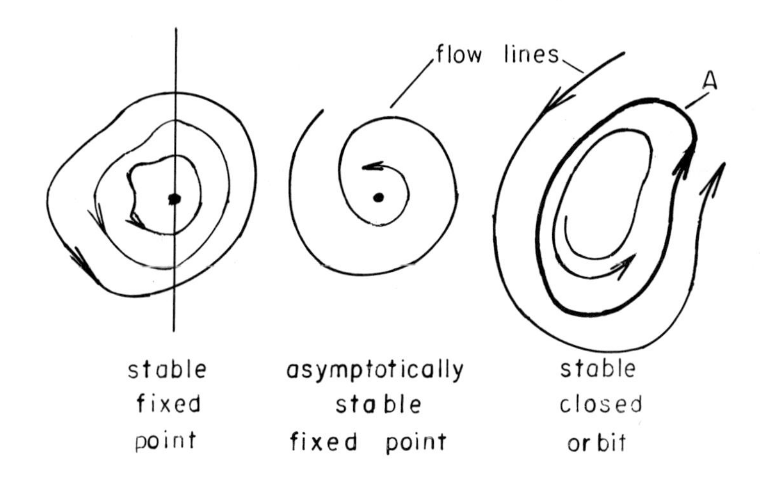

Figure 1.3

(1.2) Exercise. Show that in the ball in the hoop example, the bottom of the hoop is an attracting fixed point for $\omega < \omega_0 = \sqrt{g/R}$ and that for $\omega > \omega_0$ there are attracting fixed points determined by $\cos\theta = g/\omega^2 R$, where θ is the angle with the negative vertical axis, R is the radius of the hoop and g is the acceleration due to gravity.

The simplest case for which we can determine the stability of a fixed point x_0 is the finite dimensional, linear case. Let $X: \mathbb{R}^n \to \mathbb{R}^n$ be a linear map. The flow of

X is $x(x_0,t) = e^{tX}(x_0)$. Clearly, the origin is a fixed point. Let $\{\lambda_j\}$ be the eigenvalues of X. Then $\{e^{\lambda_j t}\}$ are the eigenvalues of e^{tX}. Suppose $\mathrm{Re}\,\lambda_j < 0$ for all j. Then $|e^{\lambda_j t}| = e^{\mathrm{Re}\,\lambda_j t} \to 0$ as $t \to \infty$. One can check, using the Jordan canonical form, that in this case 0 is asymptotically stable and that if there is a λ_j with positive real part, 0 is unstable. More generally, we have:

(1.3) Theorem. *Let $X: E \to E$ be a continuous, linear map on a Banach space E. The origin is a stable attracting fixed point of the flow of X if the spectrum $\sigma(X)$ of X is in the open left-half plane. The origin is unstable if there exists $z \in \sigma(X)$ such that $\mathrm{Re}(z) > 0$.*

This will be proved in Section 2A, along with a review of some relevant spectral theory.

Consider now the nonlinear case. Let P be a Banach manifold* and let X be a C^1 vector field on P. Let $X(p_0) = 0$. Then $dX(p_0): T_{p_0}(P) \to T_{p_0}(P)$ is a continuous linear map on a Banach space. Also in Section 2A we shall demonstrate the following basic theorem of Liapunov [1].

(1.4) Theorem. *Let X be a C^1 vector field on a Banach manifold P and let p_0 be a fixed point of X, i.e., $X(p_0) = 0$. Let F_t be the flow of X i.e., $\frac{\partial}{\partial t} F_t(x) = X(F_t(x))$, $F_0(x) = x$. (Note that $F_t(p_0) = p_0$ for all t.) If the spectrum of $dX(p_0)$ lies in the left-half plane; i.e., $\sigma(dX(p_0)) \subset \{z \in \mathbb{C} | \mathrm{Re}\, z < 0\}$, then p_0 is asymptotically*

---

*We shall use only the most elementary facts about manifold theory, mostly because of the convenient geometrical language. See Lang [1] or Marsden [4] for the basic ideas.

stable.

If there exists an isolated $z \in \sigma(dX(p_0))$ *such that* $\text{Re } z > 0$, p_0 *is unstable. If* $\sigma(dX(p_0)) \subset \{z | \text{Re } z \leq 0\}$ *and there is a* $z \in \sigma(dX(p_0))$ *such that* $\text{Re } z = 0$, *then stability cannot be determined from the linearized equation.*

(1.5) *Exercise.* Consider the following vector field on $\mathbb{R}^2$: $X(x,y) = (y, \mu(1-x^2)y-x)$. Decide whether the origin is unstable, stable, or attracting for $\mu < 0$, $\mu = 0$, and $\mu > 0$.

Many interesting physical problems are governed by differential equations depending on a parameter such as the angular velocity ω in the ball in the hoop example. Let $X_\mu: P \to TP$ be a (smooth) vector field on a Banach manifold P. Assume that there is a continuous curve $p(\mu)$ in P such that $X_\mu(p(\mu)) = 0$, i.e., $p(\mu)$ is a fixed point of the flow of X_μ. Suppose that $p(\mu)$ is attracting for $\mu < \mu_0$ and unstable for $\mu > \mu_0$. The point $(p(\mu_0), \mu_0)$ is then called a *bifurcation point* of the flow of X_μ. For $\mu < \mu_0$ the flow of X_μ can be described (at least in a neighborhood of $p(\mu)$) by saying that points tend toward $p(\mu)$. However, this is not true for $\mu > \mu_0$, and so the character of the flow may change abruptly at μ_0. Since the fixed point is unstable for $\mu > \mu_0$, we will be interested in finding stable behavior for $\mu > \mu_0$. That is, we are interested in finding bifurcation *above criticality to stable behavior*.

For example, several curves of fixed points may come together at a bifurcation point. (A curve of fixed points is a curve $\alpha: I \to P$ such that $X_\mu(\alpha(\mu)) = 0$ for all μ. One such curve is obviously $\mu \mapsto p(\mu)$.) There may be curves of stable fixed points for $\mu > \mu_0$. In the case of the ball in the hoop, there are two curves of stable fixed points for $\omega \geq \omega_0$, one moving up the left side of the hoop and one moving up the right side (Figure 1.2).

Another type of behavior that may occur is bifurcation to periodic orbits. This means that there are curves of the form $\alpha: I \to P$ such that $\alpha(\mu_0) = p(\mu_0)$ and $\alpha(\mu)$ is on a closed orbit γ_μ of the flow of X_μ. (See Figure 1.4). The Hopf bifurcation is of this type. Physical examples in fluid mechanics will be given shortly.

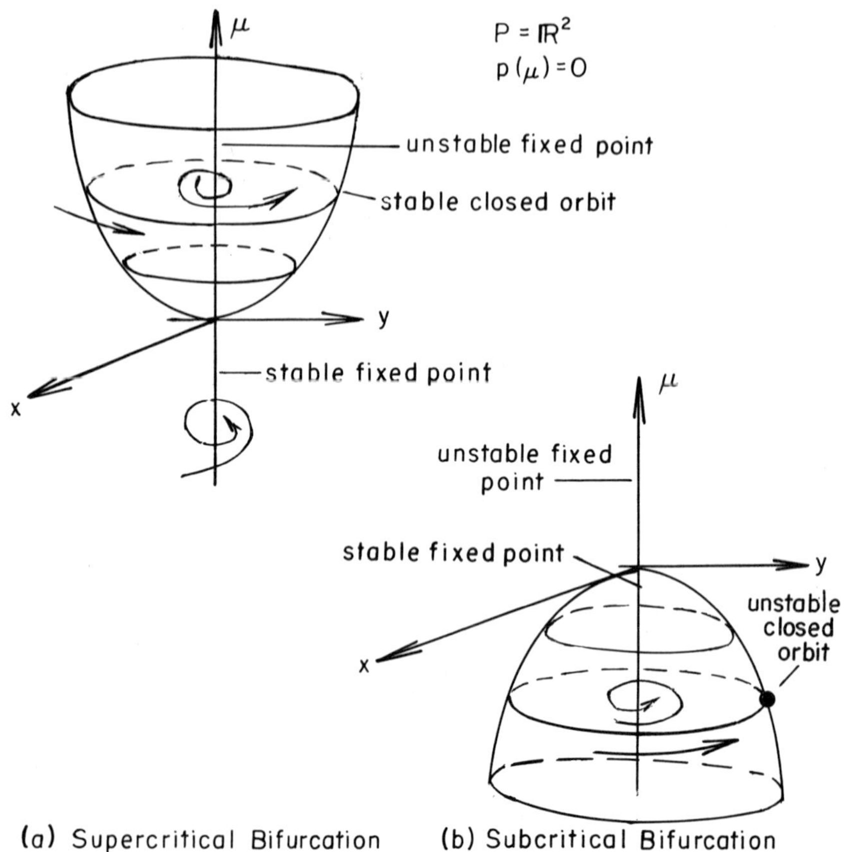

Figure 1.4
The General Nature of the Hopf Bifurcation

The appearance of the stable closed orbits (= periodic solutions) is interpreted as a "shift of stability" from the original stationary solution to the periodic one, i.e., a point near the original fixed point now is attracted to and becomes indistinguishable from the closed orbit. (See Figures 1.4 and 1.5).

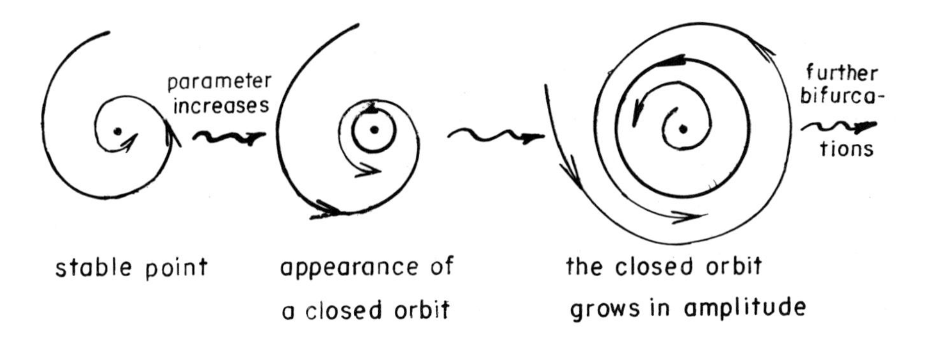

Figure 1.5
The Hopf Bifurcation

Other kinds of bifurcation can occur; for example, as we shall see later, the stable closed orbit in Figure 1.4 may bifurcate to a stable 2-torus. In the presence of symmetries, the situation is also more complicated. This will be treated in some detail in Section 7, but for now we illustrate what can happen via an example.

(1.6) <u>Example</u>: <u>The Ball in the Sphere</u>. A rigid, hollow sphere with a small ball inside it hangs from the ceiling and rotates with frequency ω about a vertical axis through its center (Figure 1.6).

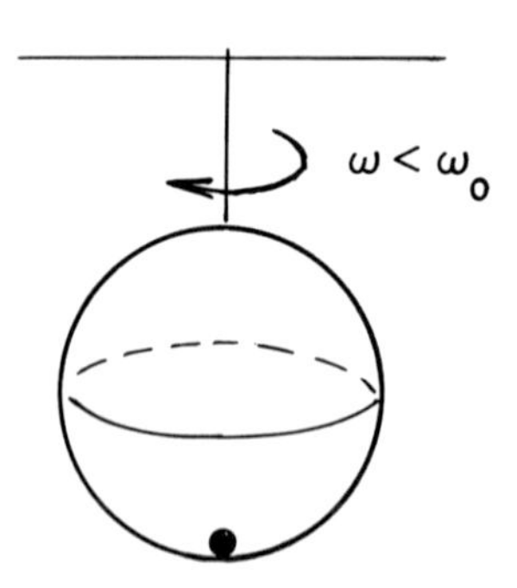

Figure 1.6

For small ω, the bottom of the sphere is a stable point, but for $\omega > \omega_0$ the ball moves up the side of the sphere to a new fixed point. For each $\omega > \omega_0$, there is a stable, invariant circle of fixed points (Figure 1.7). We get a circle of fixed points rather than isolated ones because of the symmetries present in the problem.

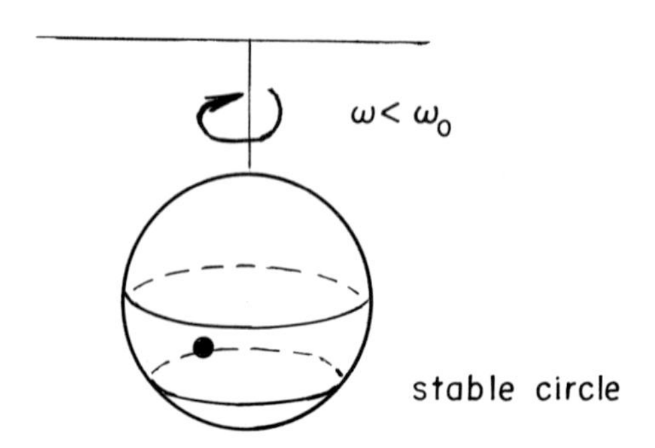

Figure 1.7

Before we discuss methods of determining what kind of bifurcation will take place and associated stability questions, we shall briefly describe the general basin bifurcation picture of R. Abraham [1,2].

In this picture one imagines a rolling landscape on which water is flowing. We picture an attractor as a basin into which water flows. Precisely, if F_t is a flow on M and A is an attractor, the basin of A is the set of all $x \in M$ which tend to A as $t \to +\infty$. (The less picturesque phrase "stable manifold" is more commonly used.)

As parameters are tuned, the landscape undulates and the flow changes. Basins may merge, new ones may form, old ones may disappear, complicated attractors may develop, etc.

The Hopf bifurcation may be pictured as follows. We begin with a simple basin of parabolic shape; i.e., a point attractor. As our parameter is tuned, a small hillock forms and grows at the center of the basin. The new attractor is, therefore, circular (viz the periodic orbit in the Hopf theorem) and its basin is the original one minus the top point of the hillock.

Notice that complicated attractors can spontaneously appear or dissappear as mesas are lowered to basins or basins are raised into mesas.

Many examples of bifurcations occur in nature, as a glance at the rest of the text and the bibliography shows. The Hopf bifurcation is behind oscillations in chemical and biological systems (see e.g. Kopell-Howard [1-6], Abraham [1,2] and Sections 10, 11), including such things as "heart flutter".*
One of the most studied examples comes from fluid mechanics, so we now pause briefly to consider the basic ideas of

---

*That "heart flutter" is a Hopf bifurcation is a conjecture told to us by A. Fischer; cf. Zeeman [2].

the subject.

The Navier-Stokes Equations

Let $D \subseteq \mathbb{R}^3$ be an open, bounded set with smooth boundary. We will consider D to be filled with an incompressible homogeneous (constant density) fluid. Let u and p be the velocity and pressure of the fluid, respectively. If the fluid is viscous and if changes in temperature can be neglected, the equations governing its motion are:

$$\frac{\partial u}{\partial t} + (u \cdot \nabla)u - \nu \Delta u = -\text{grad } p \text{ (+ external forces)} \qquad (1.3)$$

$$\text{div } u = 0 \qquad (1.4)$$

The boundary condition is $u|_{\partial D} = 0$ (or $u|_{\partial D}$ prescribed, if the boundary of D is moving) and the initial condition is that $u(x,0)$ is some given $u_0(x)$. The problem is to find $u(x,t)$ and $p(x,t)$ for $t > 0$. The first equation (1.3) is analogous to Newton's Second Law $F = ma$; the second (1.4) is equivalent to the incompressibility of the fluid.*

Think of the evolution equation (1.3) as a vector field and so defines a flow, on the space $\mathfrak{X}$ of all divergence free vector fields on D. (There are major technical difficulties here, but we ignore them for now - see Section 8.)

The Reynolds number of the flow is defined by $R = \frac{UL}{\nu}$, where U and L are a typical speed and a length associated with the flow, and ν is the fluid's viscosity. For example, if we are considering the flow near a sphere toward which fluid is projected with constant velocity $U_\infty \vec{i}$

---

*See any fluid mechanics text for a discussion of these points. For example, see Serrin [1], Shinbrot [1] or Hughes-Marsden [3].

(Figure 1.8), then L may be taken to be the radius of the sphere and $U = U_\infty$.

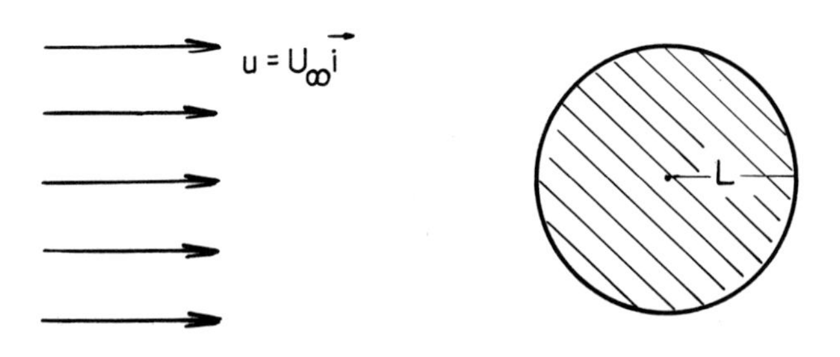

Figure 1.8

If the fluid is not viscous ($\nu = 0$), then $R = \infty$, and the fluid satisfies <u>Euler's equations</u>:

$$\frac{\partial u}{\partial t} + (u \cdot \nabla) u = -\text{grad } p \tag{1.5}$$

$$\text{div } u = 0. \tag{1.6}$$

The boundary condition becomes: $u|_{\partial D}$ is parallel to ∂D, or $u || \partial D$ for short. This sudden change of boundary condition from $u = 0$ on ∂D to $u || \partial D$ is of fundamental significance and is responsible for many of the difficulties in fluid mechanics for R very large (see footnote below).

The Reynolds number of the flow has the property that, if we rescale as follows:

$$u^* = \frac{U^*}{U} u$$

$$x^* = \frac{L^*}{L} x$$

$$t^* = \frac{T^*}{T} t$$

$$p^* = \left(\frac{U^*}{U}\right)^2 p$$

then if $T = L/U$, $T^* = L^*/U^*$ and provided $R^* = U^*L^*/\nu^* = R = UL/\nu$, u^* satisfies the same equations with respect to x^* and t^* that u satisfies with respect to x and t; i.e.,

$$\frac{\partial u^*}{\partial t^*} + (u^*\cdot\nabla^*)u^* - \nu^*\nabla u^* = -\text{grad } p^* \tag{1.7}$$

$$\text{div } u^* = 0 \tag{1.8}$$

with the same boundary condition $u^*\Big|_{\partial D} = 0$ as before. (This is easy to check and is called Reynolds' law of similarity.) Thus, the nature of these two solutions of the Navier-Stokes equations is the same. The fact that this rescaling can be done is essential in practical problems. For example, it allows engineers to test a scale model of an airplane at low speeds to determine whether the real airplane will be able to fly at high speeds.

(1.7) Example. Consider the flow in Figure 1.8. If the fluid is not viscous, the boundary condition is that the velocity at the surface of the sphere is parallel to the sphere, and the fluid slips smoothly past the sphere (Figure 1.9).

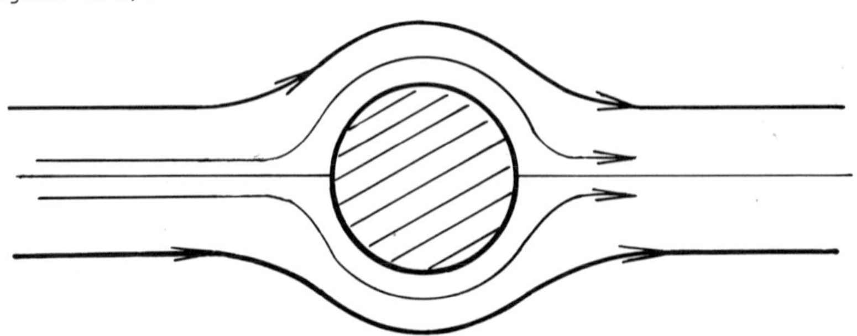

Figure 1.9

Now consider the same situation, but in the viscous case. Assume that R starts off small and is gradually increased. (In the laboratory this is usually accomplished by increasing the velocity $U_\infty \vec{i}$, but we may wish to think of it as $\nu \to 0$, i.e., molasses changing to water.) Because of the no-slip condition at the surface of the sphere, as U_∞ gets larger, the velocity gradient increases there. This causes the flow to become more and more complicated (Figure 1.10).*

For small values of the Reynolds number, the velocity field behind the sphere is observed to be stationary, or approximately so, but when a critical value of the Reynolds number is reached, it becomes periodic. For even higher values of the Reynolds number, the periodic solution loses stability and further bifurcations take place. The further bifurcation illustrated in Figure 1.10 is believed to represent a bifurcation from an attracting periodic orbit to a periodic orbit on an attracting 2-torus in $\mathfrak{X}$. These further bifurcations may eventually lead to turbulence. See Remark 1.15 and Section 9 below.

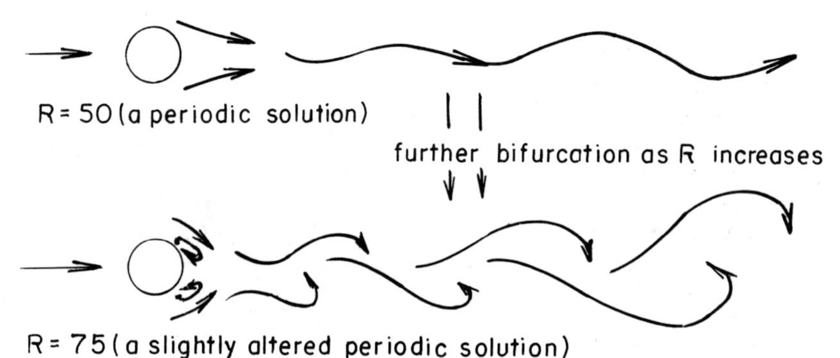

Figure 1.10

*These large velocity gradients mean that in numerical studies, finite difference techniques become useless for interesting flows. Recently A. Chorin [1] has introduced a brilliant technique for overcoming these difficulties and is able to simulate numerically for the first time, the "Karmen vortex sheet", illustrated in Figure 1.10. See also Marsden [5] and Marsden-McCracken [2].

(1.8) Example. Couette Flow. A viscous,* incompressible, homogeneous fluid fills the space between two long, coaxial cylinders which are rotating. For example, they may rotate in opposite directions with frequency ω (Figure 1.11). For small values of ω, the flow is horizontal, laminar and stationary.

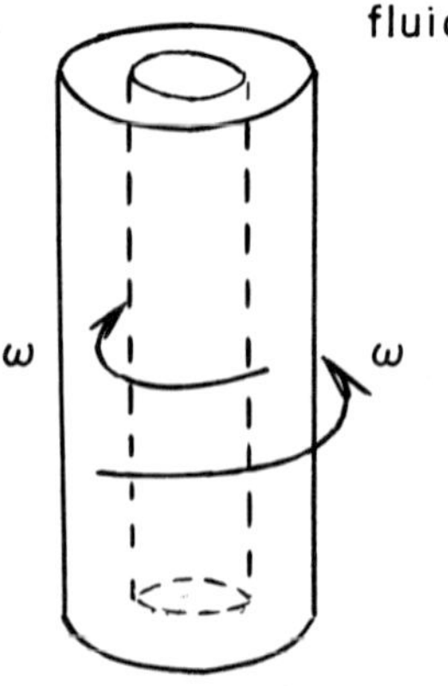

Figure 1.11

If the frequency is increased beyond some value ω_0, the fluid breaks up into what are called Taylor cells (Figure 1.12).

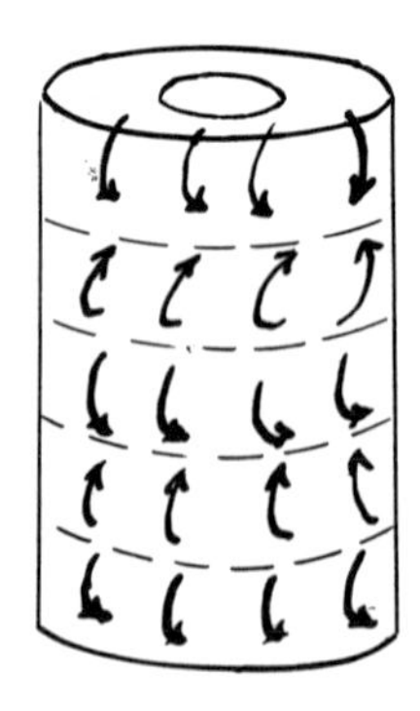

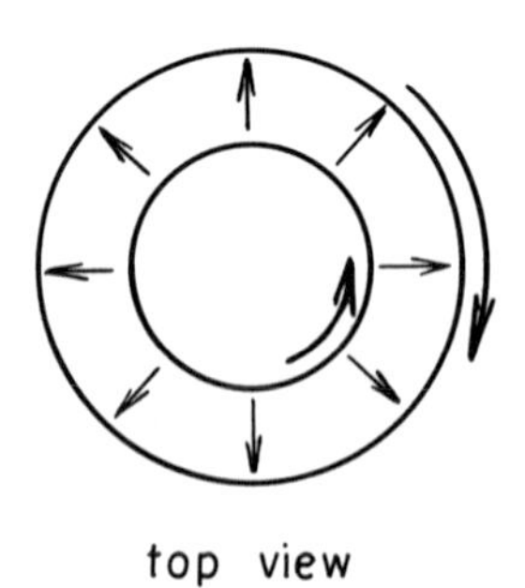

Figure 1.12

*Couette flow is studied extensively in the literature (see Serrin [1], Coles [1]) and is a stationary flow of the Euler equations as well as of the Navier-Stokes equations (see the following exercise).

Taylor cells are also a stationary solution of the Navier-Stokes equations. For larger values of ω, bifurcations to periodic, doubly periodic and more complicated solutions may take place (Figure 1.13).

helical structure

doubly periodic structure

Figure 1.13

For still larger values of ω, the structure of the Taylor cells becomes more complex and eventually breaks down completely and the flow becomes turbulent. For more information, see Coles [1] and Section 7.

(1.9) <u>Exercise</u>. Find a stationary solution $\vec{u}$ to the Navier-Stokes equations in cylindrical coordinates such that $\vec{u}$ depends only on r, $u_r = u_z = 0$, the external force $f = 0$ and the angular velocity ω satisfies $\omega|_{r=A_1} = -\rho_1$, and $\omega|_{r=A_2} = +\rho_2$ (i.e., find Couette flow). Show that $\vec{u}$ is also a solution to Euler equations.

$$\text{(Answer: } \omega = \frac{\alpha}{r^2} + \beta \text{ where } \alpha = \frac{-(\rho_1+\rho_2)A_1^2A_2^2}{A_2^2-A_1^2} \text{ and } \beta = \frac{\rho_1^2A_1^2+\rho_2^2A_2^2}{A_2^2-A_1^2}\text{).}$$

Another important place in fluid mechanics where an instability of this sort occurs is in flow in a pipe. The flow is steady and laminar (Poiseuille flow) up to Reynolds numbers around 4,000, at which point it becomes unstable and transition to chaotic or turbulent flow occurs. Actually if the experiment is done carefully, turbulence can be delayed until rather large R. It is analogous to balancing a ball on the tip of a rod whose diameter is shrinking.

Statement of the Principal Bifurcation Theorems

Let $X_\mu : P \to T(P)$ be a C^k vector field on a manifold P depending smoothly on a real parameter μ. Let F_t^μ be the flow of X_μ. Let p_0 be a fixed point for all μ, an attracting fixed point for $\mu < \mu_0$, and an unstable fixed point for $\mu > \mu_0$. Recall (Theorem 1.4) that the condition for stability of p_0 is that $\sigma(dX_\mu(p_0)) \subset \{z \mid \text{Re } z < 0\}$. At $\mu = \mu_0$, some part of the spectrum of $dX_\mu(p_0)$ crosses the imaginary axis. The nature of the bifurcation that takes place at the point (p_0,μ_0) depends on how that crossing occurs (it depends, for example, on the dimension of the generalized eigenspace* of $dX_{\mu_0}(p_0)$ belonging to the part of the spectrum that crosses the axis). If P is a finite dimensional space, there are bifurcation theorems giving necessary conditions for certain kinds of bifurcation to occur. If P is not finite dimensional, we may be able, nevertheless, to reduce the problem to a finite dimensional one via the center manifold theorem by means of the following simple but crucial suspension trick. Let ψ be the time 1 map of the flow $F_t = (F_t^\mu,\mu)$ on $P \times \mathbb{R}$. As we shall show in Section 2A, $\sigma(d\psi(p_0,\mu_0)) = e^{\sigma(dX(p_0,\mu_0))}$. That is, $\sigma(d\psi(p_0,\mu_0)) = e^{\sigma(dX_{\mu_0}(p_0))} \cup \{1\}$.

---

*The definition and basic properties are reviewed in Section 2A.

The following theorem is now applicable to ψ (see Sections 2-4 for details).

(1.10) Center Manifold Theorem (Kelley [1], Hirsch-Pugh-Shub [1], Hartman [1], Takens [2], etc.). *Let ψ be a mapping from a neighborhood of α_0 in a Banach manifold P to P. We assume that ψ has k continuous derivatives and that $\psi(\alpha_0) = \alpha_0$. We further assume that $d\psi(\alpha_0)$ has spectral radius 1 and that the spectrum of $d\psi(\alpha_0)$ splits into a part on the unit circle and the remainder, which is at a non-zero distance from the unit circle. Let Y denote the generalized eigenspace of $d\psi(\alpha_0)$ belonging to the part of the spectrum on the unit circle; assume that Y has dimension $d < \infty$. Then there exists a neighborhood V of α_0 in P and a C^{k-1} submanifold M, called a center manifold for ψ, of V of dimension d, passing through α_0 and tangent to Y at α_0, such that:*

(a) (*Local Invariance*): *If* $x \in M$ *and* $\psi(x) \in V$, *then* $\psi(x) \in M$.

(b) (*Local Attractivity*): *If* $\psi^n(x) \in V$ *for all* $n = 0,1,2,\ldots$, *then as* $n \to \infty$, $\psi^n(x) \to M$.

(1.11) Remark. It will be a corollary to the proof of the Center Manifold Theorem that if ψ is the time 1 map of F_t defined above then the center manifold M can be chosen so that properties (a) and (b) apply to F_t for all $t > 0$.

(1.12) Remark. The Center Manifold Theorem is not always true for C^∞ ψ in the following sense: since $\psi \in C^k$ for all k, we get a sequence of center manifolds M^k, but their intersection may be empty. See Remarks 2.6

regarding the differentiability of M.

We will be particularly interested in the case in which bifurcation to stable closed orbits occurs. With X_μ as before, assume that for $\mu = \mu_0$ (resp. $\mu > \mu_0$), $\sigma(dX_\mu(p_0))$ has two isolated nonzero, simple complex conjugate eigenvalues $\lambda(\mu)$ and $\overline{\lambda(\mu)}$ such that $\mathrm{Re}\,\lambda(\mu) = 0$ (resp. > 0) and such that $\left.\frac{d(\mathrm{Re}\,\lambda(\mu))}{d\mu}\right|_{\mu=\mu_0} > 0$. Assume further that the rest of $\sigma(dX_\mu(p_0))$ remains in the left-half plane at a nonzero distance from the imaginary axis. Using the Center Manifold Theorem, we obtain a 3-manifold $M \subset P$, tangent to the eigenspace of $\lambda(\mu_0), \overline{\lambda(\mu_0)}$ and to the μ-axis at $\mu = \mu_0$, locally invariant under the flow of X, and containing all the local recurrence. The problem is now reduced to one of a vector field in two dimensions $\hat{X}_\mu: \mathbb{R}^2 \to \mathbb{R}^2$. The Hopf Bifurcation Theorem in two dimensions then applies (see Section 3 for details and Figures 1.4, 1.5):

(1.13) Hopf Bifurcation Theorem for Vector Fields (Poincaré [1], Andronov and Witt [1], Hopf [1], Ruelle-Takens [1], Chafee [1], etc.). *Let* X_μ *be a* C^k $(k \geq 4)$ *vector field on* $\mathbb{R}^2$ *such that* $X_\mu(0) = 0$ *for all* μ *and* $X = (X_\mu, 0)$ *is also* C^k. *Let* $dX_\mu(0,0)$ *have two distinct, simple** *complex conjugate eigenvalues* $\lambda(\mu)$ *and* $\overline{\lambda(\mu)}$ *such that for* $\mu < 0$, $\mathrm{Re}\,\lambda(\mu) < 0$, *for* $\mu = 0$, $\mathrm{Re}\,\lambda(\mu) = 0$, *and for* $\mu > 0$, $\mathrm{Re}\,\lambda(\mu) > 0$. *Also assume* $\left.\frac{d\,\mathrm{Re}\,\lambda(\mu)}{d\mu}\right|_{\mu=0} > 0$. *Then there is a* C^{k-2} *function* $\mu: (-\varepsilon, \varepsilon) \to \mathbb{R}$ *such that*

*Simple means that the *generalized* eigenspace (see Section 2A) of the eigenvalue is one dimensional.

$(x_1, 0, \mu(x_1))$ *is on a closed orbit of period* $\approx \frac{2\pi}{|\lambda(0)|}$ *and radius growing like* $\sqrt{\mu}$, *of the flow of* X *for* $x_1 \neq 0$ *and such that* $\mu(0) = 0$. *There is a neighborhood* U *of* $(0,0,0)$ *in* $\mathbb{R}^3$ *such that any closed orbit in* U *is one of the above. Furthermore,* (c) *if* 0 *is a "vague attractor"** *for* X_0, *then* $\mu(x_1) > 0$ *for all* $x_1 \neq 0$ *and the orbits are attracting* (see Figures 1.4, 1.5).

If, instead of a pair of conjugate eigenvalues crossing the imaginary axis, a real eigenvalue crosses the imaginary axis, two stable fixed points will branch off instead of a closed orbit, as in the ball in the hoop example. See Exercise 1.16.

After a bifurcation to stable closed orbits has occurred, one might ask what the next bifurcation will look like. One can visualize an invariant 2-torus blossoming out of the closed orbit (Figure 1.14). In fact, this phenomenon can occur. In order to see how, we assume we have a stable closed orbit for F_t^μ. Associated with this orbit is a Poincaré map. To define the Poincaré map, let x_0 be a point on the orbit, let N be a codimension one manifold through x_0 transverse to the orbit. The Poincaré map P_μ takes each point $x \in U$, a small neighborhood of x_0 in N, to the next point at which $F_t^\mu(x)$ intersects N (Figure 1.15). The Poincaré map is a diffeomorphism from U to $V - P_\mu(U) \subseteq N$, with $P_\mu(x_0) = x_0$

---

*This condition is spelled out below, and is reduced to a specific hypothesis on X in Section 4A. See also Section 4C. The case in which $d \operatorname{Re} \lambda(\mu)/d\mu = 0$ is discussed in Section 3A. In Section 3B it is shown that "vague attractor" can be replaced by "asymptotically stable". For a discussion of what to expect generically, see Ruelle-Takens [1], Sotomayer [1], Newhouse and Palis [1] and Section 7.

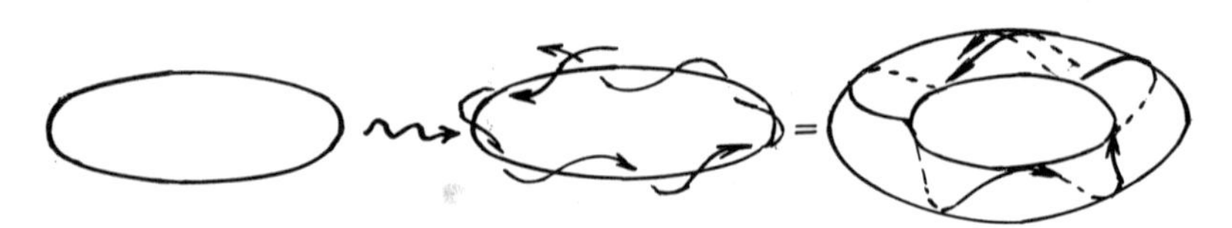

Figure 1.14

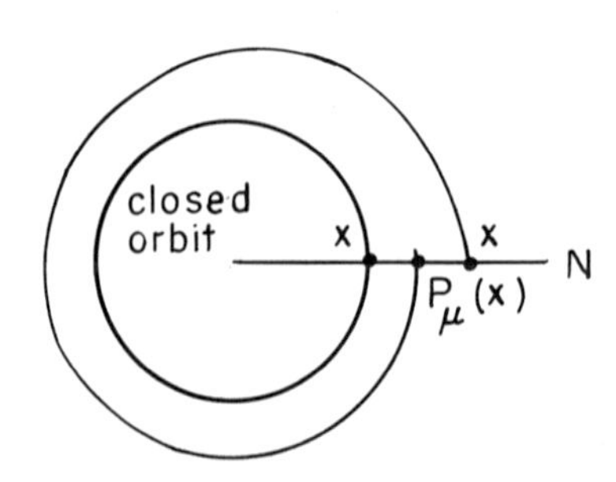

Figure 1.15

(see Section 2B for a summary of properties of the Poincaré map). The orbit is attracting if $\sigma(dP_\mu(x_0)) \subset \{z \mid |z| < 1\}$ and is not attracting if there is some $z \in \sigma(dP_\mu(x_0))$ such that $|z| > 1$.

We assume, as above, that $X_\mu: P \to TP$ is a C^k vector field on a Banach manifold P with $X_\mu(p_0) = 0$ for all μ. We assume that p_0 is stable for $\mu < \mu_0$, and that p_0 becomes unstable at μ_0, at which point bifurcation to a stable, closed orbit $\gamma(\mu)$ takes place. Let P_μ be the Poincaré map associated with $\gamma(\mu)$ and let $x_0(\mu) \in \gamma(\mu)$. We further assume that at $\mu = \mu_1$, two isolated, simple, complex conjugate eigenvalues $\lambda(\mu)$ and $\overline{\lambda(\mu)}$ of $dP_\mu(x_0(\mu))$ cross the unit circle such that $\left.\frac{d|\lambda(\mu)|}{d\mu}\right|_{\mu=\mu_1} > 0$ and such that the rest of $\sigma(dP_\mu(x_0(\mu)))$ remains inside the unit circle, at a nonzero distance from it. We then apply the Center Manifold Theorem to the map $P = (P_\mu, \mu)$ to obtain, as before, a locally invariant 3-manifold for P. The Hopf Bifurcation

Theorem for diffeomorphisms (in (1.14) below) then applies to yield a one parameter family of invariant, stable circles for P_μ for $\mu > \mu_1$. Under the flow of X_μ, these circles become stable invariant 2-tori for F_t^μ (Figure 1.16).

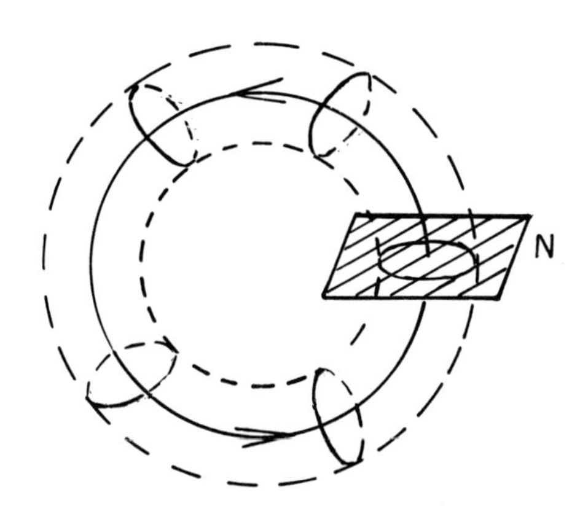

Figure 1.16

(1.14) Hopf Bifurcation Theorem for Diffeomorphisms (Sacker [1], Naimark [2], Ruelle-Takens [1]). *Let* $P_\mu: \mathbb{R}^2 \to \mathbb{R}^2$ *be a one-parameter family of* C^k $(k \geq 5)$ *diffeomorphisms satisfying*:

(a) $P_\mu(0) = 0$ *for all* μ

(b) *For* $\mu < 0$, $\sigma(dP_\mu(0)) \subset \{z \mid |z| < 1\}$

(c) *For* $\mu = 0$ $(\mu > 0)$, $\sigma(dP_\mu(0))$ *has two isolated, simple, complex conjugate eigenvalues* $\lambda(\mu)$ *and* $\overline{\lambda(\mu)}$ *such that* $|\lambda(\mu)| = 1$ $(|\lambda(\mu)| > 1)$ *and the remaining part of* $\sigma(dP_\mu(0))$ *is inside the unit circle, at a nonzero distance from it*.

(d) $\left.\dfrac{d|\lambda(\mu)|}{d\mu}\right|_{\mu=0} > 0.$

Then (under two more "vague attractor" hypotheses which will be explained during the proof of the theorem), there is a continuous one parameter family of invariant attracting circles of P_μ, one for each $\mu \in (0,\varepsilon)$ for small $\varepsilon > 0$.

(1.15) Remark. In Sections 8 and 9 we will discuss how these bifurcation theorems yielding closed orbits and invariant tori can actually be applied to the Navier-Stokes equations. One of the principal difficulties is the smoothness of the flow, which we overcome by using general smoothness results (Section 8A). Judovich [1-11], Iooss [1-6], and Joseph and Sattinger have used Hopf's original method for these results. Ruelle and Takens [1] have speculated that further bifurcations produce higher dimensional stable, invariant tori, and that the flow becomes turbulent when, as an integral curve in the space of all vector fields, it becomes trapped by a "strange attractor" (strange attractors are shown to be abundant on k-tori for $k \geq 4$); see Section 9. They can also arise spontaneously (see 4B.8 and Section 12). The question of how one can explicitly follow a fixed point through to a strange attractor is complicated and requires more research. Important papers in this direction are Takens [1,2], Newhouse [1] and Newhouse and Peixoto [1].

(1.16) Exercise. (a) Prove the following:

Theorem. Let $\mathbb{H}$ be a Hilbert space (or manifold) and $\Phi_\mu: \mathbb{H} \to \mathbb{H}$ a map defined for each $\mu \in \mathbb{R}$ such that the map $(\mu,x) \mapsto \Phi_\mu(x)$ is a C^k map, $k \geq 1$, from $\mathbb{R} \times \mathbb{H}$ to $\mathbb{H}$, and for all $\mu \in \mathbb{R}$, $\Phi_\mu(0) = 0$. Define $L_\mu = D\Phi_\mu(0)$ and suppose the spectrum of L_μ lies inside the unit circle for $\mu < 0$. Assume further there is a real, simple, isolated eigenvalue $\lambda(\mu)$ of L_μ such that $\lambda(0) = 1$, $(d\lambda/d\mu)(0) > 0$, and L_0^*

has the eigenvalue 1 (Figure 1.17); then there is a C^{k-1} curve l of fixed points of $\Phi: (x,\mu) \mapsto (\Phi_\mu(x),\mu)$ near $(0,0) \in \mathbb{H} \times \mathbb{R}$. The curve is tangent to $\mathbb{H}$ at $(0,0)$ in $\mathbb{H} \times \mathbb{R}$ (Figure 1.18). These points and the points of $(0,\mu)$ are the only fixed points of Φ in a neighborhood of $(0,0)$.

(b) Show that the hypotheses apply to the ball in the hoop example (see Exercise 1.2).

Hint: Pick an eigenvector $(z,0)$ for $(L_0,0)$ in $\mathbb{H} \times \mathbb{R}$ with eigenvalue 1. Use the center manifold theorem

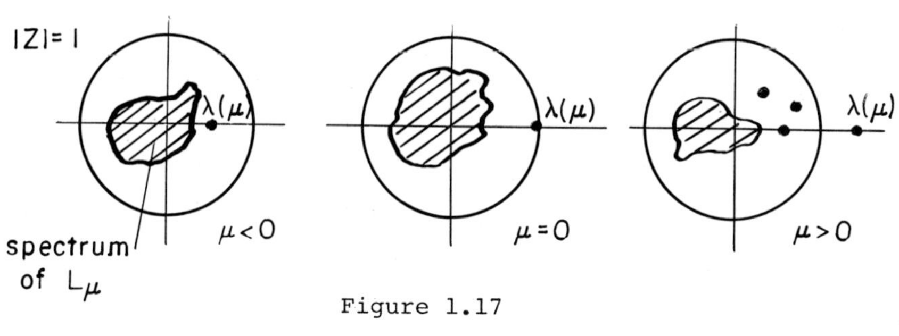

Figure 1.17

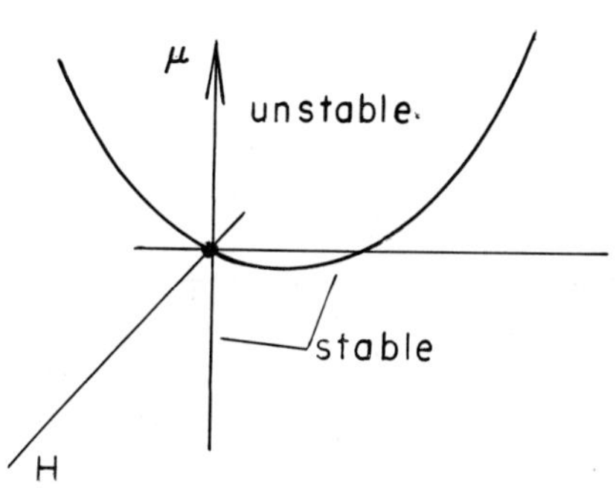

Figure 1.18

to obtain an invariant 2-manifold C tangent to $(z,0)$ and the μ axis for $\Phi(x,\mu) = (\Phi_\mu(x),\mu)$. Choose coordinates (α,μ) on C where α is the projection to the normalized eigenvector $z(\mu)$ for L_μ. Set $\Phi(x,\mu) = (f(\alpha,\mu),\mu)$ in these coordinates. Let $g(\alpha,\mu) = \frac{f(\alpha,\mu)}{\alpha} - 1$ and we use the implicit function theorem to get a curve of zeros of g in C. (See Ruelle-Takens [1, p. 190]).

(1.17) Remark. The closed orbits which appear in the Hopf theorem need not be globally attracting, nor need they persist for large values of the parameter μ. See remarks (3A.3).

(1.18) Remark. The reduction to finite dimensions using the center manifold theorem is analogous to the reduction to finite dimensions for stationary bifurcation theory of elliptic type equations which goes under the name "Lyapunov-Schmidt" theory. See Nirenberg [1] and Vainberg-Trenogin [1,2].

(1.19) Remark. Bifurcation to closed orbits can occur by other mechanisms than the Hopf bifurcation. In Figure 1.19 is shown an example of S. Wan.

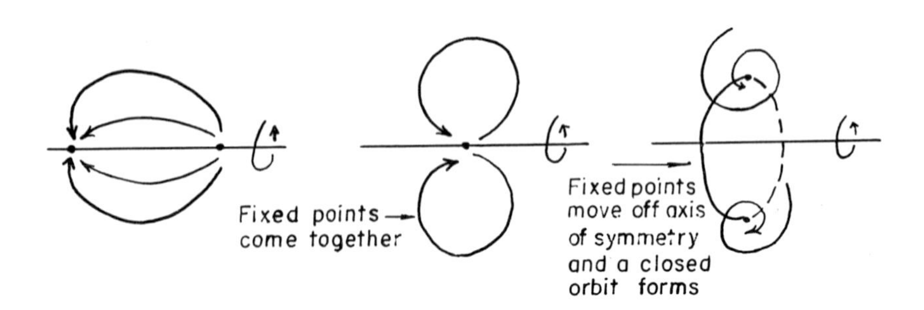

Figure 1.19

SECTION 2
THE CENTER MANIFOLD THEOREM

In this section we will start to carry out the program outlined in Section 1 by proving the center manifold theorem. The general invariant manifold theorem is given in Hirsch-Pugh-Shub [1]. Most of the essential ideas are also in Kelley [1] and a treatment with additional references is contained in Hartman [1]. However, we shall follow a proof given by Lanford [1] which is adapted to the case at hand, and is direct and complete. We thank Professor Lanford for allowing us to reproduce his proof.

The key job of the center manifold theorem is to enable one to reduce to a finite dimensional problem. In the case of the Hopf theorem, it enables a reduction to two dimensions without losing any information concerning stability. The outline of how this is done was presented in Section 1 and the details are given in Sections 3 and 4.

In order to begin, the reader should recall some results about basic spectral theory of bounded linear operators by consulting Section 2A. The proofs of Theorems 1.3

and 1.4 are also found there.

Statement and Proof of the Center Manifold Theorem

We are now ready for a proof of the center manifold theorem. It will be given in terms of an invariant manifold for a map Ψ, not necessarily a local diffeomorphism. Later we shall use it to get an invariant manifold theorem for flows. Remarks on generalizations are given at the end of the proof.

(2.1) Theorem. Center Manifold Theorem. Let Ψ be a mapping of a neighborhood of zero in a Banach space Z into Z. We assume that Ψ is C^{k+1}, $k \geq 1$ and that $\Psi(0) = 0$. We further assume that $D\Psi(0)$ has spectral radius 1 and that the spectrum of $D\Psi(0)$ splits into a part on the unit circle and the remainder which is at a non-zero distance from the unit circle.* Let Y denote the generalized eigenspace of $D\Psi(0)$ belonging to the part of the spectrum on the unit circle; assume that Y has dimension $d < \infty$.

Then there exists a neighborhood V of 0 in Z and a C^k submanifold M of V of dimension d, passing through 0 and tangent to Y at 0, such that

a) (Local Invariance): If $x \in M$ and $\Psi(x) \in V$, then $\Psi(x) \in M$

b) (Local Attractivity): If $\Psi^n(x) \in V$ for all $n = 0, 1, 2, \ldots$, then, as $n \to \infty$, the distance from $\Psi^n(x)$ to M goes to zero.

We begin by reformulating (in a slightly more general way) the theorem we want to prove. We have a mapping Ψ of a neighborhood of zero in a Banach space Z into Z, with

*This holds automatically if Z is finite dimensional or, more generally, if $D\Psi(0)$ is compact.

$\Psi(0) = 0$. We assume that the spectrum of $D\Psi(0)$ splits into a part on the unit circle and the remainder, which is contained in a circle of radius strictly less than one, about the origin. The basic spectral theory discussed in Section 2A guarantees the existence of a spectral projection P of Z belonging to the part of the spectrum on the unit circle with the following properties:

i) P commutes with $D\Psi(0)$, so the subspaces PZ and $(I-P)Z$ are mapped into themselves by $D\Psi(0)$.

ii) The spectrum of the restriction of $D\Psi(0)$ to PZ lies on the unit circle, and

iii) The spectral radius of the restriction of $D\Psi(0)$ to $(I-P)Z$ is strictly less than one.

We let X denote $(I-P)Z$, Y denote PZ, A denote the restriction of $D\Psi(0)$ to X and B denote the restriction of $D\ (0)$ to Y. Then $Z = X \oplus Y$ and

$$\Psi(x,y) = (Ax+X^*(x,y),\ \ By + Y^*(x,y)),$$

where

A is bounded linear operator on X with spectral radius strictly less than one.

B is a bounded operator on Y with spectrum on the unit circle. (All we actually need is that the spectral radius of B^{-1} is no larger than one.)

X^* is a C^{k+1} mapping of a neighborhood of the origin in $X \oplus Y$ into X with a second-order zero at the origin, i.e. $X(0,0) = 0$ and $DX(0,0) = 0$, and

Y is a C^{k+1} mapping of a neighborhood of the origin in $X \oplus Y$ into Y with a second-order zero at the origin.

We want to find an invariant manifold for Ψ which is tangent to Y at the origin. Such a manifold will be the graph of a mapping u which maps a neighborhood of the origin in Y into X, with $u(0) = 0$ and $Du(0) = 0$.

In the version of the theorem we stated in 2.1, we assumed that Y was finite-dimensional. We can weaken this assumption, but not eliminate it entirely.

(2.2) Assumption. There exists a C^{k+1} real-valued function ϕ on Y which is 1 on a neighborhood of the origin and zero for $||y|| > 1$. Perhaps surprisingly, this assumption is actually rather restrictive. It holds trivially if Y is finite-dimensional or if Y is a Hilbert space; for a more detailed discussion of when it holds, see Bonic and Frampton [1].

We can now state the precise theorem we are going to prove.

(2.3) Theorem. *Let the notation and assumptions be as above. Then there exist* $\varepsilon > 0$ *and a* C^k*-mapping* u^* *from* $\{y \in Y: ||y|| < \varepsilon\}$ *into* X, *with a second-order zero at zero, such that*

a) *The manifold* $\Gamma_{u^*} = \{(x,y) \mid x = u^*(y)$ and $||y|| < \varepsilon\} \subset X \oplus Y$, i.e. *the graph of* u^* *is invariant for* Ψ *in the sense that, if* $||y|| < \varepsilon$ *and if* $\Psi(u^*(y),y) = (x_1,y_1)$ *with* $||y_1|| < \varepsilon$ *then* $x_1 = u^*(y_1)$.

b) *The manifold* Γ_{u^*} *is locally attracting for* Ψ *in the sense that, if* $||x|| < \varepsilon$, $||y|| < \varepsilon$, *and if* $(x_n,y_n) = \Psi^n(x,y)$ *are such that* $||x_n|| < \varepsilon$, $||y_n|| < \varepsilon$ *for all* $n > 0$, *then*

$$\lim_{n\to\infty} ||x_n - u^*(y_n)|| = 0.$$

Proceeding with the proof, it will be convenient to assume that $||A|| < 1$ and that $||B^{-1}||$ is not much greater than 1. This is not necessarily true but we can always make it true by replacing the norms on X, Y by equivalent norms. (See Lemma 2A.4). We shall assume that we have made this change of norm. It is unfortunately a little awkward to explicitly set down exactly how close to one $||B^{-1}||$ should be taken. We therefore carry out the proof as if $||B^{-1}||$ were an adjustable parameter; in the course of the argument, we shall find a finite number of conditions on $||B^{-1}||$. In principle, one should collect all these conditions and impose them at the outset.

The theorem guarantees the existence of a function u^* defined on what is perhaps a very small neighborhood of zero. Rather than work with very small values of x,y, we shall scale the system by introducing new variables x/ε, y/ε (and calling the new variables again x and y). This scaling does not change A,B, but, by taking ε very small, we can make X^*, Y^*, together with their derivatives of order $\leq k+1$, as small as we like on the unit ball. Then by multiplying $X^*(x,y)$, $Y^*(x,y)$ by the function $\phi(y)$ whose existence is asserted in the assumption preceding the statement of the theorem, we can also assume that $X^*(x,y)$, $Y^*(x,y)$ are zero when $||y|| > 1$. Thus, if we introduce

$$\lambda = \sup_{\substack{||x||\leq 1 \\ y \text{ unrestricted}}} \; \sup_{\substack{j_1,j_2 \\ j_1+j_2\leq k+1}} \{||D_x^{j_1}D_y^{j_2}X^*(x,y)|| + ||D_x^{j_1}D_y^{j_2}Y^*(x,y)||\},$$

we can make λ as small as we like by choosing ε very small. The only use we make of our technical assumption on Y is to arrange things so that the supremum in the definition of λ may be taken over all y and not just over a bounded set.

Once we have done the scaling and cutting off by ϕ, we can prove a global center manifold theorem. That is, we shall prove the following.

(2.4) *Lemma*. *Keep the notation and assumptions of the center manifold theorem. If* λ *is sufficiently small (and if* $||B^{-1}||$ *is close enough to one), there exists a function* u^*, *defined and* k *times continuously differentiable on all of* Y, *with a second-order zero at the origin, such that*

a) *The manifold* $\Gamma_{u^*} = \{(x,y) | x = u^*(y), y \in Y\}$ *is invariant for* Ψ *in the strict sense*.

b) *If* $||x|| < 1$, *and* y *is arbitrary then* $\lim_{n\to\infty} ||x_n - u^*(y_n)|| = 0$ (*where* $(x_n, y_n) = \Psi^n(x,y)$).

As with $||B^{-1}||$, we shall treat λ as an adjustable parameter and impose the necessary restrictions on its size as they appear. It may be worth noting that λ depends on the choice of norm; hence, one must first choose the norm to make $||B^{-1}||$ close to one, then do the scaling and cutting off to make λ small. To simplify the task of the reader who wants to check that all the required conditions on $||B^{-1}||$ can be satisfied simultaneously, we shall note these conditions with a * as with (2.3)* on p. 34.

The strategy of proof is very simple. We start with a manifold M of the form $\{x = u(y)\}$ (this stands for the graph of u); we let M denote the image of M under Ψ. With some mild restrictions on u, we first show that the manifold

ΨM again has the form

$$\{x = \hat{u}(y)\}$$

for a new function $\hat{u}$. If we write $\mathscr{F}u$ for $\hat{u}$ we get a (non-linear) mapping

$$u \mapsto \mathscr{F}u$$

from functions to functions. The manifold M is invariant if and only if $u = \mathscr{F}u$, so we must find a fixed point of $\mathscr{F}$. We do this by proving that $\mathscr{F}$ is a contraction on a suitable function space (assuming that λ is small enough).

More explicitly, the proof will be divided into the following steps:

I) Derive heuristically a "formula" for $\mathscr{F}$.

II) Show that the formula obtained in I) yields a well-defined mapping of an appropriate function space U into itself.

III)† Prove that $\mathscr{F}$ is a contraction on U and hence has a unique fixed point u^*.

IV) Prove that b) of Lemma (2.4) holds for u^*.

We begin by considering Step I).

I) To construct $u(y)$, we should proceed as follows

i) Solve the equation

$$y = B\tilde{y} + Y^*(u(\tilde{y}),\tilde{y}) \tag{2.1}$$

for $\tilde{y}$. This means that y is the Y-component of $\Psi(u(\tilde{y}),\tilde{y})$.

ii) Let $u(y)$ be the X-component of $\Psi(u(\tilde{y}),\tilde{y})$, i.e.,

$$u(y) = Au(\tilde{y}) + X^*(u(\tilde{y}),\tilde{y}). \tag{2.2}$$

II) We shall somewhat arbitrarily choose the space of functions u we want to consider to be

---

†One could use the implicit function theorem at this step. For this approach, see Irwin [1].

$U = \{u: Y \to X | D^{k+1}u$ continuous; $||D^j u(y)|| < 1$ for $j = 0,1,\ldots,k+1$, all y; $u(0) = Du(0) = 0\}$.

We must carry out two steps:

i) Prove that, for any given $u \in U$, equation (1) has a unique solution $\tilde{y}$ for each $y \in Y$. And

ii) Prove that $\mathcal{F}u$, defined by (2.2) is in U.

To accomplish (i), we rewrite (2.1) as a fixed-point problem:

$$\tilde{y} = B^{-1}y - B^{-1}Y^*(u(\tilde{y}),\tilde{y}).$$

It suffices, therefore, to prove that the mapping

$$\tilde{y} \mapsto B^{-1}y - B^{-1}Y^*(u(\tilde{y}),\tilde{y})$$

is a contraction on Y. We do this by estimating its derivative:

$$||D_{\tilde{y}}[B^{-1}y - B^{-1}Y^*(u(\tilde{y}),\tilde{y})]|| \le ||B^{-1}|| \quad ||D_1Y^*(u(\tilde{y}),\tilde{y})Du(\tilde{y}) + D_2Y^*(u(\tilde{y}),\tilde{y})|| \le 2\lambda||B^{-1}||$$

by the definitions of λ and U. If we require

$$2\lambda||B^{-1}|| < 1, \qquad (2.3)^*$$

equation (2.1) has a unique solution $\tilde{y}$ for each y. Note that $\tilde{y}$ is a function of y, depending also on the function u. By the inverse function theorem, $\tilde{y}$ is a C^{k+1} function of y.

Next we establish (ii). By what we have just proved, $\mathcal{F}u \in C^{k+1}$. Thus to show $\mathcal{F}u \in U$, what we must check is that

$$||D^j\mathcal{F}u(y)|| \le 1 \text{ for all } y, \; j = 0,1,2,\ldots,k+1 \qquad (2.4)$$

and $\mathcal{F}u(0) = 0$, $D\mathcal{F}u(0) = 0$. (2.5)

First take $j = 0$:

$$||\mathcal{F}u(y)|| \le ||A||\cdot||u(\tilde{y})|| + ||X^*(u(\tilde{y}),\tilde{y})|| \le ||A|| + \lambda,$$

so if we require

$$||A|| + \lambda \le 1, \qquad (2.6)^*$$

then $||\mathcal{F}u(y)|| \le 1$ for all y.

To estimate $D\mathscr{F}u$ we must first estimate $D\tilde{y}(y)$. By differentiating (2.1), we get

$$I = [B+DY^u(\tilde{y})]D\tilde{y}, \tag{2.7}$$

where $Y^u: Y \to Y$ is defined by

$$Y^u(y) = Y^*(u(y),y).$$

By a computation we have already done,

$$||DY^u(\tilde{y})|| \leq 2\lambda \quad \text{for all} \quad \tilde{y}.$$

Now $B + DY^u = B[I+B^{-1}DY^u]$ and since $2\lambda||B^{-1}|| < 1$ (by (2.3)*), $B + DY^u$ is invertible and

$$||(B+DY^u)^{-1}|| \leq ||B^{-1}||(1-2\lambda||B^{-1}||)^{-1}.$$

The quantity on the right-hand side of this inequality will play an important role in our estimates, so we give it a name:

$$\gamma \equiv ||B^{-1}||(1-\lambda||B^{-1}||)^{-1}. \tag{2.8}$$

Note that, by first making $||B^{-1}||$ very close to one and then by making λ small, we can make γ as close to one as we like.

We have just shown that

$$||Dy(\tilde{y})|| \leq \gamma \quad \text{for all} \quad y. \tag{2.9}$$

Differentiating the expression (2.2) for $\mathscr{F}u(y)$ yields

$$\left.\begin{array}{l} D\mathscr{F}u(y) = [A\ Du(\tilde{y}) + DX^u(\tilde{y})]\,D\tilde{y}(y); \\ \text{and} \\ (X^u(\tilde{y}) = X^*(u(\tilde{y}),\tilde{y})). \end{array}\right\} \tag{2.10}$$

Thus

$$||D\mathscr{F}u(y)|| \leq (||A|| + 2\lambda)\cdot\gamma, \tag{2.11}$$

so if we require

$$(||A|| + 2\lambda)\gamma \leq 1, \tag{2.12}*$$

we get

$$||D\mathscr{F}u(y)|| \leq 1 \quad \text{for all} \quad y.$$

We shall carry the estimates just one step further. Differentiating (2.7) yields

$$0 = (B+DY^u(\tilde{y}))D^2\tilde{y} + D^2Y^u(D\tilde{y})^2.$$

By a straightforward computation,

$$||D^2Y^u(\tilde{y})|| \leq 5\lambda \quad \text{for all} \quad \tilde{y},$$

so

$$||D^2y(\tilde{y})|| = ||(B + DY^u(\tilde{y}))^{-1}D^2Y^u(D\tilde{y})^2||$$

$$\leq \gamma \cdot 5\lambda \cdot \gamma^2 = 5\lambda\gamma^3.$$

Now, by differentiating the formula (2.10) for $D\mathscr{F}u$, we get

$$D^2\mathscr{F}u(y) = [A\ D^2u(\tilde{y}) + D^2X^u(\tilde{y})](D\tilde{y})^2 + [A\ Du(\tilde{y}) + DX^u(\tilde{y})]D^2\tilde{y},$$

so

$$||D^2\mathscr{F}u(y)|| \leq (||A|| + 5\lambda)\gamma^2 + (||A|| + 2\lambda)\cdot 5\lambda\gamma^3.$$

If we require

$$(||A|| + 5\lambda)\gamma^2 + (||A|| + 2\lambda)\cdot 5\lambda\gamma^3 \leq 1, \tag{2.13}*$$

we have

$$||D^2\mathscr{F}u(y)|| \leq 1 \quad \text{for all} \quad y.$$

At this point it should be plausible by imposing a sequence of stronger and stronger conditions on $\gamma,\lambda,$ that we can arrange

$$||D^j\mathscr{F}(y)|| \leq 1 \quad \text{for all} \quad y, \quad j = 3,4,\ldots,k+1.$$

The verification that this is in fact possible is left to the reader.

To check (2.5), i.e. $\mathscr{F}u = 0$, $D\mathscr{F}u = 0$ (assuming $u = 0$, $Du = 0$) we note that

$$\begin{aligned}
\tilde{y}(0) &= 0 \quad \text{since } 0 \text{ is a solution of } 0 = B\tilde{y} + Y(u(\tilde{y}),\tilde{y}) \\
\mathscr{F}u(0) &= Au(0) + X(u(0),0) = 0 \quad \text{and} \\
D\mathscr{F}u(0) &= [A\ Du(0) + D_1X(0,0)Du(0) + D_2X(0,0)]\cdot D\tilde{y}(0) \\
&= [A\cdot 0 + 0 + 0]\cdot D\tilde{y}(0) = 0.
\end{aligned}$$

This completes step II). Now we turn to III)

III) We show that $\mathscr{F}$ is a contraction and apply the contraction mapping principle. What we actually do is slightly more complicated.

i) We show that $\mathscr{F}$ is a contraction in the supremum norm. Since U is not complete in the supremum norm, the contraction mapping principle does not imply that $\mathscr{F}$ has a fixed point in U, but it does imply that $\mathscr{F}$ has a fixed point in the completion of U with respect to the supremum norm.

ii) We show that the completion of U with respect to the supremum norm is contained in the set of functions u from Y to X with Lipschitz-continuous k^{th} derivatives and with a second-order zero at the origin. Thus, the fixed point u^* of $\mathscr{F}$ has the differentiability asserted in the theorem.

We proceed by proving i).

i) Consider $u_1, u_2 \in U$, and let $||u_1-u_2||_0 = \sup_y ||u_1(y)-u_2(y)||$. Let $\tilde{y}_1(y)$, $\tilde{y}_2(y)$ denote the solution of

$$y = B\tilde{y}_i + Y(u_i(\tilde{y}_i),\tilde{y}_i) \quad i = 1,2.$$

We shall estimate successively $||\tilde{y}_1-\tilde{y}_2||_0$, and $||\mathscr{F}u_1-\mathscr{F}u_2||_0$. Subtracting the defining equations for $\tilde{y}_1, \tilde{y}_2$, we get

$$B(\tilde{y}_1-\tilde{y}_2) = Y(u_2(\tilde{y}_2),\tilde{y}_2) - Y(u_1(\tilde{y}_1),\tilde{y}_1),$$

so that

$$||\tilde{y}_1-\tilde{y}_2|| \leq ||B^{-1}||\cdot\lambda\cdot[||u_2(\tilde{y}_2)-u_1(\tilde{y}_1)|| + ||\tilde{y}_2-\tilde{y}_1||]. \tag{2.14}$$

Since $||Du_1||_0 \leq 1$, we can write

$$\begin{aligned}||u_2(\tilde{y}_2) - u_1(\tilde{y}_1)|| &\leq ||u_2(\tilde{y}_2) - u_1(\tilde{y}_2)|| + \\ ||u_1(\tilde{y}_2) - u_1(\tilde{y}_1)|| &\leq ||u_2-u_1||_0 + ||\tilde{y}_2-\tilde{y}_1||.\end{aligned} \tag{2.15}$$

Inserting (2.15) in (2.14) and rearranging, yields

$$\left.\begin{aligned}(1-2\lambda\cdot||B^{-1}||)||\tilde{y}_1-\tilde{y}_2|| &\leq \lambda\cdot||B^{-1}||\cdot||u_2-u_1||_0,\\ \text{i.e.} \qquad ||\tilde{y}_1-\tilde{y}_2||_0 &\leq \lambda\cdot\gamma\cdot||u_2-u_1||.\end{aligned}\right\} \tag{2.16}$$

Now insert estimates (2.15) and (2.16) in

$$\begin{aligned}\mathscr{F}u_1(y) - \mathscr{F}u_2(y) &= A[u_1(\tilde{y}_1) - u_2(\tilde{y}_2)]\\ &\quad + [X(u_1(\tilde{y}_1),\tilde{y}_1) - X(u_2(\tilde{y}_2),\tilde{y}_2)]\end{aligned}$$

to get

$$\begin{aligned}||\mathscr{F}u_1 - \mathscr{F}u_2||_0 &\leq ||A||[||u_2-u_1||_0 + ||\tilde{y}_2-\tilde{y}_1||_0]\\ &\quad + \lambda[||u_2-u_1||_0 + 2\cdot||\tilde{y}_2-\tilde{y}_1||_0]\\ &\leq ||u_2-u_1||_0\{||A||(1+\gamma\lambda) + \lambda(1+2\gamma\lambda)\}.\end{aligned}$$

If we now require

$$\alpha = ||A||(1+\gamma\lambda) + \lambda(1+2\gamma\lambda) < 1, \tag{2.17)*$$

$\mathscr{F}$ will be a contraction in the supremum norm.

ii) The assertions we want all follow directly from the following general result.

(2.5) Lemma. *Let (u_n) be a sequence of functions on a Banach space Y with values on a Banach space X. Assume that, for all n and $y \in Y$,*

$$||D^j u_n(y)|| \leq 1 \qquad j = 0,1,2,\ldots,k,$$

and that each $D^k u_n$ is Lipschitz continuous with Lipschitz constant one. Assume also that for each y, the sequence $(u_n(y))$ converges weakly (i.e., in the weak topology on X) to a unit vector $u(y)$. Then

a) *u has a Lipschitz continuous k^{th} derivative with Lipschitz constant one.*

b) *$D^j u_n(y)$ converges weakly to $D^j u(y)$* for all y and $j = 1,2,\ldots,k$.*

If X, Y are finite dimensional, all the Banach space technicalities in the statement of the proposition disappear, and the proposition becomes a straightforward consequence of the Arzela-Ascoli Theorem. We postpone the proof for a moment, and instead turn to step IV).

IV) We shall prove the following: Let $x \in X$ with $||x|| \leq 1$ and let $y \in Y$ be arbitrary. Let $(x_1,y_1) = \Psi(x,y)$.

---

*This statement may require some interpretation. For each $n,y,D^j u_n(y)$ is a bounded symmetric j-linear map from Y^j to X. What we are asserting is that, for each $y,y_1,\ldots,y_j$, the sequence $(D^j u_n(y)(y_1,\ldots,y_j))$ of elements of X converges in the weak topology on X to $D^j u(y)(y_1,\ldots,y_j)$.

Then

$$||x_1|| \leq 1$$

and

$$||x_1 - u^*(y_1)|| \leq \alpha \cdot ||x - u^*(y)||, \tag{2.18}$$

where α is as defined in (2.17). By induction,

$$||x_n - u^*(y_n)|| \leq \alpha^n ||x - u^*(y)|| \to 0 \quad \text{as} \quad n \to \infty,$$

as asserted.

To prove $||x_1|| \leq 1$, we first write

$$x_1 = Ax + X(x,y), \quad \text{so that}$$

$$||x_1|| \leq ||A|| \cdot ||X|| + \lambda \leq ||A|| + \lambda \leq 1 \quad \text{by} \quad (2.6)$$

To prove (2.18), we essentially have to repeat the estimates made in proving that $\mathcal{F}$ is a contraction. Let $\tilde{y}_1$ be the solution of

$$y_1 = B\tilde{y}_1 + Y(u^*(\tilde{y}_1), \tilde{y}_1).$$

On the other hand, by the definition of y_1 we have

$$y_1 = By + Y(x,y).$$

Subtracting these equations and proceeding exactly as in the derivation of (2.16), we get

$$||\tilde{y}_1 - y|| \leq \lambda \cdot \gamma \cdot ||u^*(y) - x||.$$

Next, we write

$$u^*(y_1) = \mathcal{F}u^*(y_1) = Au^*(y_1) + X(u^*(\tilde{y}_1), \tilde{y}_1)$$

$$x_1 = Ax + X(x,y).$$

Subtracting and making the same estimates as before, we get

$$||x_1-u^*(y_1)|| \leq \alpha\cdot||x-u^*(y)||$$

as desired. This completes step IV).

Let us finish the argument by supplying the details for Lemma (2.5).

<u>Proof of Lemma (2.5)</u>

We shall give the argument only for $k = 1$; the generalization to arbitrary k is a straightforward induction argument.

We start by choosing $y_1, y_2 \in Y$ and $\phi \in X^*$ and consider the sequence of real-valued functions of a real variable

$$t \to \phi(u_n(y_1+ty_2)) \equiv \psi_n(t).$$

From the assumptions we have made about the sequence (u_n), it follows that

$$\lim_{n\to\infty} \psi_n(t) = \phi(u(y_1+ty_2)) \equiv \psi(t)$$

for all t, that $\psi_n(t)$ is differentiable, that

$$|\psi_n'(t)| \leq ||\phi|| \cdot||y_1|| \quad \text{for all } n, t$$

and that

$$|\psi_n'(t_1) - \psi_n'(t_2)| \leq ||\phi||\cdot||y_2||^2|t_1-t_2|$$

for all n, t_1, t_2. By this last inequality and the Arzela-Ascoli Theorem, there exists a subsequence $\psi_{n_j}'(t)$ which converges uniformly on every bounded interval. We shall temporarily denote the limit of this subsequence by $X(t)$. We have

$$\psi_{n_j}(t) = \psi_{n_j}(0) + \int_0^t \psi_{n_j}'(\tau)\, d\tau;$$

hence, passing to the limit $j \to \infty$, we get

$$\psi(t) = \psi(0) + \int_0^t X(\tau)\, d\tau,$$

which implies that $\psi(t)$ is continuously differentiable and that

$$\psi'(t) = X(t).$$

To see that

$$\lim_{n\to\infty} \psi_n'(t) = \psi'(t)$$

(i.e., that it is not necessary to pass to a subsequence), we note that the argument we have just given shows that <u>any</u> subsequence of $(\psi_n'(t))$ has a subsequence converging to $\psi'(t)$; this implies that the original sequence must converge to this limit.

Since

$$\psi_n'(0) = \phi(Du_n(y_1)(y_2)),$$

we conclude that the sequence

$$Du_n(y_1)(y_2)$$

converges in the weak topology on X^{**} to a limit, which we shall denote by $Du(y_1)(y_2)$; this notation is at this point only suggestive. By passage to a limit from the corresponding property of $Du_n(y_1)(y_2)$, we see that

$$y_2 \mapsto Du_n(y_1)(y_2)$$

is a bounded linear mapping of norm ≤ 1 from Y to X^{**} for each y_1. We denote this linear operator by $Du(y_1)$. Since

$$||Du_n(y_1)(y_2) - Du_n(y_1')(y_2)|| \leq ||y_1 - y_1'||\cdot||y_2||,$$

we have

$$||Du(y_1) - Du(y_1')|| \leq ||y_1 - y_1'||,$$

i.e., the mapping $y \mapsto Du(y)$ is Lipschitz continuous from Y to $L(Y,X^{**})$.

The next step is to prove that

$$u(y_1+y_2) - u(y_1) = \int_0^1 Du(y_1+\tau y_2)(y_2)\, d\tau; \tag{2.19}$$

this equation together with the norm-continuity of $y \mapsto Du(y)$ will imply that u is (Fréchet)-differentiable. The integral in (2.19) may be understood as a vector-valued Riemann integral. By the first part of our argument,

$$\phi(u(y_1+y_2)) - \phi(u(y_1)) = \int_0^1 \phi(Du(y_1+\tau y_2)(y_2))\, d\tau,$$

for all $\phi \in X^{**}$, and taking Riemann integrals commutes with continuous linear mappings, so that

$$\phi([u(y_1+y_2) - u(y_1) - \int_0^1 Du(y_1+\tau y_2)(y_2)])\, d\tau = 0$$

for all $\phi \in X^*$. Therefore (2.19) is proved.

The situation is now as follows: We have shown that, if we regard u as a mapping into X^{**}, which contains X, then it is Fréchet differentiable with derivative Du. On the other hand, we know that u actually takes values in X and want to conclude that it is differentiable as a mapping into X. This is equivalent to proving that $Du(y_1)(y_2)$ belongs to X for all y_1, y_2. But

$$Du(y_1)(y_2) = \text{norm} \lim_{t \to 0} \frac{u(y_1+ty_2)-u(y_1)}{t};$$

the difference quotients on the right all belong to X, and X is norm closed in X^{**}. Thus, $Du(y_1)(y_2)$ is in X and the proof is complete. □

(2.6) Remarks on the Center Manifold Theorem

1. It may be noted that we seem to have lost some differentiability in passing from Ψ to u^*, since we assumed that Ψ is C^{k+1} and only concluded that u^* is C^k. In fact, however, the u^* we obtain has a Lipschitz continuous k^{th} derivative, and our argument works just as well if we only assume that Ψ has a Lipschitz continuous k^{th} derivative, so in this class of maps, no loss of differentiability occurs. Moreover, if we make the weaker assumption that the kth derivative of Ψ is uniformly continuous on some neighborhood of zero, we can show that the same is true of u^*. (Of course, if X and Y are finite dimensional, continuity on a neighborhood of zero implies uniform continuity on a neighborhood of zero, but this is no longer true if X or Y is infinite dimensional).

2. As C. Pugh has pointed out, if Ψ is infinitely differentiable, the center manifold <u>cannot</u>, in general, be taken to be infinitely differentiable. It is also <u>not</u> true that, if Ψ is analytic there is an analytic center manifold. We shall give a counterexample in the context of equilibrium points of differential equations rather than fixed points of maps; cf. Theorem 2.7 below. This example, due to Lanford, also shows that the center manifold is not unique; cf. Exercise 2.8.

Consider the system of equations:

$$\frac{dy_1}{dt} = -y_2, \quad \frac{dy_2}{dt} = 0, \quad \frac{dx}{dt} = -x + h(y_1), \tag{2.20}$$

where h is analytic near zero and has a second-order zero at zero. We claim that, if h is not analytic in the whole complex plane, there is no function $u(y_1,y_2)$, analytic in a neighborhood of $(0,0)$ and vanishing to second order at $(0,0)$, such that the manifold

$$\{x = u(y_1,y_2)\}$$

is locally invariant under the flow induced by the differential equation near $(0,0)$. To see this, we assume that we have an invariant manifold with

$$u(y_1,y_2) = \sum_{\substack{j_1,j_2 \\ j_1+j_2\geq 1}} c_{j_1j_2} y_1^{j_1} y_2^{j_2} .$$

Straightforward computation shows that the expansion coefficients c_{j_1,j_2} are uniquely determined by the requirement of invariance and that

$$c_{j_1,j_2} = \frac{(j_1+j_2)!}{(j_1)!} h_{j_1+j_2},$$

where

$$h(y_1) = \sum_{j\geq 2} h_j y_1^j .$$

If the series for h has a finite radius of convergence, the series for $u(o,y_2)$ diverges for all non-zero y_2.

The system of differential equations has nevertheless many infinitely differentiable center manifolds. To construct one, let $\tilde{h}(y_1)$ be a bounded infinitely differentiable function agreeing with h on a neighborhood of zero. Then the manifold defined by

$$u(y_1,y_2) = \int_{-\infty}^{0} e^{\sigma}\tilde{h}(y_1-\sigma y_2)\, d\sigma \tag{2.21}$$

is easily verified to be globally invariant for the system

$$\frac{dy_1}{dt} = -y_2, \quad \frac{dy_2}{dt} = 0, \quad \frac{dx}{dt} = -x + \tilde{h}(y_1) \tag{2.22}$$

and hence locally invariant at zero for the original system.

(To make the expression for u less mysterious, we sketch its derivation. The equations for y_1, y_2 do not involve x and are trivial to solve explicitly. A function u defining an invariant manifold for the modified system (2.22) must satisfy

$$\frac{d}{dt} u(y_1 - ty_2, y_2) = -u(y_1 - ty_2) + \tilde{h}(y_1 - ty_2)$$

for all t, y_1, y_2. The formula (2.21) for u is obtained by solving this ordinary differential equation with a suitable boundary condition at $t = -\infty$.)

3. Often one wishes to replace the fixed point 0 of Ψ by an invariant manifold V and make spectral hypotheses on a normal bundle of V. We shall need to do this in Section 9. This general case follows the same procedure; details are found in Hirsch-Pugh-Shub [1].

The Center Manifold Theorem for Flows

The center manifold theorem for maps can be used to prove a center manifold theorem for flows. We work with the time t maps of the flow rather than with the vector fields themselves because, in preparation for the Navier Stokes equations, we want to allow the vector field generating the flow to be only densely defined, but since we can often prove that the time t-maps are C^∞ this is a reasonable hypothesis for many partial differential equations (see Section 8A for details).

(2.7) Theorem. Center Manifold Theorem for Flows. Let Z be a Banach space admitting a C^∞ norm* away from 0 and let F_t be a C^0 semiflow defined in a neighborhood of zero for $0 \leq t \leq \tau$. Assume $F_t(0) = 0$, and that for $t > 0$, $F_t(x)$ is C^{k+1} jointly in t and x. Assume that the spectrum of the linear semigroup $DF_t(0): Z \to Z$ is of the form $e^{t(\sigma_1 \cup \sigma_2)}$ where $e^{t\sigma_1}$ lies on the unit circle (i.e. σ_1 lies on the imaginary axis) and $e^{t\sigma_2}$ lies inside the unit circle a nonzero distance from it, for $t > 0$; i.e. σ_2 is in the left half plane. Let Y be the generalized eigenspace corresponding to the part of the spectrum on the unit circle. Assume $\dim Y = d < \infty$.

Then there exists a neighborhood V of 0 in Z and a C^k submanifold $M \subset V$ of dimension d passing through 0 and tangent to Y at 0 such that

(a) If $x \in M$, $t > 0$ and $F_t(x) \in V$, then $F_t(x) \in M$.

(b) If $t > 0$ and $F_t^n(x)$ remains defined and in V for all $n = 0,1,2,\ldots$, then $F_t^n(x) \to M$ as $n \to \infty$.

This way of formulating the result is the most convenient for it applies to ordinary as well as to partial differential equations, the reason is that we do not need to worry about "unboundedness" of the generator of the flow. Instead we have used a smoothness assumption on the flow.

The center manifold theorem for maps, Theorem 2.1, applies to each F_t, $t > 0$. However, we are claiming that V and M can be chosen independent of t. The basic reason

---

*Notice that this assumption on Z was not required above, but it is needed here. The reason will be evident below. Such a Banach space is often called "smooth".

for this is that the maps $\{F_t\}$ commute: $F_s \circ F_t = F_{t+s} = F_t \circ F_s$, where defined. However, this is somewhat oversimplified. In the proof of the center manifold theorem we would require the F_t to remain globally commuting after they have been cut off by the function ϕ. That is, we need to ensure that in the course of proving lemma 2.4, λ can be chosen small (independent of t) and the F_t's are globally defined and commute.

The way to ensure this is to first cut off the F_t in Z outside a ball B in such a way that the F_t are not disturbed in a small ball about 0, $0 \leq t \leq \tau$, and are the identity outside of B. This may be achieved by joint continuity of F_t and use of a C^∞ function f which is one on a neighborhood of 0 and is 0 outside B. Then defining

$$G_t(x) = F_\tau(x) \qquad \text{where} \quad \tau = \int_0^t f(F_s(x))ds, \tag{2.23}$$

it is easy to see that G_t extends to a global semiflow[†] on Z which coincides with F_t, $0 \leq t \leq \tau$ on a neighborhood of zero, and which is the identity outside B. Moreover, G_t remains a C^{k+1} semiflow. (For this to be true we required the smoothness of the norm on Z and that for $t > 0$ F_t has smoothness in t and x jointly[*]).

Now we can rescale and chop off <u>simultaneously</u> the G_t outside B as in the above proof. Since this does not affect F_t on a small neighborhood of zero, we get our result.

---

*In linear semigroup theory this corresponds to analyticity of the semigroup; it holds for the heat equation for instance. For the Navier Stokes eqations, see Sections 8,9.

†See Renz [1] for further details.

(2.8) Exercise. (nonuniqueness of the center manifold for flows). Let $X(x,y) = (-x,y^2)$. Solve the equation $\frac{d(x,y)}{dt} = X(x,y)$ and draw the integral curves.

Show that the flow of X satisfies the conditions of Theorem 2.7 with Y the y-axis. Show that the y-axis is a center manifold for the flow. Show that each integral curve in the lower half plane goes toward the origin as $t \to \infty$ and that the curve becomes parallel to the y-axis as $t \to -\infty$. Show that the curve which is the union of the positive y-axis with any integral curve in the lower half plane is a center manifold for the flow of X. (see Kelley [1], p. 149).

(2.9) Exercise. (Assumes a knowledge of linear semi-group theory).

Consider a Hilbert space H (or a "smooth" Banach space) and let A be the generator of an analytic semigroup. Let $K : H \to H$ be a C^{k+1} mapping and consider the evolution equations

$$\frac{dx}{dt} = Ax + K(x), \quad x(0) = x_0 \tag{2.24}$$

(a) Show that these define a local semiflow $F_t(x)$ on H which is C^{k+1} in (t,x) for $t > 0$. (Hint: Solve the Duhamel integral equation $x(t) = e^{At}x_0 + \int_0^t e^{(t-s)A}K(x(s))ds$ by iteration or the implicit function theorem on a suitable space of maps (references: Segal [1], Marsden [1], Robbin [1]).

(b) Assume $K(0) = 0$, $DK(0) = 0$. Show the existence of invariant manifolds for (2.24) under suitable spectral hypotheses on A by using Theorem 2.7.

SECTION 2A

SOME SPECTRAL THEORY

In this section we recall quickly some relevant results in spectral theory. For details, see Rudin [1] or Dunford-Schwartz [1,2]. Then we go on and use these to prove Theorems 1.3 and 1.4.

Let $T: E \to E$ be a bounded linear operator on a Banach space E and let $\sigma(T)$ denote its spectrum, $\sigma(T) = \{\lambda \in \mathbb{C} | T - \lambda I$ is not invertible on the complexification of $E\}$. Then $\sigma(T)$ is non-empty, is compact, and for $\lambda \in \sigma(T), |\lambda| \leq ||T||$. Let $r(T)$ denote its spectral radius defined by $r(T) = \sup\{|\lambda| \; |\lambda \in \sigma(T)\}$. $r(t)$ is also given by the spectral radius formula:

$$r(T) = \lim_{n\to\infty} ||T^n||^{1/n}.$$

(The proof is analogous to the formula for the radius of convergence of a power series and can be supplied without difficulty; cf. Rudin [1, p.355].)

The following two lemmas are also not difficult and are

proven in the references given:

(2A.1) Lemma. *Let* $f(z) = \sum_0^\infty a_n z^n$ *be an entire function and define* $\tilde{f}(T) = \sum_0^\infty a_n T^n$, *then* $\sigma(\tilde{f}(T)) = f(\sigma(T))$.

(2A.2) Lemma. *Suppose* $\sigma(T) = \sigma_1 \cup \sigma_2$ *where* $d(\sigma_1,\sigma_2) > 0$, *then there are unique* T-*invariant subspaces* E_1 *and* E_2 *such that* $E = E_1 \oplus E_2$ *and if* $T_i = T|_{E_i}$, *then* $\sigma(T_i) = \sigma_i$. E_i *is called the generalized eigenspace of* σ_i.

Lemma 2A.2 is done as follows: Let γ_1 be a closed curve with σ_1 in its interior and σ_2 in its exterior, then $T_1 = \frac{1}{2\pi i}\int_{\gamma_1} \frac{dz}{zI-T}$.

Note that the eigenspace of an eigenvalue λ is not always the same as the generalized eigenspace of λ. In the finite dimensional case, the generalized eigenspace of λ is the subspace corresponding to all the Jordan blocks containing λ in the Jordan canonical form.

(2A.3) Lemma. *Let* T, σ_1, *and* σ_2 *be as in Lemma 2A.2 and assume that* $d(e^{\sigma_1},e^{\sigma_2}) > 0$. *Then for the operator* e^{tT}, *the generalized eigenspace of* e^{tT_i} *is* E_i.

Proof. Write, according to Lemma 2A.2, $E = E_1 \oplus E_2$. Thus,

$$e^{tT}(e_1,e_2) = \sum_0^\infty \frac{t^nT^n}{n!}(e_1,e_2) = \left(\sum_0^\infty \frac{t^nT^n}{n!}e_1, \frac{t^nT^n}{n!}e_2\right)$$
$$= \left(\sum_0^\infty \frac{t^nT_1^n}{n!}e_1, \frac{t^nT_2^n}{n!}e_2\right) = \left(e^{tT_1}e_1, e^{tT_2}e_2\right).$$

From this the result follows easily. □

(2A.4) Lemma. *Let* $T\colon E \to E$ *be a bounded, linear operator on a Banach space* E. *Let* r *be any number greater than* $r(T)$, *the spectral radius of* T. *Then there is a norm* $|\ |$ *on* E *equivalent to the original norm such that* $|T| \leq r$.

Proof. We know that $r(t)$ is given by $r(T) = \lim ||T^n||^{1/n}$. Therefore, $\sup\limits_{n\geq 0} \frac{||T^n||}{r^n} < \infty$. If we define $|x| = \sup\limits_{n\geq 0} \frac{||T^n(x)||}{r^n}$, we see that $|\ |$ is a norm and that $||x|| \leq |x| \leq \left(\sup\limits_{n\geq 0} \frac{||T^n||}{r^n}\right) ||x||$. Hence $|T(x)| = \sup\limits_{n\geq 0} \frac{||T^{n+1}(x)||}{r^n} = r \sup\limits_{n\geq 0} \frac{||T^{n+1}(x)||}{r^{n+1}} \leq r|x|$. □

(2A.5) *Lemma*. *Let* $A: E \to E$ *be a bounded operator on* E *and let* $r > \sigma(A)$ (i.e. *if* $\lambda \in \sigma(A)$, Re $\lambda < r$). *Then there is an equivalent norm* $|\ |$ *on* E *such that for* $t \geq 0$,

$$|e^{tA}| \leq e^{rt}.$$

Proof. Note that (see Lemma 2A.1) e^{rt} is $\geq$ spectral radius of e^{tA}; i.e. $e^{rt} \geq \lim\limits_{n\to\infty} ||e^{ntA}||^{1/n}$. Set

$$|x| = \sup_{\substack{n\geq 0\\ t\geq 0}} \frac{||e^{ntA}(x)||}{e^{rnt}}$$

and proceed as in Lemma 2A.4. □

There is an analogous lemma if $r < \sigma(A)$, giving $|e^{tA}| \geq e^{rt}$. With this machinery we now turn to Theorems 1.3 and 1.4 of Section 1:

(1.3) *Theorem*. *Let* $T: E \to E$ *be a bounded linear operator*. *Let* $\sigma(T) \subset \{z | \text{Re } z < 0\}$ (*resp*. $\sigma(T) \supset \{z | \text{Re } z > 0\}$), *then the origin is an attracting* (*resp. repelling*) *fixed point for the flow* $\phi_t = e^{tT}$ of T.

Proof. This is immediate from Lemma 2A.5 for if $\sigma(T) \supset \{z \mid \text{Re } z < 0\}$, there is an $r < 0$ with $\sigma(T) < r$, as $\sigma(T)$ is compact. Thus $|e^{tA}| \leq e^{rt} \to 0$ as $t \to +\infty$. □

Next we prove the first part of Theorem 1.4 from Section 1.

(1.4) Theorem. *Let X be a C^1 vector field on a Banach manifold P and be such that $X(p_0) = 0$. If $\sigma(dX(p_0)) \subset \{z \in \mathbb{C} \mid \text{Re } z < 0\}$, then p_0 is asymptotically stable.*

Proof. We can assume that P is a Banach space E and that $p_0 = 0$. As above, renorm E and find $\varepsilon > 0$ such that $||e^{tA}|| \leq e^{-\varepsilon t}$, where $A = dX(0)$.

From the local existence theory of ordinary differential equations*, there is a r-ball about 0 for which the time of existence is uniform if the initial condition x_0 lies in this ball. Let

$$R(x) = X(x) - DX(0)\cdot x.$$

Find $r_2 \leq r_1$ such that $||x|| \leq r_2$ implies $||R(x)|| \leq \alpha||x||$ where $\alpha = \varepsilon/2$.

Let B be the open $r_2/2$ ball about 0. We shall show that if $x_0 \in B$, then the integral curve starting at x_0 remains in B and $\to 0$ exponentially as $t \to +\infty$. This will prove the result.

Let $x(t)$ be an integral curve of X starting at x_0. Suppose $x(t)$ remains in B for $0 \leq t < T$. Then noting that

---

*See for instance, Hale [1], Hartman [1], Lang [1], etc.

$$x(t) = x_0 + \int_0^t X(x(s))\, ds$$

$$= x_0 + \int_0^t A(x(s)) + R(x(s))\, ds$$

gives (the Duhamel formula; Exercise 2.9),

$$x(t) = e^{tA}x_0 + \int_0^t e^{(t-s)A}R(x(s))\, ds$$

and so

$$||x(t)|| \leq e^{-t\varepsilon}x_0 + \alpha \int_0^t e^{-(t-s)\varepsilon}||x(s)||ds.$$

Thus, letting $f(t) = e^{t\varepsilon}||x(t)||$,

$$f(t) \leq x_0 + \alpha \int_0^t f(s)\, ds$$

and so

$$f(t) \leq ||x_0||e^{\alpha t}.$$

(This elementary inequality is usually called Gronwall's inequality cf. Hartman [1].)

Thus

$$||x(t)|| \leq ||x_0||e^{(\alpha-\varepsilon)t} = ||x_0||e^{-\varepsilon t/2}.$$

Hence $x(t) \in B$, $0 \leq t < T$ so $x(t)$ may be indefinitely extended in t and the above estimate holds. It shows $x(t) \to 0$ as $t \to +\infty$. □

The instability part of Theorem 1.4 requires use of the invariant manifold theorems, splitting the spectrum into two parts in the left and right half planes. The above analysis shows that on the space corresponding to the spectrum in the right half plane, p_0 is repelling, so is unstable.

(2A.6) Remarks. Theorem 1.3 is also true by the same

proof if T is an unbounded operator, provided we know $\sigma(e^{tT}) = e^{t\sigma(T)}$ which usually holds for decent T, (e.g.: if the spectrum is discrete) but need not always be true (See Hille-Phillips [1]). This remark is important for applications to partial differential equations. Theorem 1.4's proof would require $R(x) = o(||x||)$ which is unrealistic for partial differential equations. However, the following holds:

Assume $\sigma(DF_t(0)) = e^{t\sigma(DX(0))}$ and we have a C^0 flow F_t and each F_t is C^1 with derivative strongly continuous in t (See Section 8A), then the conclusion of 1.4 is true.

This can be proved as follows: in the notation above, we have, by Taylor's theorem:

$$F_t(x) - 0 = F_t(x) - F_t(0) = DF_t(0)\cdot x + R(t,x)$$

where $R(t,x)$ is the remainder. Now we will have

$$||R(t,x)|| = 0(||x||)e^{-\varepsilon t},$$

as long as x remains in a small neighborhood of 0; this is because $||DF_t(0)|| \leq e^{-2\varepsilon t}$ and hence $||DF_t(x)|| \leq e^{-\varepsilon t}$ for some $\varepsilon > 0$ and x in some neighborhood of $0 \cdots$ remember $R(t,x) = \int_0^1 \{DF_t(sx)\cdot x - DF_t(0)\cdot x\}\, ds$. Therefore, arguing as in Theorem 1.4, we can conclude that x remains close to 0 and $F_t(x) \to 0$ exponentially as $t \to \infty$. □

(2A.7) Exercise. Let E be a Banach space and $F: E \to E$ a C^1 map with $F(0) = 0$ and the spectrum of $DF(0)$ inside the unit circle. Show that there is a neighborhood U of 0 such that if $x \in U$, $F^{(n)}(x) \to 0$ as $n \to \infty$; i.e. 0 is a stable point of F.

SECTION 2B

THE POINCARÉ MAP

We begin by recalling the definition of the Poincaré map. In doing this, one has to prove that the mapping exists and is differentiable. In fact, one can do this in the context of C^0 flows $F_t(x)$ such that for each t, F_t is C^k, as was the case for the center manifold theorem for flows, but here with the additional assumption that $F_t(x)$ is smooth in t as well for $t > 0$. Again, this is the appropriate hypothesis needed so that the results will be applicable to partial differential equations. However, let us stick with the ordinary differential equation case where F_t is the flow of a C^k vector field X at first.

First of all we recall that a closed orbit γ of a flow F_t on a manifold M is a non-constant integral curve $\gamma(t)$ of X such that $\gamma(t+\tau) = \gamma(t)$ for all $t \in \mathbb{R}$ and some $\tau > 0$. The least such τ is the period of γ. The image of γ is clearly diffeomorphic to a circle.

(2B.1) Definition. Let γ be a closed orbit, let

$m \in \gamma$, say $m = \gamma(0)$ and let S be a local transversal section; i.e. a submanifold of codimension one transverse to γ(i.e. $\gamma'(0)$ is not tangent to S). Let $\mathscr{D} \subset M \times \mathbb{R}$ be the domain (assumed open) on which the flow is defined.

A Poincaré map of γ is a mapping $P: W_0 \to W_1$ where:

(PM 1) $W_0, W_1 \subset S$ are open neighborhoods of $m \in S$, and P is a C^k diffeomorphism;

(PM 2) there is a function $\delta: W_0 \to \mathbb{R}$ such that for all $x \in W_0$, $(x, \tau-\delta(x)) \in \mathscr{D}$ and $P(x) = F(x, \tau-\delta(x))$; and finally,

(PM 3) if $t \in (0, \tau-\delta(x))$, then $F(x,t) \notin W_0$ (see Figure 2B.1).

(2B.2) *Theorem* (*Existence and uniqueness of Poincaré maps*).

(i) *If* X *is a* C^k *vector field on* M, *and* γ *is a closed orbit of* X, *then there exists a Poincaré map of* γ.

(ii) *If* $P: W_0 \to W_1$ *is a Poincaré map of* γ (*in a local transversal section* S *at* $m \in \gamma$) *and* P' *also* (*in* S' *at* $m' \in \gamma$), *then* P *and* P' *are locally conjugate*. *That is, there are open neighborhoods* W_2 *of* $m \in S$, W_2' *of* $m' \in S'$, *and a* C^k *diffeomorphism* $H: W_2 \to W_2'$, *such that* $W_2 \subset W_0 \cap W_1$, $W_2' \subset W_0'$ *and the diagram*

$$\begin{array}{ccc} \Theta^{-1}(W_2) \cap W_2 & \xrightarrow{\ \Theta\ } & W_2 \cap \Theta W_2 \\ \Big\downarrow H & & \Big\downarrow H \\ W_2' & \xrightarrow{\ \Theta'\ } & S' \end{array}$$

commutes.

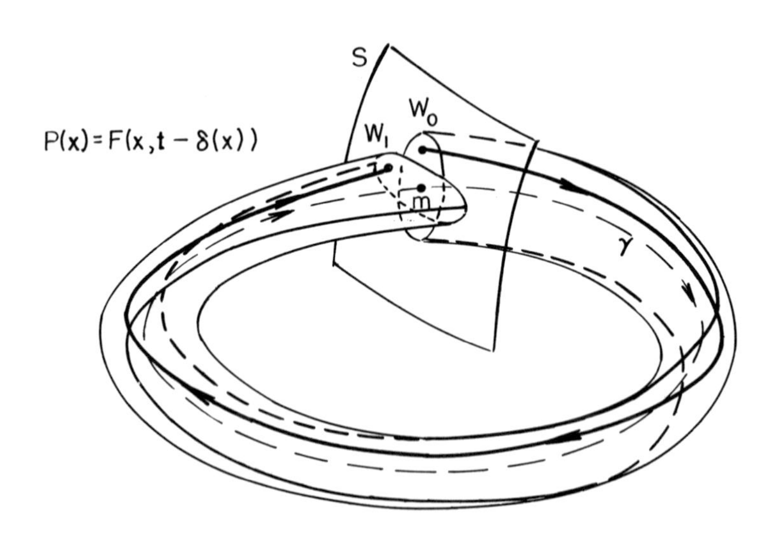

Figure 2B.1

Here is the idea of the proof of existence of P (for further details see Abraham-Marsden [1]). Choose S arbitrarily. By continuity, F_τ is a homeomorphism of a neighborhood U_0 of m to another neighborhood U_2 of m. By assumption, $F_{\tau+t}(x)$ is t-differentiable at $t = 0$ and is transverse to S at $x = m$ and hence also in a neighborhood of m. Therefore, there is a unique number $\delta(x)$ near zero such that $F_{\tau-\delta(x)}(x) \in S$. This is $P(x)$, and by construction P will be as differentiable as F is. The derivative of P at m is seen to be just the projection of the derivative of F_τ on T_mS (this is done below). Hence if F_τ is a diffeomorphism, P will be as well. □

For partial differential equations, F_t is often just a semi-flow i.e. defined for $t \geq 0$ (See Section 8A). In particular, this means F_t is not generally a diffeomorphism.

(For instance if F_t is the flow of the heat equation on L_2, F_t is not surjective). For the Poincaré mapping this means that we will have a P (if $F_t(x)$ is differentiable in t, x for $t > 0$), but P is just a <u>map</u>, not necessarily a diffeomorphism. This is one of the technical reasons why it is important to have the center manifold theorem for <u>maps</u> and not just diffeomorphisms.

As we outlined in Section 1, there is a Hopf theorem for <u>diffeomorphisms</u> and this is to be eventually applied to the Poincaré map P to deduce the existence of an invariant circle for P and hence an invariant torus for F_t.

For partial differential equations, this seems like a dilemma since P need not be a diffeomorphism. However, this can be overcome by a trick: first apply the center manifold theorem to reduce everything to finite dimensions--this does not require diffeomorphisms; then, as proved in Section 8A, (see 8A.9) in <u>finite</u> dimensions F_t and hence P will <u>automatically</u> become (local) diffeomorphisms. Thus the dilemma is only apparent.

Let us now prove the fundamental results concerning the derivative of P so we can relate the spectrum of P to that of F_t. It suffices to do the computation in a Banach space E, and we can let $m = 0$, the origin of E.

We begin by calculating $dP(0)$ in terms of F_t.

(2B.3) <u>Lemma</u>. <u>Let</u> $F_t\colon E \to E$ <u>be a</u> C^1 <u>flow on a Banach space</u>. <u>Let</u> 0 <u>be on a closed orbit</u> γ <u>with period</u> $\tau \neq 0$. <u>Let</u> $\left.\frac{\partial F_t}{\partial t}(0)\right|_{t=0} = v$. <u>Let</u> V <u>be the subspace generated by</u> v <u>and let</u> F <u>be a complementary subspace, so that</u> $E = F \oplus V$ <u>and</u> $F_t(x,y) = (F_t^1(x,y), F_t^2(x,y))$. <u>Let</u> $P\colon F \to F$

be the Poincaré map associated with γ *at the point* 0. *Then* $dP(0) = d_1F_\tau^2(0)$.

Proof. Let $\tau(x)$ be the time at $P(x)$ ($\tau-\delta(x)$ in the above notation). Then by definition of P,

$$P(x) = F^1_{\tau(x)}(x,0)$$

so

$$dP(x) = \frac{\partial F^1_{(x)}}{\partial t}(x,0)d\tau(x) + d_1F^1_{\tau(x)}(x,0).$$

Letting $(x,0) = 0$, we get

$$dP(0) = \frac{\partial F^1}{\partial t}(0)d\tau(0) + d_1F^1_\tau(0).$$

However $\frac{\partial F_\tau}{\partial t}(0) = \left(\frac{\partial F^1_\tau}{\partial t}(0), \frac{\partial F^2_\tau}{\partial t}(0)\right) = (0,V)$ by construction, so $\frac{\partial F^1_\tau}{\partial t}(0) = 0$. Thus, $dP(0) = d_1F^1_\tau(0)$. □

(2B.4) *Lemma.* $d_2F_\tau(0,0)V = V$.

Proof. $\frac{dF_{\tau+s}}{ds}(0,0)\Big|_{s=0} = \frac{dF_t}{dt}(0,0)\Big|_{t=\tau} = V$

$$\frac{dF_{\tau+s}}{ds}(0,0)\Big|_{s=0} = \frac{dF_\tau \circ F_s}{ds}(0,0)\Big|_{s=0} = dF_\tau(0,0)\frac{dF_s}{ds}(0,0)\Big|_{s=0}$$

$$= dF_\tau(0,0)(0,V) = (d_1F_\tau(0,0)\cdot 0 + d_2F_\tau(0,0)V) = d_2F_\tau(0,0)V.$$

So, $d_2F_\tau(0,0)V = V$. □

(2B.5) *Lemma.* $\sigma(dF_\tau(0,0)) = \sigma(P(0)) \cup \{1\}$. (*This is true of the point spectrum, too*).

Proof. The matrix of $dF_\tau(0,0)$ is

$$\begin{pmatrix} dP(0) & 0 \\ * & 1 \end{pmatrix}$$

where * indicates some unspecified matrix entry.

Let $\lambda \in \mathbb{C}$. Then $\lambda \in \sigma(dF_\tau(0,0))$ iff $dF_\tau(0,0) - \lambda I$ is not

1-1 or is not onto. But

$$(dF_\tau(0,0) - \lambda I)\binom{\alpha}{\beta} = (dP(0)\alpha,\ *(\alpha) + \beta) + (-\lambda\alpha, -\lambda\beta)$$
$$= (dP(0)\alpha - \lambda\alpha,\ *(\alpha) + \beta(1 - \lambda)). \tag{2B.1}$$

Clearly, $1 \in \sigma(dF_\tau(0,0))$. Assume $\lambda \neq 1$ and $\lambda \in \sigma(dF_\tau(0,0))$. Then either, there exists (α,β) such that the expression (2B.1) is zero or the map is not onto. The former implies $\lambda \in \sigma(dP(0))$. Assume the latter. Since $\lambda \neq 1$, for any α, we can choose β so that the second component is onto. Therefore, $\lambda \in \sigma(dP(0))$. On the other hand, let $1 \neq \lambda \in \sigma(dP(0))$. If, $dP(0) - \lambda I$ is not onto, then clearly neither is $dF_\tau(0,0) - \lambda I$. Suppose there exists α such that $dP(0)\alpha - \lambda\alpha = 0$. Then choosing $\beta = \frac{*(\alpha)}{\lambda - 1}$, we see that $\lambda \in \alpha(dF_\tau(0,0))$. □

Consult Abraham-Marsden [1] Abraham-Robbin[1] or Hartman [1, Section IV. 6, IX.10] for additional details on this and and the associated Floquet theory.

One of the most basic uses of the Poincaré map is in the proof of the following.

(2B.6) Theorem. *Let γ be a closed orbit for F_t with period τ. Let $m \in \gamma$ and suppose $dF_\tau(m)$ has spectrum inside the unit circle except for one point 1 on the unit circle. Then γ is an attracting (stable) closed orbit.*

Proof. By Lemma 2B.5, the condition on the spectrum means that the spectrum of the derivative of the Poincaré map P at m is inside the unit circle. Hence from 2A.7 m is an attracting fixed pojnt for P. It follows from the construction of P that γ is attracting for F_t. □

(2B.7) Exercise.

(a) Give the details in the last step of the above proof.

(b) If, in 2B.6, P has an attracting invariant circle, give the details of the proof that this yields an attracting invariant 2 torus for the flow.

SECTION 3
THE HOPF BIFURCATION THEOREM IN $\mathbb{R}^2$ AND IN $\mathbb{R}^n$

The center manifold theorem is used to reduce bifurcation problems to finite dimensional ones as follows.

Consider a one parameter family of maps $\psi^\mu: Z \to Z$ on a Banach space Z, where $\mu \in R$ or an interval in R containing 0. Assume $(\mu,x) \mapsto \Psi^\mu(x)$ is C^{k+1} and $\Psi^\mu(0) = 0$ for all μ. Assume that for $\mu < 0$ the spectrum of $D\Psi^\mu(0)$ is strictly inside the unit circle, for $\mu = 0$ the spectrum splits in two pieces as in the center manifold theorem and for $\mu > 0$ the spectrum has two pieces, one inside and one outside the unit circle. See Figure 3.1.

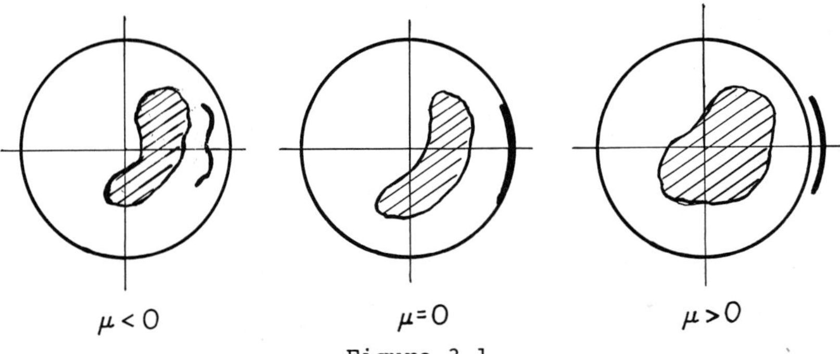

Figure 3.1

Consider $\Psi: R \times Z \to R \times Z$, $(\mu,x) \mapsto (\mu,\Psi^{\mu}(x))$. The derivative at zero is

$$D\Psi(0,0)(\nu,h) = \left(\nu, \left.\frac{\partial\Psi^{\mu}}{\partial\mu}(x)\nu\right|_{\substack{\mu=0\\x=0}} + \left.D_x\Psi^{\mu}(x)\cdot h\right|_{\substack{\mu=0\\x=0}}\right)$$

$$= (\nu, D_x\Psi^{\mu}(0)\cdot h)$$

(since $\Psi^{\mu}(0) = 0$ for all μ). Thus, the spectrum of $D\Psi(0,0)$ consists of the spectrum of $D_x\Psi^{\mu}(0)$ plus the point one. Therefore, we can apply the center manifold theorem to Ψ to produce an invariant manifold in $R \times Z$.

The μ = constant slices of this invariant manifold produces a one parameter family of invariant manifolds for Ψ^{μ}. These manifolds have the same dimensionality as the eigenspace of the piece of the spectrum crossing the unit circle, and this is often finite dimensional.

There is an entirely analogous reduction possible for flows using the center manifold theorem for flows.

It should be cautioned that the center manifold, while containing all the local recurrence, is not globally invariant nor need it be attracting in the strict sense. However, if care is exercised, and if a pair of eigenvalues crosses the unit circle or the imaginary axis if we are thinking in terms of the vector field, then we are in the two dimensional case. (We remark--see Section 8A that a semiflow of C^{k+1} maps on a finite dimensional space automatically has a C^k generating vector field, so that <u>after</u> the reduction is made we may usually assume that the generator of the flow is smooth.)

Hence we shall next examine, in detail, the finite dimensional case. (Details of the above reduction process are given in Section 4.)

First, we consider the two dimensional case. (The n-dimensional case is treated in Theorem 3.15 below.)

The Hopf Theorem in R^2

The following theorem is essentially due to Andronov (1930) and Hopf (1942) and was suggested in the work of Poincaré (1892)[+]

(3.1) *Theorem. Let* X_μ *be a* C^k $(k \geq 4)$ *vector field on* R^2 *such that* $X_\mu(0) = 0$ *for all* μ *and* $X = (X_\mu, 0)$ *is also* C^k. *Let* $dX_\mu(0,0)$ *have two distinct, complex conjugate eigenvalues* $\lambda(\mu)$ *and* $\overline{\lambda(\mu)}$ *such that for* $\mu > 0$ $\text{Re}\,\lambda(\mu) > 0$. *Also, let* $\left.\frac{d(\text{Re}\,\lambda(\mu))}{d\mu}\right|_{\mu=0} > 0$. *Then*

(A) *there is a* C^{k-2}[*] *function* $\mu\colon (-\varepsilon, \varepsilon) \to R$ *such that* $(x_1, 0, \mu(x_1))$ *is on a closed orbit of period* $\approx$ $2\pi/|\lambda(0)|$ *and radius growing like* $\sqrt{\mu}$ *of* X *for* $x_1 \neq 0$ *and such that* $\mu(0) = 0$.

(B) *There is a neighborhood* U *of* $(0,0,0)$ *in* R^3 *such that any closed orbit in* U *is one of those above. Furthermore, if* 0 *is a "vague attractor" for* X_0, *then*

(C) $\mu(x_1) > 0$ *for all* $x_1 \neq 0$ *and the orbits are attracting.*

The meaning of "vague attractor" will be spelled out as we go along and the detailed calculations involved in this condition are worked out in Section 4 (see also Section 5A). In Section 3A it is shown that "vague attractor" can be

---

[+] The present version of Theorem 3.1 is due to Ruelle and Takens [1]. The n dimensional case is due to Hopf. See Section 5A for comparisons.

[*] If X is analytic, then μ will also be analytic (Hopf [1]; see Section 5).

weakened to "attractor" in the usual Liapunov sense. In any case, this condition is usually not obvious in examples and will be discussed extensively below.

Our proof follows Ruelle-Takens [1], with the details included. At the end of the section we shall discuss what happens if $d(\mathrm{Re}\,\lambda(\mu))/d\mu = 0$ (see Section 3A).

<u>Proof</u>. The essence of the proof is an application of the implicit function theorem. We show that for small μ, there is a C^{k-1} function which takes the point $(x_1,0,\mu)$ to the first intersection $(P(x_1,\mu),0,\mu)$ of the orbit of $(x_1,0,\mu)$ under the flow of X with the x_1-axis such that x_1 and $P(x_1,\mu)$ have the same sign (Figure 3.2). Let $V(x_1,\mu) = P(x_1,\mu) - x_1$. V is a displacement function.
We use the implicit function theorem to get

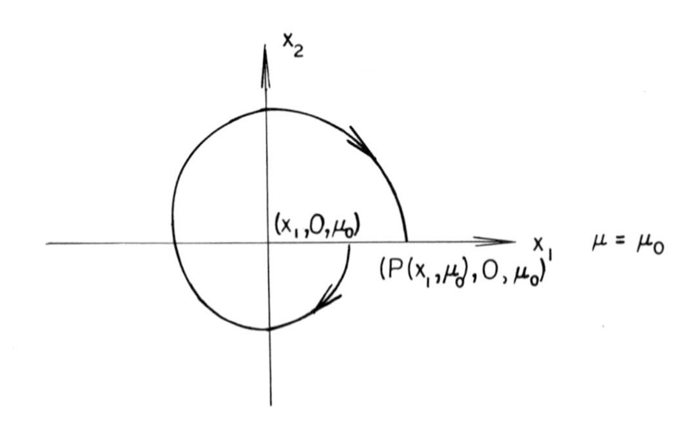

Figure 3.2

a curve $(x_1,0,\mu(x_1))$ of zeros of V, i.e., a curve of closed orbits of the flow of X. The map $(x_1,0) \mapsto (P(x_1,\mu(x_1)),0)$

is the Poincaré map associated with the closed orbit through $(x_1, \mu(x_1))$. We use standard results about Poincaré maps to find conditions under which the orbits are attracting. The uniqueness of the orbits is essentially the uniqueness of the implicitly defined function in the implicit function theorem. (Proving uniqueness of the closed orbits is more complicated in higher dimensions, see Section 5 and Section 5A, page 198.)

<u>Step 1</u>: By a μ-dependent change of basis on R^2, we may assume that $dX_\mu(0,0) = \begin{pmatrix} \mathrm{Re}\,\lambda(\mu) & \mathrm{Im}\,\lambda(\mu) \\ -\mathrm{Im}\,\lambda(\mu) & \mathrm{Re}\,\lambda(\mu) \end{pmatrix}$ where $\lambda(\mu)$ is chosen so that $\mathrm{Im}\,\lambda(\mu) > 0$. In the new coordinates, X_μ will have continuous derivatives up to order k except that $\partial^k X_\mu/\partial\mu^k$ may not exist. Furthermore, for each μ, the x_1-axis is invariant under the change of basis. (That is, the new x_1-axis is the same as the old one, and we are only changing the x_2-axis). Let us now note a few simple lemmas:

(3.2) <u>Lemma</u>. <u>Let</u> $\mu \to \begin{pmatrix} a_{11}(\mu) & a_{12}(\mu) \\ a_{21}(\mu) & a_{22}(\mu) \end{pmatrix}$ <u>be a</u> C^k <u>function from</u> $U \subseteq R \to R^4$. <u>Let the matrix have two distinct eigenvalues for all</u> $\mu \in [a,b] \subset U$. <u>Then the eigenvalues are</u> C^k <u>functions from</u> $(a,b) \subseteq R \to \mathbb{C}$.

<u>Proof</u>. By the quadratic formula, the eigenvalues are $\dfrac{a_{11} + a_{22} \pm \sqrt{(a_{11}+a_{22})^2 - 4a_{12}a_{21}}}{2}$. By assumption $(a_{11}+a_{22})^2 - 4a_{12}a_{21}$ is bounded away from zero on (a,b), so the eigenvalues are C^k functions of μ on this interval. □

(3.3) <u>Lemma</u>. <u>Let</u> $T: \mathbb{C}^2 \to \mathbb{C}^2$ <u>be a linear transformation that is real on real vectors and has no real eigenvalues</u>.

Let $v_1 + iv_2$ be an eigenvector with eigenvalue λ. There is an eigenvector $\begin{pmatrix}1\\0\end{pmatrix} + i\begin{pmatrix}\alpha\\\beta\end{pmatrix}$ which has the same eigenvalue.

Proof. Any complex multiple of $v_1 + iv_2$ is an eigenvector of T with eigenvalue λ. Thus, it is sufficient to show that there is a $z = x + iy$ such that

$$(x+iy)\left[\begin{pmatrix}v_{11}\\v_{21}\end{pmatrix} + i\begin{pmatrix}v_{12}\\v_{22}\end{pmatrix}\right] = \begin{pmatrix}1\\0\end{pmatrix} + i\begin{pmatrix}\alpha\\\beta\end{pmatrix}.$$ This is equivalent to solving the pair of equations:

$$xv_{11} - yv_{12} = 1$$
$$xv_{21} - yv_{22} = 0$$

i.e.,

$$\begin{pmatrix}v_{11} & -v_{12}\\v_{21} & -v_{22}\end{pmatrix}\begin{pmatrix}x\\y\end{pmatrix} = \begin{pmatrix}1\\0\end{pmatrix}.$$

The columns of this matrix are independent over $\mathbb{R}$ since if $v_2 = cv_1$, then $v_1 + iv_2 = (1+ic)v_1$. Therefore, $v_1 = (1+iv)^{-1}(v_1+iv_2)$ is a real eigenvector, which cannot be. Thus, the equation can be solved. □

(3.4) Lemma. Let T be as in the previous lemma. Then $Tv_1 = \text{Re}\,\lambda v_1 - \text{Im}\,\lambda v_2$ and $Tv_2 = \text{Im}\,\lambda v_1 + \text{Re}\,\lambda v_2$.

Proof. $Tv_1 = \text{Re}\,T(v_1+iv_2)$ because T is real. $Tv_1 = \text{Re}[\lambda(v_1+iv_2)] = \text{Re}\,\lambda v_1 - \text{Im}\,\lambda v_2$. $Tv_2 = \text{Im}[\lambda(v_1+iv_2)] = \text{Im}\,\lambda v_1 + \text{Re}\,\lambda v_2$. □

Using the preceding lemmas, we see that if $\begin{pmatrix}1\\0\end{pmatrix} + i\begin{pmatrix}\alpha(\mu)\\\beta(\mu)\end{pmatrix}$ is an eigenvector of $dX_\mu(0,0)$ with eigenvalue $\lambda(\mu)$, then $\begin{pmatrix}1\\0\end{pmatrix}$ and $\begin{pmatrix}\alpha(\mu)\\\beta(\mu)\end{pmatrix}$ are independent vectors such that the matrix of $dX_\mu(0,0)$ with respect to $\begin{pmatrix}1\\0\end{pmatrix}$ and $\begin{pmatrix}\alpha(\mu)\\\beta(\mu)\end{pmatrix}$ is

$\begin{pmatrix}\text{Re }\lambda(\mu) & \text{Im }\lambda(\mu)\\-\text{Im }\lambda(\mu) & \text{Re }\lambda(\mu)\end{pmatrix}$. We now show that the vector $\begin{pmatrix}\alpha(\mu)\\\beta(\mu)\end{pmatrix}$ is

a C^{k-1} function of μ. Let $dX_\mu(0,0) = \begin{bmatrix}a_{11}(\mu) & a_{12}(\mu)\\a_{21}(\mu) & a_{22}(\mu)\end{bmatrix}$.

We solve the equation $\begin{pmatrix}a_{11}(\mu) & a_{12}(\mu)\\a_{22}(\mu) & a_{22}(\mu)\end{pmatrix}\begin{pmatrix}1+i\alpha(\mu)\\i\beta(\mu)\end{pmatrix} =$

$= \lambda(\mu)\begin{pmatrix}1+i\alpha(\mu)\\i\beta(\mu)\end{pmatrix}$. From this we obtain the equations

$a_{11}(\mu) = \text{Re }\lambda(\mu) - \text{Im }\lambda(\mu)\alpha(\mu)$, $a_{21}(\mu) = -\text{Im }\lambda(\mu)\beta(\mu)$.

Therefore,

$$\alpha(\mu) = \frac{\text{Re }\lambda(\mu) - a_{11}(\mu)}{\text{Im }\lambda(\mu)}, \qquad \beta(\mu) = \frac{-a_{21}(\mu)}{\text{Im }\lambda(\mu)}.$$

Because the change of coordinates is linear for each μ and because α and β are C^{k-1} functions of μ, in the new coordinates X will have continuous k^{th} partials except that $\frac{\partial^k X}{\partial_\mu k}$ may not exist. In particular, $\frac{\partial X}{\partial x_1}$ and $\frac{\partial X}{\partial x_2}$ are C^{k-1} functions in the new coordinates. From now on, we will assume that the coordinate change has been made, i.e., that

$$dX_\mu(0,0) = \begin{pmatrix}\text{Re }\lambda(\mu) & \text{Im }\lambda(\mu)\\-\text{Im }\lambda(\mu) & \text{Re }\lambda(\mu)\end{pmatrix}.$$

Step 2: There is a unique C^{k-1} vector field $\tilde{X}_\mu$ on R^2 such that $\psi_*\tilde{X} = X_\mu$, where $\psi: R^2 \to R^2$ is the polar coordinate map $\psi(r,\theta) = (r\cos\theta, r\sin\theta)$, and ψ_* is the differential of ψ.

Let $\tilde{X} = \tilde{X}_r \frac{\partial}{\partial r} + \tilde{X}_\theta \frac{\partial}{\partial \theta}$ be any vector field* on R^2.

*Note that $\tilde{X}_\theta$ is the "angular velocity" of $\tilde{X}$, and not the component of $\tilde{X}$ along a unit vector e_θ in the $\partial/\partial\theta$ direction which is what $\tilde{X}_\theta$ often stands for.

Then

$$\psi_*(\tilde{X}) = \begin{pmatrix} \cos\theta & -r\sin\theta \\ \sin\theta & r\cos\theta \end{pmatrix} \begin{pmatrix} \tilde{X}_r \\ \tilde{X}_\theta \end{pmatrix}.$$

$\psi_*(\tilde{X}_\mu) = X_\mu$ implies

$$\begin{pmatrix} \tilde{X}_{\mu r} \\ \tilde{X}_{\mu\theta} \end{pmatrix} = \begin{pmatrix} \cos\theta & \sin\theta \\ \dfrac{-\sin\theta}{r} & \dfrac{\cos\theta}{r} \end{pmatrix} \begin{pmatrix} X_{\mu 1} \\ X_{\mu 2} \end{pmatrix}, \quad r \neq 0.$$

Thus, if the vector field $\tilde{X}_\mu$ can be extended to be C^{k-1} on all of R^2, it is clearly unique. $\tilde{X}_{\mu r} = \cos\theta\, X_{\mu 1} + \sin\theta\, X_{\mu 2}$, which is C^{k-1} for all (r,θ). Consider

$$\tilde{X}_{\mu\theta}(r,\theta) = \frac{-\sin\theta}{r}\tilde{X}_{\mu 1}(r\cos\theta, r\sin\theta)$$

$$+ \frac{\cos\theta}{r}\tilde{X}_{\mu 2}(r\cos\theta, r\sin\theta), \qquad r \neq 0.$$

Then

$$\lim_{r\to 0}\tilde{X}_{\mu\theta}(r,\theta) = -\sin\theta \lim_{r\to 0}\frac{X_{\mu 1}(r\cos\theta, r\sin\theta) - X_{\mu 1}(0,0)}{r}$$

$$+ \cos\theta \lim_{r\to 0}\frac{X_{\mu 2}(r\cos\theta, r\sin\theta) - X_{\mu 2}(0,0)}{r}$$

because $X_\mu(0,0) = 0$. Thus

$$\lim_{r\to 0}\tilde{X}_{\mu\theta}(r,\theta) = (-\sin\theta)dX_{\mu 1}(0,0)(\cos\theta, \sin\theta)$$

$$+ (\cos\theta)dX_{\mu 2}(0,0)(\cos\theta, \sin\theta)$$

$$= (-\sin\theta)(\mathrm{Re}\,\lambda(\mu)\cos\theta + \mathrm{Im}\,\lambda(\mu)\sin\theta)$$

$$+ (\cos\theta)(-\mathrm{Im}\,\lambda(\mu)\cos\theta + \mathrm{Re}\,\lambda(\mu)\sin\theta)$$

$$= -\mathrm{Im}\,\lambda(\mu).$$

We therefore define

$$\tilde{X}_\mu(r,\theta) = \begin{cases} (\cos\theta\, X_{\mu 1}(r\cos\theta,\, r\sin\theta) \\ \qquad + \sin\theta\, X_{\mu 2}(r\cos\theta,\, r\sin\theta))\frac{\partial}{\partial r} \\ + (\frac{-\sin\theta}{r} X_{\mu 2}(r\cos\theta,\, r\sin\theta) \\ \qquad + \frac{\cos\theta}{r} X_{\mu 2}(r\cos\theta,\, r\sin\theta))\frac{\partial}{\partial\theta}, \quad r \neq 0 \\ -\operatorname{Im}\lambda(\mu)\frac{\partial}{\partial\theta}, \quad r = 0 \end{cases}$$

$$\tilde{X}_{\mu\theta}(r,\theta) = \begin{cases} \frac{-\sin\theta}{r} X_{\mu 1}(r\cos\theta,\, r\sin\theta) + \\ \qquad + \frac{\cos\theta}{r} X_{\mu 2}(r\cos\theta,\, r\sin\theta), \quad r \neq 0 \\ -\operatorname{Im}\lambda(\mu), \quad r = 0 \end{cases}$$

To see that $\tilde{X}_{\mu\theta}(r,\theta)$ is C^{k-1}, we show that the functions $\frac{1}{r} X_{\mu 1}(r\cos\theta,\, r\sin\theta)$ and $\frac{1}{r} X_{\mu 2}(r\cos\theta,\, r\sin\theta)$ are C^{k-1} when extended as above.

(3.5) Lemma. Let $A\colon R^2 \to R$ be C^k. Then

$$A(x,y) - A(0,0) = \int_0^1 \frac{\partial A(tx,ty)}{\partial x} x + \frac{A(tx,ty)}{\partial y} y\, dt.$$ Let

$$A_1(x,y) = \int_0^1 \frac{\partial A(tx,ty)}{\partial x} dt \quad \text{and} \quad A_2(x,y) = \int_0^1 \frac{\partial A(tx,ty)}{\partial y} dt.$$

$$A_1(0,0) = \frac{\partial A(0,0)}{\partial x} \quad \text{and} \quad A_2(0,0) = \frac{\partial A(0,0)}{\partial y}.$$

Proof. The first statement is true by Taylor's theorem and the second statement is easy to prove by induction. By the lemma, $\frac{1}{r} X_{\mu j}(r\cos\theta,\, rt\sin\theta) = \cos\theta \int_0^1 \frac{\partial X_{\mu j}}{\partial x}(rt\cos\theta,\, rt\sin\theta)\, dt$ $+ \sin\theta \int_0^1 \frac{\partial X_{\mu j}}{\partial y}(rt\cos\theta,\, rt\sin\theta)\, dt$ for $j = 1,2$. Since all k^{th}

partials of X are continuous except $\frac{\partial X_k}{\partial \mu^k}$, the functions under the integral sign are C^{k-1} and so the integrals are C^{k-1}, too. □

Step 3: The Poincaré Map (see Section 2B for a discussion of Poincaré maps). Let the flows of $\tilde{X}$ and X be $\tilde{\Phi}_t$ and Φ_t respectively. It is elementary to see that $\psi \circ \tilde{\Phi}_t = \Phi_t \circ \psi$. Consider the vector field X. Since $X(0,\theta) = -\text{Im}\,\lambda(\mu)\frac{\partial}{\partial\theta}$, we have $\tilde{\Phi}_{\mu t}(0,\theta) = (0,\theta - \text{Im}\,\lambda(\mu)t,\mu)$. $\tilde{\Phi}_{02\pi/|\lambda(0)|}(0,0)$ $= (0,0-|\lambda(0)|2\pi/|\lambda(0)|,0) = (0,-2\pi,0)$ (Figure 3.3).

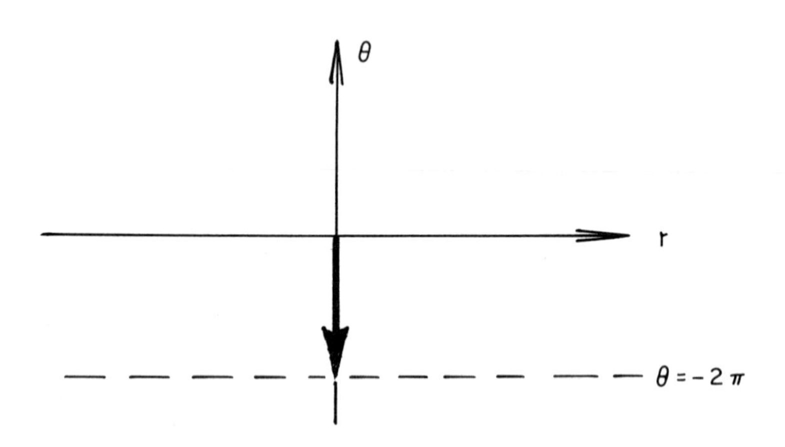

Trajectory of $(0,0,0)$ under $\tilde{\Phi}_{0t}$

Figure 3.3

Because $\tilde{X}$ is periodic with period 2π, it is a C^{k-1} vector field on a thick cylinder and the orbit of the origin is closed. We can associate a Poincaré map P with this orbit (Figure 3.4).

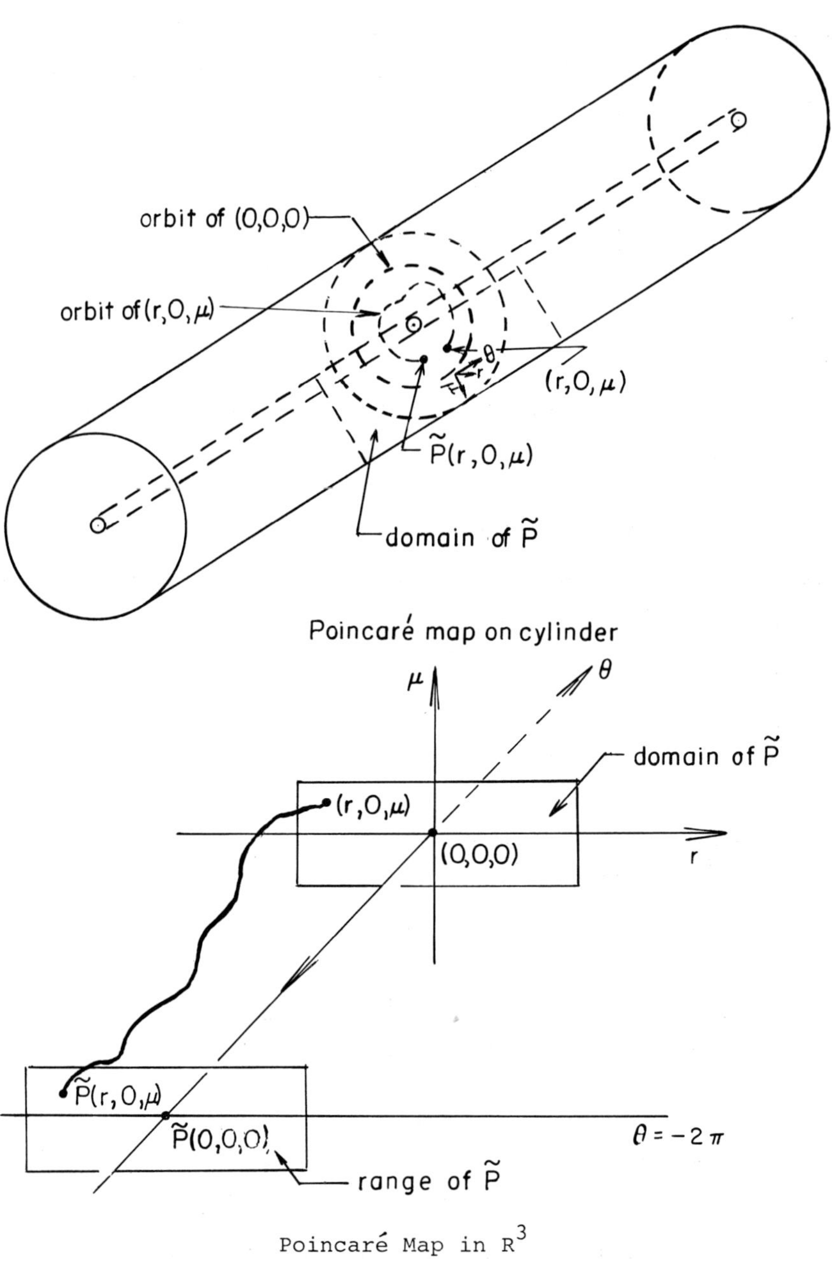

Poincaré Map in R^3

Figure 3.4

That is, there is a neighborhood $U = \{(r,0,\mu) \mid r \in (-\varepsilon,\varepsilon)$ and $\mu \in (-\varepsilon,\varepsilon)\}$ such that the map $\tilde{P}(r,0,\mu) = (\hat{P}(r,\mu),-2\pi,\mu)$, where $\hat{P}(r,\mu)$ is the r coordinate of the first intersection of the orbit of $(r,0,\mu)$ with the line $\theta = -2\pi$, is defined. This map is C^{k-1}. The map $T(r,\mu)$, which is the time t when $\tilde{\Phi}_t(r,0,\mu) = \tilde{P}(r,0,\mu)$ is also C^{k-1}. Note that under ψ, the r-axis becomes the x_1-axis. Therefore, the displacement map $(x_1,0,\mu) \mapsto (x_1+V(x_1,\mu),0,\mu)$ is defined and C^{k-1} on the neighborhood $\theta = \{(x_1,0,\mu) \mid x_1 \in (-\varepsilon,\varepsilon)$ and $\mu \in (-\varepsilon,\varepsilon)\}$ $(x_1+V(x_1,\mu),0,\mu)$ is the first intersection of the orbit of $(x_1,0,\mu)$ with the x_1-axis such that the sign of x_1 and the sign of $P(x_1,\mu) = x_1 + V(x_1,\mu)$ are the same.

(3.6) Remark. Using uniform continuity and the fact that $\tilde{\Phi}_t$ is θ-periodic, it is easy to see that there is a neighborhood $\tilde{N} = \{(r,\theta,\mu) \mid r^2 + \mu^2 < \delta\}$ such that no point of $\tilde{N}$ is a fixed point of $\tilde{\phi}_t$. Thus, the only fixed points of ϕ_t in $N = \{(x_1,x_2,\mu) \mid x_1^2 + x_2^2 + \mu^2 < \delta\}$ are the points $(0,0,\mu)$.

(3.7) Lemma. $$\left.\frac{\partial P(x_1,\mu)}{\partial x_1}\right|_{\substack{x_1=0\\ \mu=\mu}} = e^{2\pi(\mathrm{Re}\,\lambda(\mu))/\mathrm{Im}\,\lambda(\mu)}$$

Proof. Let $\Phi_{\mu t}(x_1,x_2) = (a_{\mu t}(x_1,x_2),\ b_{\mu t}(x_1,x_2))$. The flow satisfies the following equation:

$$V(x_1,\mu) = P(x_1,\mu) - x_1 = \int_0^{T(x_1,\mu)} X_{1\mu}(a_{\mu t}(x_1,0),\ b_{\mu t}(x_1,0))dt.$$

We differentiate this equation with respect to x_1 to get the desired result; the differentiation proceeds along standard lines as follows

$$\frac{V(x_1+\Delta x_1,\mu) - V(x_1,\mu)}{\Delta x_1}$$

$$= \frac{1}{\Delta x_1}\left[\int_0^{T(x_1+\Delta x_1,\mu)} X_{1\mu}(a_{\mu t}(x_1+\Delta x_1,0),\ b_{\mu t}(x_1+\Delta x_1,0))dt - \int_0^{T(x_1,\mu)} X_{1\mu}(a_{\mu t}(x_1,0),\ b_{\mu t}(x_1,0))dt\right]$$

$$= \int_0^{T(x_1,\mu)} \frac{X_{1\mu}(a_{\mu t}(x_1+\Delta x_1,0),b_{\mu t}(x_1+\Delta x_1,0)) - X_{1\mu}(a_{\mu t}(x_1,0),b_t(x_1,0))}{\Delta x_1}\,dt$$

$$+ \frac{1}{\Delta x_1}\int_{T(x_1,\mu)}^{T(x_1+\Delta x_1,\mu)} X_{1\mu}(a_{\mu t}(x_1+\Delta x_1,0),\ b_{\mu t}(x_1+\Delta x_1,0))dt.$$

From this we see that

$$\frac{\partial V(x_1,\mu)}{\partial x_1} = \int_0^{T(x_1,\mu)} \frac{\partial X_{1\mu}}{\partial x_1}(a_{\mu t}(x_1,0),\ b_{\mu t}(x_1,0))dt$$

$$+ \frac{\partial T(x_1,\mu)}{\partial x_2} X_{1\mu}(a_{\mu T(x_1,\mu)}(x_1,0),\ b_{\mu T(x_1,\mu)}(x_1,0)).$$

In case $x_1 = 0$, we can evaluate this expression. Since $a_{\mu t}(0,0) = b_{\mu t}(0,0) = 0$ and since $X_{1\mu}(0,0) = 0$, the second term on the right-hand side vanishes. Recall that $T(0,\mu) = 2\pi/\mathrm{Im}\,\lambda(\mu)$. By the chain rule, $\frac{\partial X_{1\mu}}{\partial x_1}(a_{\mu t}(0,0),\ b_{\mu t}(0,0)) = \frac{\partial X_1}{\partial a}(0,0)\,\frac{\partial a_{\mu t}}{\partial x_1}(0,0) + \frac{\partial X_{1\mu}}{\partial x_1}(0,0) + \frac{\partial X_{1\mu}}{\partial b}\,\frac{\partial b_{\mu t}}{\partial x_1}(0,0)$. Since $\frac{\partial X_{1\mu}}{\partial a}(0,0) = \mathrm{Re}\,\lambda(\mu)$ and $\frac{\partial X_{1\mu}}{\partial b}(0,0) = \mathrm{Im}\,\lambda(\mu)$ and because $(0,0)$ is a fixed point of $\Phi_{\mu t}$, we can evaluate the derivatives of the flow here.

$$d\Phi_{\mu t}(0,0) = \exp[tdX_\mu(0,0)] = \exp\left[t\begin{pmatrix} \text{Re }\lambda(\mu) & \text{Im }\lambda(\mu) \\ -\text{Im }\lambda(\mu) & \text{Re }\lambda(\mu) \end{pmatrix}\right]$$

$$\begin{pmatrix} e^{t \text{ Re } \lambda(\mu)}\cos \text{ Im } \lambda(\mu)t & e^{t \text{ Re } \lambda(\mu)}\sin \text{ Im } \lambda(\mu)t \\ -e^{t \text{ Re } \lambda(\mu)}\sin \text{ Im } \lambda(\mu)t & e^{t \text{ Re } \lambda(\mu)}\cos \text{ Im } \lambda(\mu)t \end{pmatrix}$$

Therefore, $\frac{\partial a_{\mu t}}{\partial x_1}(0,0) = e^{t, \text{Re } \lambda(\mu)}\cos \text{ Im } \lambda(\mu)t$ and $\frac{\partial b_{\mu t}}{\partial x_1}(0,0) = -e^{t \text{ Re } \lambda(\mu)t}\sin \text{ Im } \lambda(\mu)t$. So we have

$$\frac{\partial V}{\partial x_1}(0,\mu)$$

$$= \int_0^{2\pi/\text{Im } \lambda(\mu)} e^{t \text{ Re } \lambda(\mu)}(\text{Re } \lambda(\mu)\cos \text{ Im}\lambda(\mu)t - \text{Im } \lambda(\mu)\sin \text{ Im } \lambda(\mu)t)dt$$

$$= e^{2\pi(\text{Re } \lambda(\mu))/\text{Im } \lambda(\mu)} - 1. \quad \square$$

Step 4. Use of the Implicit Function Theorem to Find Closed Orbits

The most obvious way to try to find closed orbits of Φ_t is to try to find zeros of V. Since $V(0,0) = 0$, if either $\frac{\partial V}{\partial x_1}(0,0)$ or $\frac{\partial V}{\partial \mu}(0,0)$ were not equal to 0, the conditions for the implicit function theorem would be satisfies and we could have a curve of the form $(x_1(\mu),\mu)$ or $(x_1,\mu(x_1))$ such that $V = 0$ along the curve. Unfortunately, $\frac{\partial V}{\partial \mu}(0,0) = 0 = \frac{\partial V}{\partial x_1}(0,0)$. Instead of V we use the function

$$\tilde{V}(x_1,\mu) = \begin{cases} \dfrac{V(x_1,\mu)}{x_1} & x_1 \neq 0 \\ \dfrac{\partial V}{\partial x_1}(0,\mu) & x_1 = 0 \end{cases}$$

(3.8) Lemma. $\tilde{V}$ is C^{k-2}.

Proof. Recall that V is C^{k-1}. $V(x_1,\mu) = \int_0^1 \frac{\partial V}{\partial x_1}(tx_1,\mu)x_1 dt$ because $V(0,\mu) = 0$, $\frac{V(x_1,\mu)}{\partial x_1} = \int_0^1 \frac{\partial V}{\partial x_1}(tx_1,\mu)dt$, $x_1 \neq 0$. The function $\int_0^1 \frac{\partial V}{\partial x_1}(tx_1,\mu)dt$ is easily seen to be C^{k-2} by induction. □

(3.9) Lemma. $V(0,0) = 0$. $\frac{\partial \tilde{V}}{\partial \mu}(0,0) \neq 0$. Therefore, there are neighborhoods N_1 and N_2 of 0 and a unique function $\mu\colon N_1 \to N_2$ such that $\mu(0) = 0$ and such that $V(x_1,\mu(x_1)) = 0$.

Proof. $\tilde{V}(0,0) = \frac{\partial \tilde{V}}{\partial x_1}(0,0) = e^{2\pi(\mathrm{Re}\,\lambda(0))/\mathrm{Im}\,\lambda(0)} - 1 = 0$ since $\mathrm{Re}\,\lambda(0) = 0$. $\frac{\partial \tilde{V}}{\partial \mu}(0,0) = \lim_{\mu\to 0} \frac{\tilde{V}(0,\mu) - \tilde{V}(0,0)}{\mu} =$

$$\lim_{\mu\to 0} \frac{1}{\mu}\left(\frac{\partial V}{\partial x_1}(0,\mu) - \frac{\partial V}{\partial x_1}(0,0)\right) = \frac{\partial^2 V}{\partial \mu \partial x_1}(0,0) =$$

$$\frac{d}{d\mu}\left(e^{2\pi(\mathrm{Re}\,\lambda(\mu))/\mathrm{Im}\,\lambda(\mu)} - 1\right)\Bigg|_{\mu=0} = \frac{2\pi}{\mathrm{Im}\,\lambda(0)} \left.\frac{d(\mathrm{Re}\,\lambda(\mu))}{d\mu}\right|_{\mu=0} \neq 0.$$

(Note that this is where the hypothesis that the eigenvalues cross the imaginary axis with nonzero speed is used.) The rest of the lemma follows from the implicit function theorem. □

Step 5. Conditions for Stability

Now let us assemble results on the derivatives of μ and V at zero.

(3.10) Lemma. $\mu'(0) = 0$.

Proof. By the way that the domain of V was chosen, we know that if $V(x_1,\mu(x_1)) = 0$, then the orbit through

$(x_1,\mu(x_1))$ crosses the x_1-axis at a point $(x_1,0,\mu(x_1))$ such that x_1 and $\hat{x}_1$ have opposite sign. (In polar coordinates, this corresponds to the fact that the orbit of $(x_1,0,\mu(x_1))$ crosses the line $\theta = -\pi$). Choose a sequence of points $x_n \downarrow 0$. Then for each x_n, there is a y_n such that $y_n < 0$ and $\mu(x_n) = \mu(y_n)$. By continuity of Φ, $y_n \to 0$. (To show this, one uses the fact that $T(x_1,\mu)$ is bounded in a neighborhood of $(0,0)$ and the fact that Φ is uniformly continuous on bounded sets.) Therefore, since $\mu(0) = 0$ and $\frac{\mu(x_n)}{x_n}$ has opposite sign to $\frac{\mu(y_n)}{y_n}$, we have $\mu'(0) = 0$. □

(3.11) Lemma. $V(0,0) = \frac{\partial V}{\partial x_1}(0,0) = \frac{\partial^2 V}{\partial x_1^2}(0,0) = 0$.

Proof. We already know that $V(0,0) = \frac{\partial V}{\partial x_1}(0,0) = 0$. To see that $\frac{\partial^2 V}{\partial x_1^2}(0,0) = 0$, we differentiate the equation $V(x_1,\mu(x_1)) = 0$. Thus, $\left.\frac{\partial V}{\partial x_1}\right|_{(x_1,\mu(x_1))} + \left.\frac{\partial V}{\partial \mu}\right|_{(x_1,\mu(x_1))} \mu'(x_1) = 0$

and $\left.\frac{\partial^2 V}{\partial x_1^2}\right|_{(x_1,\mu(x_1))} + 2\left.\frac{\partial^2 V}{\partial x_1}\right|_{(x_1,\mu(x_1))} \mu'(x_1)$

$+ \left.\frac{\partial^2 V}{\partial \mu^2}\right|_{(x_1,\mu(x_1))} \mu'(x_1)^2 + \left.\frac{\partial V}{\partial \mu}\right|_{(x_1,\mu(x_1))} \mu''(x_1) = 0$. If $x_1 = 0$, then $\mu(x_1) = 0$ and we get the equation $\left.\frac{\partial^2 V}{\partial x_1^2}\right|_{(0,0)} = 0$. □

(3.12) Definition. $(0,0)$ is a vague attractor for X_0 if $\frac{\partial^3 V}{\partial x_1^3}(0,0) < 0$.*

---

*This condition is computable; see Section 4.

(3.13) Lemma. If $(0,0)$ is a vague attractor for X_0, then the orbits through $(x_1,\mu(x_1))$ are attracting and $\mu(x_1) > 0$ for small $x_1 \neq 0$.*

Proof. To show that $\mu(x_1) > 0$ for small $x_1 \neq 0$, we show that $\mu''(0) > 0$. Since $\mu(0) = \mu'(0) = 0$, this shows that μ has a local minimum at $x_1 = 0$. Again we differentiate the equation $V(x_1,\mu(x_1)) = 0$. Having done this three times and evaluated the result at $x_1 = 0$, we get $\mu''(0) = \frac{-\partial^3 V}{\partial x_1^3}(0,0) \Big/ 3\frac{\partial^2 V}{\partial\mu\partial x_1}(0,0)$. Recall that $\frac{\partial^2 V}{\partial\mu\partial x_1}(0,0) = \frac{2\pi}{\operatorname{Im}\lambda(0)} \left.\frac{d\operatorname{Re}\lambda(\mu)}{d\mu}\right|_{\mu=0} > 0$. Therefore, $\mu''(0) > 0$. To show that the orbit through $(x_1,0,\mu(x_1))$ is attracting, we must show that the eigenvalues of the derivative of the Poincaré map associated with this orbit are less than 1 in absolute value (see Section 2B). Clearly, the Poincaré map associated with the orbit through $(x_1,0,\mu(x_1))$ is $P_{\mu(x_1)}(x_1') = P(x_1',\mu(x_1))$. The derivative of $P_\mu(x_1)$ at the point x_1 is $\left.\frac{\partial P}{\partial x_1}\right|_{(x_1,\mu(x_1))}$. Because $\left.\frac{\partial P}{\partial x_1}\right|_{(0,0)} = 1$, there is a neighborhood of $(0,0)$ in which $\frac{\partial P}{\partial x_1} > -1$. Thus, we need only show that for $x_1 \neq 0$, $\left.\frac{\partial P}{\partial x_1}\right|_{(x_1,\mu(x_1))} < 1$, i.e., for $\left.\frac{\partial V}{\partial x_1}\right|_{(x_1,\mu(x_1))} < 0$. We show that the function $f(x_1) = \left.\frac{\partial V}{\partial x_1}\right|_{(x_1,\mu(x_1))}$ experiences a local maximum at $x_1 = 0$. We

*This is not the most general possible statement of the theorem; see Section 3B below for a generalization.

already know that $f(0) = \left.\frac{\partial V}{\partial x_1}\right|_{(0,0)} = 0$. $f'(x_1) = \left.\frac{\partial^2 V}{\partial x_1^2}\right|_{(x_1,\mu(x_1))} +$ $\mu'(x_1) \left.\frac{\partial^2 V}{\partial \mu \partial x_1}\right|_{(x_1,\mu(x_1))}$. Thus, $f'(0) = 0$. $f''(x_1) = \left.\frac{\partial^3 V}{\partial x_1^3}\right|_{(x_1,\mu(x_1))} +$

$$2 \left.\frac{\partial^3 V}{\partial x_1^2 \partial \mu}\right|_{(x_1,\mu(x_1))} \mu'(x_1) + \mu'(x) \left.\frac{\partial^3 V}{\partial x_1 \partial \mu^2}\right|_{(x_1,\mu(x_1))} + \mu''(x_1) \left.\frac{\partial^2 V}{\partial \mu \partial x_1}\right|_{(x_1,\mu(x_1))}.$$

Therefore, $f''(0) = \frac{\partial^3 V}{\partial x_1^3}(0,0) + \mu''(0) \frac{\partial^2 V}{\partial \mu \partial x_1}(0,0) = \frac{\partial^3 V}{\partial x_1^3}(0,0) -$

$$\frac{\partial^2 V}{\partial \mu \partial x_1}(0,0) \left(\frac{\partial^3 V}{\partial x_1^3}(0,0) \Big/ 3 \frac{\partial^2 V}{\partial x_1 \partial \mu}(0,0) \right) = \frac{2}{3} \frac{\partial^3 V}{\partial x_1^3}(0,0) < 0.$$

Thus, $f(x_1)$ experiences a local maximum at $x_1 = 0$ and the orbits are attracting. □

Step 6. The Uniqueness of the Closed Orbits

(3.14) Lemma. There is a neighborhood N of $(0,0,0)$ such that any closed orbit in N of the flow of X passes through one of the points $(x_1,0,\mu(x_1))$.

Proof. There is a neighborhood N_ε of $(0,0,0)$ such that if $(x_1,x_2,\mu) \in N_\varepsilon$, then the orbit of Φ_t through (x_1,x_2,μ) crosses the x_1-axis at a point $(\hat{x}_1,0,\mu)$ such that $|\hat{x}_1| < \varepsilon$. This is true by the same argument that was used to show the existence of $P(x_1,\mu)$. We choose $N = \{(x_1,x_2,\mu) | (x_1,\mu) \in \text{Domain }(P)$ and $\left.\frac{\partial P}{\partial x_1}\right|_{(x_1,\mu)} > 0$ and $T(x_1,\mu) > \varepsilon > 0$ and μ is small enough so that $V(x_1,\mu) = 0$ for $\mu \in N$ iff $\mu = \mu(x_1)$. Assume that the $(x_1,0,\mu) \in \gamma$, is a closed orbit of Φ_t in N. If $V(x_1,\mu) = 0$, then $\mu = \mu(x_1)$ and there is nothing to prove. Suppose $V(x_1,\mu) \neq 0$. Then $P(x_1,\mu) > x_1$,

$(P(x_1,\mu) < x_1)$. Since $\gamma \subseteq N$, $P^n(x_1,\mu)$ is defined for all $n \geq 0$. $P^n(x_1,\mu) - P^{n-1}(x_1,\mu) = \frac{\partial P}{\partial x_1}(\xi,\mu)$, $(P^{n-1}(x_1,\mu) - P^{n-2}(x_1,\mu))$. Thus, $P^n(x_1,\mu) - P^{n-1}(x_1,\mu)$ has the same sign as $P^{n-1}(x_1,\mu) - P^{n-2}(x_1,\mu)$. By induction $P^n(x_1,\mu) > P^{n-1}(x_1,\mu)$, $(P^n(x_1,\mu) < P^{n-1}(x_1,\mu))$ for all n. Because there is a nonzero lower bound on $T(x_1,\mu)$ for $(x_1,0,\mu) \in N$, this shows that $(x_1,0,\mu)$ is not on a closed orbit of Φ_t. □

The Hopf Theorem in $\mathbb{R}^n$

Now let us consider the n-dimensional case. Reference is made to Theorem 3.1, p. .

(3.15) Theorem. Let X_μ be a C^{k+1}, $k \geq 4$, vector field on $\mathbb{R}^n$, with all the assumptions of Theorem 3.1 holding except that we assume that the rest of the spectrum is distinct from the two assumed simple eigenvalues $\lambda(\mu),\overline{\lambda(\mu)}$. Then conclusion (A) is true. Conclusion (B) is true if the rest of the spectrum remains in the left half plane as μ crosses zero. Conclusion (C) is true if, relative to $\lambda(\mu),\overline{\lambda(\mu)}$, 0 is a "vague attractor" in the same sense as in Theorem 3.1 and if when coordinates are chosen so that

$$dX_0(0) = \begin{pmatrix} 0 & |\lambda(0)| & d_3X^1(0) \\ -|\lambda(0)| & 0 & d_3X^2(0) \\ 0 & 0 & d_3X^3(0) \end{pmatrix}, \quad \lambda(0) \notin \sigma(d_3X^3(0)).$$

(3.16) Remarks. (1) The condition $\lambda(0) \notin \sigma(d_3X^3(0))$ is independent of the way R^n is split into a space corresponding to the $\lambda(0),\overline{\lambda(0)}$ space and a complementary one since choosing a different complementing subspace will only replace $d_3X^3(0)$ by a conjugate operator: $Cd_3X^3(0)C^{-1}$ as is easily seen.

(2) The condition $\lambda(0) \notin \sigma(d_3X^3(0))$ is automatic if $n = 3$ since the matrix $dX_0(0)$ is real.

(3) Further details concerning this theorem are given in Section 4.

The proof of Theorem 3.15 is obtained by combining the center manifold theorem with Theorem 3.1; i.e., we find a center manifold tangent to the eigenspace of $\lambda(\mu)$ and $\overline{\lambda(\mu)}$ and apply Theorem 3.1 to this. One important point is that in (B) of Theorem 3.1 we concluded stability of the orbit within the center manifold. Here we are, in (B), claiming it in a whole R^n neighborhood of the orbit. The reason for this is that we will be able to reduce our problem to one in which the center manifold is the x_1,x_2-plane and that plane is invariant under the flow. If (x,μ) is on a closed orbit with period τ,

$$d\phi_{\tau,\mu}(x) = \begin{pmatrix} a_{11} & a_{12} & d_3\phi_\tau^1(x) \\ a_{21} & a_{22} & d_3\phi_\tau^2(x) \\ 0 & 0 & d_3\phi_\tau^3(x) \end{pmatrix} .$$

The two-dimensional theorem will imply that the spectrum of the upper block transverse to the closed orbit is in $\{z \mid |z| < 1\}$ and our assumptions plus continuity will imply the same for $\sigma(d_3\phi_\tau^3(x))$. Since the spectrum of the Poincaré map is the spectrum of $d\phi_{\tau,\mu}(x)$ restricted to a subspace transverse to the closed orbit, this shows $\sigma(dP(x)) \subset \{z \mid |z| < 1\}$ and the orbit is attracting. □

(3.17) <u>Exercise</u>. Show that the vector field $X_\mu(x_1,x_2) = (x_2, \mu(1-x_1^2)x_2-x_1)$ satisfies the conditions of Theorem 3.1.

(3.18) <u>Exercise</u>. All equations are given in polar coordinates. Match each set of equations to the appropriate picture and state which hypotheses of the Hopf bifurcation theorem are violated. (If you get stuck, come back to this problem after reading Section 3A.)

1. $\dot{r} = -r(r+\mu)^2$, $\dot{\theta} = 1$
2. $\dot{r} = r(\mu-r^2)(2\mu-r^2)^2$, $\dot{\theta} = 1$
3. $\dot{r} = r(r+\mu)(r-\mu)$, $\dot{\theta} = 1$
4. $\dot{r} = \mu r(r+\mu)^2$, $\dot{\theta} = 1$
5. $\dot{r} = -\mu^2 r(r+\mu)^2(r-\mu)^2$, $\dot{\theta} = 1$

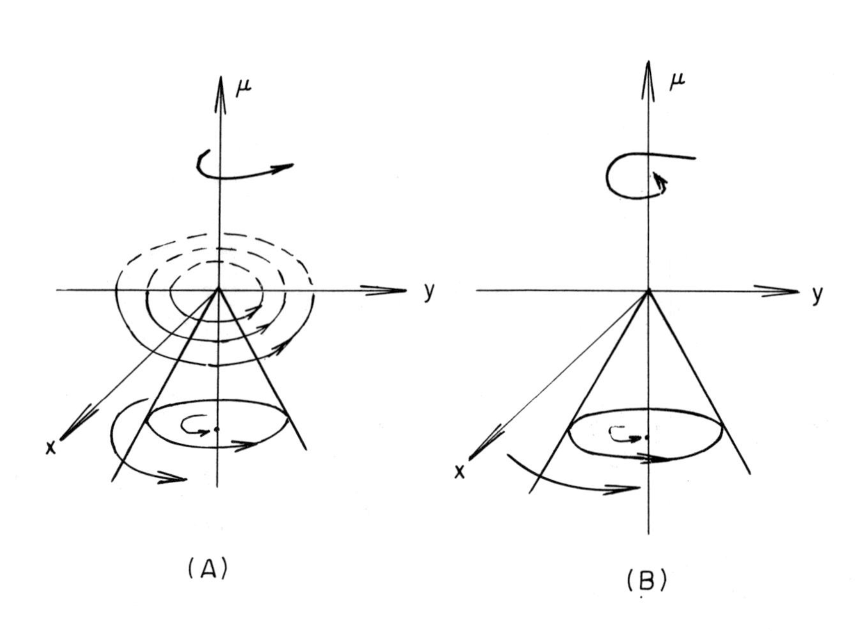

(A) (B)

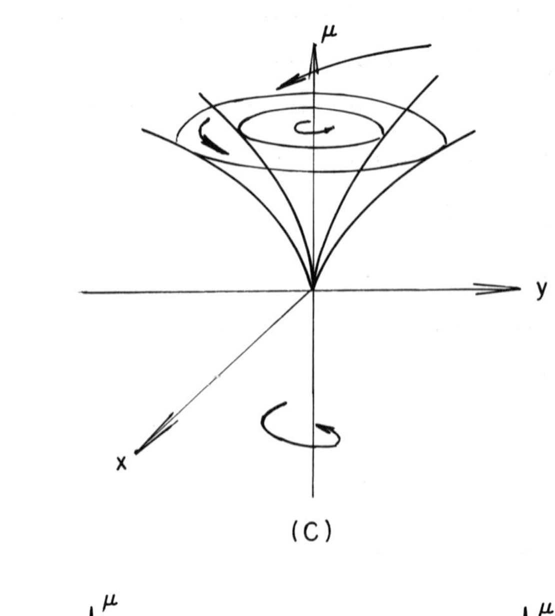

(C)

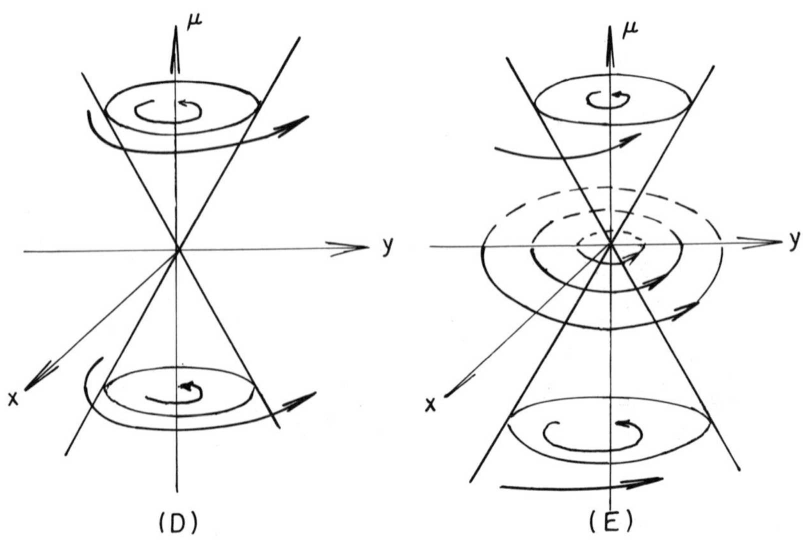

(D)

(E)

SECTION 3A

OTHER BIFURCATION THEOREMS

Several authors have published generalizations of the Hopf Bifurcation. In particular, Chafee [1] has eliminated the condition that the eigenvalue $\lambda(\mu)$ cross the imaginary axis with nonzero speed. In this case, bifurcation to periodic orbits occurs, but it is not possible to predict from eigenvalue conditions exactly how many families of periodic orbits will bifurcate from the fixed point. Chafee's result gives a good description of the behavior of the flow of the vector field near the bifurcation point. See also Bautin [1] and Section 3C.

Chafee's Theorem

We consider an autonomous differential equation of the form

$$\dot{x} = P(\varepsilon)x + X(x,\varepsilon), \qquad (3A.1)$$

where x and X vary in real Euclidean space R^n $(n > 2)$, $\varepsilon \geq 0$ is a small parameter (called μ in previous sections), and P is a real $n \times n$ matrix. We assume the following hypotheses.

(H_1) There exist numbers $r_0 > 0$ and $\varepsilon_0 > 0$ such that P is continuous on the closed interval $[0,\varepsilon_0]$ and X is continuous on the domain $B^n(r_0) \times [0,\varepsilon_0]$.

(H_2) For each ε in $[0,\varepsilon_0]$ we have $X(0,\varepsilon) = 0$ so that the origin $x = 0$ is an equilibrium point of (3A.1).

(H_3) For each r in $[0,r_0]$ there exists a $k(r) > 0$ such that on the domain $B^n(r) \times [0,\varepsilon_0]$ the function X is uniformly Lipschitzian in x with Lipschitz constant $k(r)$; moreover, $k(r) \to 0$ as $r \to 0$.

(H_4) For each ε in $[0,\varepsilon_0]$ the matrix $P(\varepsilon)$ has a complex-conjugate pair of eigenvalues $a(\varepsilon) \pm ib(\varepsilon)$ whose real and imaginary parts satisfy the conditions

$$a(0) = 0, \qquad a(\varepsilon) > 0 \qquad (0 < \varepsilon \leq \varepsilon_0),$$
$$b(\varepsilon) > 0 \qquad (0 \leq \varepsilon \leq \varepsilon_0).$$

The other eigenvalues $\lambda_1(\varepsilon)$, $\lambda_2(\varepsilon),\ldots,\lambda_{n-2}(\varepsilon)$ of $P(\varepsilon)$ have their real parts negative for all ε in $[0,\varepsilon_0]$.

(H_5) For $\varepsilon = 0$ the equilibrium point of (3A.1) at the origin is asymptotically stable in the sense of Liapunov (Lefschetz [1], p. 89, and Section 1 above).

Hypotheses (H_1) and (H_3) are sufficient to guarantee the usual properties of existence, uniqueness, and continuity in initial conditions for solutions of (3A.1). In that which follows the solution of (3A.1) assuming a given initial value x_0 at $t = 0$ will be denoted by $x(t,x_0,\varepsilon)$. In connection with this notation we should mention the well-known autonomous property of (3A.1): the solution of (3A.1) assuming a given initial value x_0 at a specified value of t, say t_0, is given by $x(t - t_0, x_0, \varepsilon)$. The hypothesis (H_5) replaces the "vague attractor" hypothesis considered earlier.

(3A.1) Theorem. *Let* (3A.1) *satisfy the hypotheses* (H_1) *through* (H_5) *and let* r_0 *and* ε_0 *be as in* (H_1). *Then, there exist numbers* r_1, r_2, *and* ε_1 *such that* $0 < r_2 \leq r_1 \leq r_0$, $0 < \varepsilon_1 \leq \varepsilon_0$, *and such that the following assertions are true*.

(i) *For each* $\varepsilon \in (0, \varepsilon_1]$ *there exist for Equation* (3A.1) *two closed orbits* $\gamma_1(\varepsilon)$ *and* $\gamma_2(\varepsilon)$ (*not necessarily distinct*) *which lie inside a neighborhood of the form* $B^n(r(\varepsilon))$, *where* $0 < r(\varepsilon) \leq r_2$ *and* $r(\varepsilon) \to 0$ *as* $\varepsilon \to 0+$. *Moreover,* $\gamma_1(\varepsilon)$ *and* $\gamma_2(\varepsilon)$ *lie on a local integral manifold* $M^2(\varepsilon)$ *homeomorphic to an open disk in* R^2 *and containing the origin* $x = 0$. *Regarded as closed Jordan curves in* $M^2(\varepsilon)$, $\gamma_1(\varepsilon)$ *and* $\gamma_2(\varepsilon)$ *are concentric about the origin with, say,* $\gamma_1(\varepsilon)$ *inside* $\gamma_2(\varepsilon)$ *when these curves are distinct*.

(ii) *For each* $\varepsilon \in (0, \varepsilon_1]$ *that part of* $M^2(\varepsilon)$ *which lies inside* $\gamma_1(\varepsilon)$ *is filled by solutions of* (3A.1) *which approach the origin as* $t \to -\infty$ *and which, except for the equilibrium point* ar $x = 0$, *approach* $\gamma_1(\varepsilon)$ *as* $t \to +\infty$. *No other solutions of* (3A.1) *remain in* $B^n(r_1)$ *for all* $t < 0$.

(iii) *For each* $\varepsilon \in (0, \varepsilon_1]$ *that part of* $M^2(\varepsilon)$ *lying outside* $\gamma_2(\varepsilon)$ *but contained in* $B^n(r_2)$ *is filled by solutions of* (3A.1) *which remain in* $M^2(\varepsilon) \cap B^n(r_1)$ *for all* $t > 0$ *and which approach* $\gamma_2(\varepsilon)$ *as* $t \to +\infty$.

(iv) *For each* $\varepsilon \in (0, \varepsilon_1]$ *there exist solutions of* (3A.1) *which approach the origin* $x = 0$ *as* $t \to +\infty$ *and these solutions fill a local integral manifold* (= *invariant manifold*) $M^{n-2}(\varepsilon)$ *homeomorphic to an open ball in* R^{n-2} *and containing* the origin $x = 0$.

(v) If for a given $\varepsilon \in (0,\varepsilon_0]$, $x(t,x_0,\varepsilon)$ is a solution of (3A.1) for which $x_0 \in B^n(r_2)$, then $x(t,x_0,\varepsilon)$ remains in $B^n(r_1)$ for all $t > 0$. Moreover, if $x(t,x_0,\varepsilon) \not\to 0$ as $t \to +\infty$ (see (iv)) then as $t \to +\infty$, $x(t,x_0,\varepsilon)$ approaches the closed invariant set $\Omega(\varepsilon)$ consisting of those points in $M^2(\varepsilon)$ which lie on $\gamma_1(\varepsilon)$ or $\gamma_2(\varepsilon)$ or between them. The solutions which approach $\Omega(\varepsilon)$ contain in their positive-limiting sets one or more closed orbits (which may or may not coincide with $\gamma_1(\varepsilon)$ or $\gamma_2(\varepsilon)$). See Figure 3A.1.

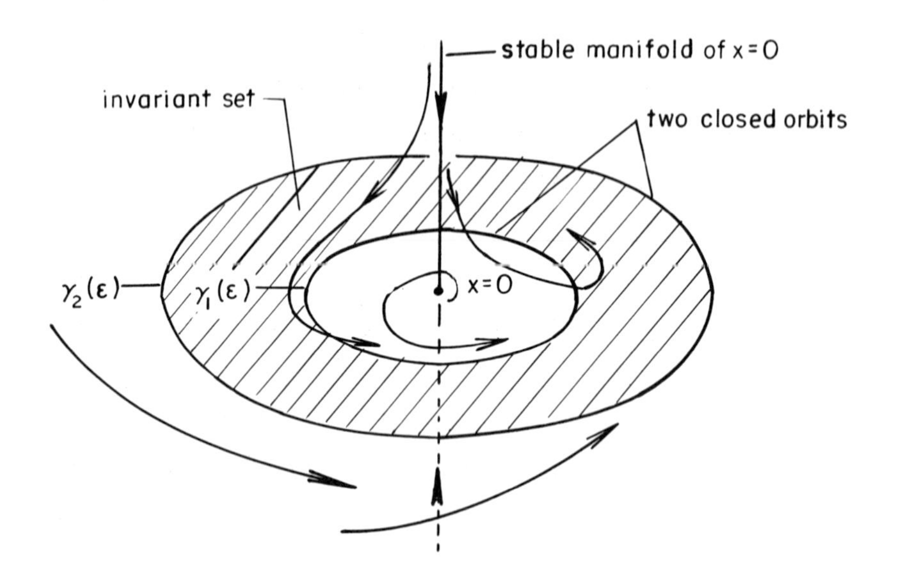

Figure 3A.1

Chafee [2] has also proved a theorem parallel to 3A.1 for the case in which the vector field at time t depends on the flow at time $t - \alpha$ for some $\alpha \geq 0$; i.e., $\dot{x} = F(t,x_t)$ where $x_t = x(t-\alpha)$.

The following example (see Chafee [1]) shows that one cannot predict the number of distinct families of closed orbits to which the flow bifurcates. In cylindrical coordinates, let

$$\left.\begin{aligned}\frac{dr}{dt} &= r(\varepsilon-r^2)^{n_1}(2\varepsilon-r^2)^{n_2}(3\varepsilon-r^2)^{n_3}\cdots(m\varepsilon-r^2)^{n_m} = rf(r,\varepsilon)\\ \frac{d\theta}{dt} &= 1\\ \frac{dz}{dt} &= -z\end{aligned}\right\}\qquad(3A.2)$$

where $n_1+\cdots+n_m$ is odd.

This equation is C^∞ in rectangular coordinates. Furthermore, it has m distinct families of closed orbits which bifurcate from the origin at $\varepsilon = 0$ (i.e., at $r^2 = j\varepsilon$, $z = 0$). In rectangular coordinates, the derivative of the vector field at the origin is

$$\begin{pmatrix}\alpha & -1 & 0\\ 1 & \alpha & 0\\ 0 & 0 & -1\end{pmatrix}$$

where $\alpha = (\varepsilon^{\sum_j n_j}) \prod_j (j^{n_j})$. The eigenvalues are -1 and $\alpha \pm i$. By varying m and the n_j's one can vary the number of distinct closed orbits independently of the order to which α vanishes at 0. For example,

$$\frac{dr}{dt} = r(\varepsilon-r^2)(2\varepsilon-r^2)(3\varepsilon-r^2) \quad\text{or}\quad \frac{dr}{dt} = r(\varepsilon-r^2)^3 \quad\text{or}$$

$$\frac{dr}{dt} = r(\varepsilon-r^2)(2\varepsilon-r^2)^2.$$

Chafee has shown us another example proving that differentiability with respect to the parameter is necessary to insure uniqueness of the closed orbits. In polar coordinates, let

$$\frac{dr}{dt} = r(r - \sqrt[3]{\varepsilon})^2(2\sqrt[3]{\varepsilon} - r)$$

$$\frac{d\theta}{dt} = 1$$

(3A.2) Exercise: Show that although $\left.\frac{d \operatorname{Re} \lambda(\varepsilon)}{d}\right|_{\varepsilon=0}$ $= 2 > 0$, bifurcation the two distinct periodic orbits (at $r = \varepsilon^{1/3}, 2\varepsilon^{1/3}$) occurs.

(3A.3) Remarks. In the paper of Jost and Zehnder [1], the situation where X depends on more than one parameter is considered. See also Takens [1].

Some interesting recent results of Joseph give another proof of Chafee's result that one does not need $V'''(0) \neq 0$, but only that the fixed point at $\mu = 0$ is stable. (See also Section 3B). Joseph also is able to deal with the case in which a finite amplitude periodic orbit arises. See Joseph-Nield [1] for details, and Joseph [2]. For the case of more than one parameter, Takens [1] also obtains finite amplitude bifurcations.

Alexander and Yorke [1] prove, roughly speaking, that if a vector field X_μ has a closed orbit $\gamma\mu$, then as μ increases either (i) $\gamma\mu$ remains a closed orbit; (ii) the period of $\gamma\mu$ becomes infinite or (iii) $\gamma\mu$ shrinks to a fixed point. This is done without regard to stability of $\gamma\mu$. For another proof, see Izé [1]. (We thank L. Nirenberg for bringing this to our attention.)

SECTION 3B

MORE GENERAL CONDITIONS FOR STABILITY

Here we shall prove that the least n for which $\frac{\partial^n V}{\partial x_1^n}(0,0) \neq 0$ is odd, and that if this coefficient is negative, the periodic orbits obtained in Theorem 3.1 by use of the implicit function theorem are attracting and occur for $\mu > 0$ (we assume we have enough differentiability so that V is C^n).

We also show that if the origin is attracting in the sense of Liapunov for the flow of X_0, then the periodic orbits obtained in Theorem 3.15 (conclusions (A) and (B)) are attracting and occur for $\mu > 0$.

(3B.1) Lemma. *Let the vector field* X *be* C^{2k} *for* $k \geq 2$. *Then the function* $\mu(x_1)$ *is* $C^{2(k-1)}$. *Assume that there is a* $j \leq 2(k-1)$ *such that* $\mu^{(j)}(0) \neq 0$. *Then the least* j *for which this is true is even*.

Proof. Let n be the least integer j such that $\mu^{(j)}(0) \neq 0$. Assume $\mu^{(n)}(0) > 0$ and choose $\varepsilon > 0$ such that for all x_1 with $|x_1| < \varepsilon$, $\mu^{(n)}(x_1) > 0$. Then by the mean value theorem we have $\mu(x_1) = \mu^{(n)}(\alpha_n)x_1\alpha_1\cdots\alpha_{n-1}$ where $0 < \alpha_j < x_1$ (or $x_1 < \alpha_j < 0$) for all j. Suppose n is odd, then $\text{sign}(x_1) = \text{sign}(x_1\alpha_1\cdots\alpha_{n-1})$. Therefore, for all x_1 with $|x_1| < \varepsilon$, if $x_1 > 0$, then $\mu(x_1) > 0$ and if $x_1 < 0$, then $\mu(x_1) < 0$. However, we know that this cannot occur. For let $x_1^n \downarrow 0$ as $n \to \infty$ then for each x_1^n there is a $y_1^n < 0$ such that $\mu(x_1^n) = \mu(y_1^n)$ and $y_1^n \to 0$ as $n \to \infty$. The same argument shows that if $\mu^{(n)}(x_1) < 0$, n must be even. Therefore, n is even. This also shows that bifurcation occurs above or below criticality. □

(3B.2) <u>Lemma</u>. <u>Let</u> X <u>be</u> C^{2k} <u>for</u> $k \geq 2$. <u>Assume that there is a</u> $j \leq 2(k-1)$ <u>such that</u> $\mu^{(j)}(0) \neq 0$, <u>and let</u> n <u>be as in the preceding lemma</u>. <u>In this case</u> $\frac{\partial^j V}{\partial x_1^j}(0,0) = 0$ <u>for all</u> $j \leq n$ <u>and</u> $\frac{\partial^{n+1} V}{\partial x_1^{n+1}}(0,0) = -3\frac{\partial^2 V}{\partial x_1 \partial \mu}(0,0)\mu^{(n)}(0)$.

<u>Proof</u>. Upon differentiating the equation $V(x_1,\mu(x_1)) = 0$ j times and evaluating at $x_1 = 0$, we get $\frac{\partial^j V}{\partial x_1^j}(0,0) + (*) = 0$. $(*)$ is a sum of terms of the form $A_\ell \mu^{(\ell)}(0)$ for $\ell \leq j$ and $A_j = \frac{\partial V}{\partial \mu}(0,0) = 0$ because $V(0,\mu) = 0$ for all μ. Therefore, if $j \leq n$, $\frac{\partial^j V}{\partial x_1^j}(0,0) = 0$. To find $\frac{\partial^{n+1} V}{\partial x_1^{n+1}}(0,0)$, we must find the coefficient of $\mu^{(n)}(0)$ in the equation $\frac{\partial^{n+1} V}{\partial x_1^{n+1}}(0,0) + (*) = 0$. This coefficient is easily seen to be $3\frac{\partial^2 V}{\partial x_1 \partial \mu}(0,0) + 3\frac{\partial^2 V}{\partial \mu^2}(0,0)\mu'(0)$. Therefore, $\frac{\partial^{n+1} V}{\partial x_1^{n+1}}(0,0) = -3\frac{\partial^2 V}{\partial x_1 \partial \mu}(0,0)\mu^{(n)}(0)$.

(3B.3) Theorem. Let X be C^{2k} for $k \geq 2$. Then the function $\mu(x_1)$ is $C^{2(k-1)}$. Assume that there is an integer $j \leq 2(k-1)$ such that $\mu^{(j)}(0) \neq 0$. Let n be the least such integer. If $\frac{\partial^{n+1} V}{\partial x_1^{n+1}}(0,0) < 0$, then $\mu(x_1) > 0$ for all $x_1 \neq 0$ and sufficiently small. Furthermore, the periodic orbits obtained from the implicit function theorem are attracting.

Proof. We have already seen that $\mu(x_1) > 0$ for all small x_1 such that $x_1 \neq 0$. To show that the periodic orbits are stable, we must show that the function

$f(x_1) = \left.\frac{\partial V}{\partial x_1}\right|_{(x_1,\mu(x_1))}$ has a local maximum at $x_1 = 0$ (See Step 5 above). $f^{(j)}(0) = 0$ for all $j < n$ because $f^{(j)}(0) = \frac{\partial^{j+1} V}{\partial x_1^{j+1}}(0,0) + (*)$ where $(*)$ is a sum of terms of the form $A_\ell \mu^{(\ell)}(0)$ for $\ell \leq j$. $f^{(n)}(0) = \frac{\partial^{n+1} V}{\partial x_1^{n+1}}(0,0) + \frac{\partial^2 V}{\partial x_1 \partial \mu}(0,0)\mu^{(n)}(0) = \frac{2}{3}\frac{\partial^{n+1} V}{\partial x_1^{n+1}}(0,0) < 0$. Recall that n is even. Therefore, the mean value theorem shows that $f(x_1)$ has a local maximum at $x_1 = 0$. □

(3B.4) Theorem. Let the conditions of Theorem 3.15 be satisfied so that conclusions (A) and (B) hold. Furthermore, let the origin be Liapunov attracting for the flow of X_0. Then the periodic orbits obtained from Theorem 3.15 are attracting and occur for $\mu > 0$.

Proof. Under these conditions, Chafee's Theorem (page 85) holds. Therefore, the periodic orbits occur for $\mu > 0$.

Since conclusion (B) of Theorem 3.15 holds, the orbits are unique, that is $\gamma_1 = \gamma_2$. Under these circumstances, Chafee's theorem implies that the orbits are attracting. $\square$

SECTION 3C

HOPF'S BIFURCATION THEOREM AND THE CENTER THEOREM OF LIAPUNOV

by

Dieter S. Schmidt

Introduction

In recent years numerous papers have dealt with the bifurcation of periodic orbits from an equilibrium point. The starting point for most investigations is the Liapunov Center Theorem [1] or the Hopf Bifurcation Theorem [1]. Local results concerning these theorems were published by Chafee [1], Henrard [1] Schmidt and Sweet [1] among many others, noted in previous sections, whereas Alexander and Yorke [1] discussed the global problem of the bifurcation of periodic orbits. They showed in their paper that Liapunov's Center Theorem can be derived as a consequence of Hopf's bifurcation theorem.

J. A. Yorke suggested that one should show also on the local level that Liapunov's theorem can be obtained from the one of Hopf. For this we provide an analytic proof of Hopf's theorem based on the alternative method as outlined in Berger's article in Antman-Keller [1] which is general enough to include the center theorem as a corollary. In addition our proof of Hopf's theorem is simple enough to allow the discussion of some exceptional cases.

The Hopf Bifurcation Theorem

We consider the n-dimensional autonomous system of differential equations given by

$$\dot{x} = F(x,\mu) \tag{3C.1}$$

which depends on the real parameter μ. We assume that (3C.1) possesses an analytic family $x = x(\mu)$ of equilibrium points; that is $F(x(\mu),\mu) = 0$. Without loss of generality we assume that this family is given by $x \equiv 0$, that is $F(0,\mu) = 0$. We suppose that for a certain value of μ, say $\mu = 0$, the matrix $F_x(0,\mu)$ has two purely imaginary eigenvalues $\pm i\beta$ and no other eigenvalue of $F_x(0,0)$ is an integral multiple of $i\beta$. If $\alpha(\mu) + i\beta(\mu)$ is the continuation of the eigenvalue $i\beta$ then we assume that $\alpha'(0) \neq 0$.

(3C.1) Theorem (Hopf). Under the above conditions there exist continuous functions $\mu = \mu(\varepsilon)$ and $T = T(\varepsilon)$ depending on a parameter ε with $\mu(0) = 0$ $T(0) = 2\pi\beta^{-1}$ such that there are nonconstant periodic solutions $x(t,\varepsilon)$ of (3C.1) with period $T(\varepsilon)$ which collapse into the origin as $\varepsilon \to 0$.

(3C.2) Remark. Our assumptions for the Hopf theorem are slightly less restrictive than they are usually stated as we do not require the other eigenvalues to be non imaginary. Furthermore $F(x,\mu)$ does not have to be analytic for the proof to hold but a certain degree of differentiability is required.

Proof. Through a linear change of coordinates of the form $y = S(\mu)x$ and by a change of the independent variable $\tau = \beta(\mu)t$ we can bring equation (3C.1) into the following form

$$\begin{aligned} \dot{y}_1 &= (u(\mu) + i)y_1 + \phi_1(y_1, y_2, \tilde{y}, \mu) \\ \dot{y}_2 &= (u(\mu) - i)y_2 + \phi_2(y_1, y_2, \tilde{y}, \mu) \\ \tilde{y} &= B(\mu)\tilde{y} + \tilde{\phi}(y_1, y_2, \tilde{y}, \mu) \end{aligned} \tag{3C.2}$$

y_1 and y_2 are the first two complex components of the vector y. Real solutions are only given if $y_1 = \bar{y}_2$. The remaining $n - 2$ components of the vector y are real and denoted by $\tilde{y}$. $B(\mu)$ is a real $n - 2$ square matrix not necessarily in normal form and the functions ϕ_1, ϕ_2, and $\tilde{\phi}$ are at least quadratic in the components of the vector y.

We introduce now the following polar coordinates

$$y_1 = r\,e^{i\theta} \qquad y_2 = r\,e^{-i\theta} \qquad \tilde{y} = r\eta$$

and arrive at the following system

$$\begin{aligned} \dot{r} &= u(\mu)r + \mathrm{Re}\{e^{-i\theta}\,\phi_1\} \\ \dot{\theta} &= 1 + \frac{1}{r}\,\mathrm{Im}\,\{e^{-i\theta}\,\phi_1\} \\ \dot{\eta} &= B(\mu)\eta - u(\mu)\eta + \frac{1}{r}\,(\phi - \mathrm{Re}\{e^{-i\theta}\,\phi_1\}\eta). \end{aligned} \tag{3C.3}$$

Into this system we introduce the scale factor ε by $r = \varepsilon\rho$ $\mu = \varepsilon\mu_1$.

Because of $\dot{\theta} = 1 + O(\varepsilon)$ we can use θ as a new independent variable to overcome the autonomous character of the given system. The resulting differential equations have the following form

$$\frac{d\rho}{d\theta} = \varepsilon\, R(\theta,\rho,\eta,\varepsilon)$$
$$\frac{d\eta}{d\theta} = B(0) \quad + \quad V(\theta,\rho,\eta,\varepsilon) \tag{3C.4}$$

and we are searching for 2π - periodic solutions of this system. By our assumptions on the eigenvalues of $B(0)$ we find that for $\varepsilon = 0$ the only 2π - periodic solution is $\rho = \rho_0$ $=$ const, $\eta = 0$. This solution persists for $\varepsilon \neq 0$ if the following bifurcation equation holds (see Berger [1])

$$\int_0^{2\pi} R(\theta,\rho,\eta,\varepsilon)d\theta = 0.$$

In this expression ρ and η represent the 2π periodic solution of the given system (3C.1), but the terms of order ε^0 are already known and we can evaluate the term of the same order in the bifurcation equation. This leads to the equation

$$\mu_1\, u'(0)\rho_0 + 0(\varepsilon) = 0$$

which can be solved uniquely to yield $\mu_1 = \mu_1(\varepsilon) = 0(\varepsilon)$ by the implicit function theorem, since $u'(0) \neq 0$ by assumption and $\rho_0 \neq 0$ because we are looking for nontrivial solutions. Therefore the function $\mu = \mu(\varepsilon) = \varepsilon\, \mu_1(\varepsilon) = \theta(\varepsilon^2)$ has been found. The period of the solution in the original x-coordinate system as found from the expression for $\frac{d\theta}{dt}$ is

$$T = T(\varepsilon) = \frac{2\pi}{\beta(0)}\,(1 + 0(\varepsilon^2)). \quad \square$$

<u>The Liapunov Center Theorem</u>

(3C.3) <u>Theorem (Liapunov)</u>. Consider the system

$$\dot{x} = Ax + f(x) \tag{3C.5}$$

where f is a smooth function which vanishes along with its first partial derivatives at $x = 0$. Assume that the system admits a first integral of the form $I(x) = \frac{1}{2} x^T S x + \cdots$ where $S = S^T$ and $\det S \neq 0$. Let A have eigenvalues $\pm i\beta, \lambda_3, \ldots, \lambda_n$ $\beta \neq 0$. Then if $\lambda_j/i\beta \neq$ integer for $j = 3, \cdots, n$ the above system has a one parameter family of periodic solutions emanating from the origin starting with period $2\pi/\beta$.

For the usual proof, see Kelley [1].

Proof. As announced in the introduction we will show that this theorem is a consequence of the Hopf bifurcation theorem. To this end we consider the modified system

$$\dot{x} = Ax + f(x) + \mu \text{ grad } I(x) \tag{3C.6}$$

and we will show that all conditions of Hopf's theorem are met and that the nonstationary periodic orbits can only occur for $\mu = 0$.

The second part is easily done by evaluating dI/dt along solutions of (3C.6), which gives

$$\begin{aligned}\frac{dI}{dt} &= \langle \text{ grad } I(x), Ax + f(x) + \mu \text{ grad } I(x) \rangle \\ &= \mu |\text{grad } I(x)|^2.\end{aligned}$$

The second equality holds because $I(x)$ is an integral for (3C.5). Therefore $\frac{1}{\mu}\frac{dI}{dt}$ is monotonically increasing unless $\text{grad } I(x(t)) = 0$, which gives $x(t) = x(0)$ that is a stationary point.

In order to apply the theorem of the previous section we only have to verify the condition concerning the real part of the eigenvalue near $i\beta$. Again through a linear change we will bring the linear part of system (3C.5) into a normal form. We assume that this has been done already and the matrix A

has thus the following real form

$$A = \begin{pmatrix} 0 & \beta & 0 \\ -\beta & 0 & 0 \\ 0 & 0 & \tilde{A} \end{pmatrix}$$

where $\tilde{A}$ is a real $n-2$ square matrix. It follows from $A^T S + SA = 0$ that the matrix S in the integral has the form

$$S = \begin{pmatrix} a & 0 & 0 \\ 0 & a & 0 \\ 0 & 0 & \tilde{S} \end{pmatrix}$$

with $a \neq 0$ since $\det S \neq 0$.

From the matrix

$$A + \mu S = \begin{pmatrix} \mu a & \beta & 0 \\ -\beta & \mu a & 0 \\ 0 & 0 & \tilde{A} + \mu\tilde{S} \end{pmatrix}$$

it follows at once that the eigenvalue near $i\beta$ has real part $\alpha(\mu) = a\mu$ and therefore $\alpha'(0) = a \neq 0$. $\square$

An Exceptional Case for the Hopf Bifurcation Theorem

Our proof of Hopf's theorem is easy enough to allow us to discuss the case where the real part of the eigenvalue does not satisfy the condition $\alpha'(0) \neq 0$, but instead the second derivative is nonzero $\alpha''(0) \neq 0$. The term with μ_1 in the bifurcation equation is zero and we will have to evaluate some higher order terms.

We use the same normal form as given earlier in equation (3C.2) and for simplicity we assume that ϕ_1 and ϕ_2 are analytic functions in their variables. We assume that in a preliminary nonlinear transformation mixed quadratic terms involving y_1 or y_2 and a component of $\tilde{y}$ have been eliminated. This can be achieved with a method similar to the one used in the Birkhoff normalization of Hamiltonian systems, that is by a transformation of the form (cf. Section 6A).

$$y_1 \to y_1 + y_1\, \alpha^T\tilde{y} + y_2\beta^T\tilde{y}$$

$$y_2 \to y_2 + y_1\, \bar{\beta}^T\, \tilde{y} + y_2\, \bar{\alpha}^T\tilde{y}$$

$$\tilde{y} \to \tilde{y}.$$

The $n-2$ dimensional complex valued vectors α and β can be determined uniquely to eliminate the terms under question, because the matrix $B(\mu)$ in system (3C.2) does neither have 0 nor $2i$ as eigenvalue for small μ.

In the function ϕ_1 we need to know the quadratic and cubic terms made up of y_1 and y_2. They are

$$\phi_1 = ay_1^2 + by_1y_2 + cy_2^2 +\cdots+ \alpha y_1^3 + \beta y_1^2y_2 + \lambda y_1y_2^2 + \delta y_2^3 +\cdots .$$

The terms not written down either only involve the $\tilde{y}$ variables or are of higher order. The coefficients depend of course on the parameter μ and we write $a = a(\mu) = a_0 + a_1\mu + O(\mu^2)$ and similarly for the other coefficients.

The differential equation of interest in the θ, r, η variables is the one for r which reads

$$\frac{dr}{d\theta} = \frac{u(\mu)r + \mathrm{Re}\{e^{-i\theta}\phi_1\}}{1 + r^{-1}\ \mathrm{Im}\{e^{-i\theta}\phi_1\}} .$$

This time we scale by

$$r = \varepsilon^2\rho \qquad \mu = \varepsilon\,\mu_1$$

and obtain

$$\frac{d\rho}{d\theta} = \varepsilon^2(R_0+\varepsilon^2R_1+\varepsilon\mu_1R_2+\mu_1^2R_3+\cdots)$$

$$\frac{d\eta}{d\theta} = B(0)\eta + O(\varepsilon^2)$$

with

$$R_0 = \rho^2 \operatorname{Re}\{a_0 e^{i\theta} + b_0 e^{-i\theta} + c_0 e^{-3i\theta} + \cdots\}$$

$$R_1 = \rho^3 \,(\operatorname{Re}\{\alpha_0 e^{2i\theta} + \beta_0 + \gamma_0 e^{-2i\theta} + \delta_0 e^{-4i\theta} + \cdots\}$$
$$- \operatorname{Re}\{a_0 e^{i\theta} + b_0 e^{-i\theta} + c_0 e^{-3i\theta} + \cdots\} \operatorname{Im}\{a_0 e^{i\theta} + b_0 e^{i\theta}$$
$$+ c_0 e^{-3i\theta} + \cdots\})$$

$$R_2 = \rho^2 \operatorname{Re}\{a_1 e^{i\theta} + b_1 e^{-i\theta} + c_1 e^{-3i\theta} + \cdots\}$$

$$R_3 = \frac{1}{2} u''(0)\rho.$$

The dots in the functions R_0, R_1 and R_2 stand for terms involving the η variables. Because $\eta = 0(\varepsilon^2)$ for 2π periodic solutions those terms will be insignificant in evaluating the bifurcation equation, which has the same form as earlier and is given by

$$\int_0^{2\pi} (R_0 + \varepsilon^2 R_1 + \varepsilon\mu_1 R_2 + \mu_1^2 R_3 + \cdots)\, d\theta = 0.$$

In evaluating this integral it is seen at once that there is no constant term. Nevertheless care has to be exercised in integrating R_0 because it will contribute to the ε^2 term due to the form of the solution of ρ which is

$$\rho = \rho_0 + \varepsilon^2 \rho_0^2 \operatorname{Re}\{a_0 i(1-e^{i\theta}) + b_0 i(e^{-i\theta}-1) + \frac{c_0 i}{3}(e^{-3i\theta}-1)\} + 0(\varepsilon^3)$$

Due to our preliminary transformation the η variables appear quadratic on R_0 and therefore they will only contribute to higher order terms in ε and μ_1. The integration leads to the following bifurcation equation

$$2\pi(\varepsilon^2 \rho_0^3 \operatorname{Re}\{\beta_0 + i\, a_0 b_0\} + \tfrac{1}{2}\mu_1^2 u''(0)\rho_0 + \text{h.o.t}) = 0.$$

The implicit function theorem allows us to state the following result: If $u''(0) \operatorname{Re}\{\beta_0 + i\, a_0 b_0\} < 0$ there are two distinct solutions of the above bifurcation equation of the form $\mu_1 = 0(\varepsilon)$. The solutions correspond to two families of periodic orbits emanating from the equilibrium. In case $u''(0)\ \mathrm{Re}\{\beta_0 + i\ a_0 b_0\} > 0$ there are no such solutions. Finally, if the discriminant is equal to zero higher order terms are needed to decide what is happening. In the case $u(0) = u'(0) = \cdots = u^{(n-1)} = 0$ $u^{(n)}(0) \neq 0$ we scale by $r = \varepsilon^n \rho$ $\mu = \varepsilon\, \mu_1$ and after identical computations we arrive at the bifurcation equation

$$\frac{\rho_0}{n!}\, u^{(n)}(0)\, \mu_1^n + \rho_0^3\, \mathrm{Re}\{\beta_0 + i\ a_0 b_0\}\varepsilon^n + \text{h.o.t} = 0.$$

Call $D = u^{(n)}(0)\ \mathrm{Re}\{\beta_0 + i\ a_0 b_0\}$. If n is odd and $D \neq 0$ there is always a solution of the form $\mu_1 = \mu_1(\varepsilon) = 0(\varepsilon)$ for ε small to the above bifurcation equation. If n is even then there are two such solutions if $D < 0$ and none if $D > 0$.

(3C.4) Theorem. Consider the differential system of equations (3C.2) put into a normal form as outlined above. Assume $u(0) = u'(0) = \cdots = u^{(n-1)}(0) = 0$ $u^{(n)}(0) \neq 0$ $n = 1,2,\cdots$ let $D = u^{(n)}(0)\ \mathrm{Re}\{\beta_0 + i\ a_0 b_0\}$. Then if n is odd and $D \neq 0$ there exists at least locally a one parameter family of periodic orbits which collapse into the origin as the parameter tends to zero and the period tends to 2π. If n is even then there are two such families in case $D < 0$ and none in case $D > 0$.

The result is very close to that of Chafee [1], discussed in Section 3A. See also Takens [1].

SECTION 4
COMPUTATION OF THE STABILITY CONDITION

Seeing if the Hopf theorem applies in any given situation is a matter of analysis of the spectrum of the linearized equations; i.e. an eigenvalue problem. This procedure is normally straightforward. (For partial differential equations, consult Section 8.)

It is less obvious how to determine the stability of the resulting periodic orbits. We would now like to develop a method which is applicable to concrete examples. In fact we give a specific computational algorithm which is summarized in Section 4A below. (Compare with similar formulas based on Hopf's method discussed in Section 5A.) The results here are derived from McCracken [1].

Reduction to Two Dimensions

We begin by examining the reduction to two dimensions in detail.

Suppose $X : N \to T(N)$ is a C^k vector field, depending smoothly on μ, on a Banach manifold N such that $X_\mu(a(\mu)) = 0$

for all μ, where $a(\mu)$ is a smooth one-parameter family of zeros of X_μ. Suppose that for $\mu < \mu_0$, the spectrum $\sigma(dX_\mu(a(\mu))) \subset \{z \mid \mathrm{Re}\, z < 0\}$, so that $a(\mu)$ is an attracting fixed point of the flow of X_μ. To decide whether the Hopf Bifurcation Theorem applies, we compute $dX_\mu(a(\mu))$. If two simple, complex conjugate nonzero eigenvalues $\lambda(\mu)$ and $\overline{\lambda(\mu)}$ cross the imaginary axis with nonzero speed at $\mu = \mu_0$ and if the rest of $\sigma(dX_\mu(a(\mu)))$ remains in the left-half plane bounded away from the imaginary axis, then bifurcation to periodic orbits occurs.

However, since unstable periodic orbits are observed in nature only under special conditions (see Section 7), we will be interested in knowing how to decide whether or not the resultant periodic orbits are stable. In order to apply Theorem 3.1, we must reduce the problem to a two dimensional one. We assume that we are working in a chart, i.e., that $N = E$, a Banach space. For notational convenience we also assume that $\mu_0 = 0$ and $a(\mu) = 0$ for all μ. Let $X_\mu = (X_\mu^1, X_\mu^2, X_\mu^3)$ where X_μ^1 and X_μ^2 are coordinates in the eigenspace of $dX_0(0)$ corresponding to the eigenvalues $\lambda(0)$ and $\overline{\lambda(0)}$, and X_μ^3 is a coordinate in a subspace F complementary to this eigenspace. We assume that coordinates in the eigenspace have been chosen so that

$$dX_0(0,0,0) = \begin{pmatrix} 0 & |\lambda(0)| & 0 \\ -|\lambda(0)| & 0 & 0 \\ 0 & 0 & d_3X_3(0,0,0) \end{pmatrix} . \qquad (4.1)$$

This can always be arranged by splitting E into the subspaces corresponding to the splitting of the spectrum of

$dX_0(0)$ into $\{z \mid \text{Re } z < 0\} \subset \{\lambda(0), \overline{\lambda(0)}\}$, as in 2A.2. By the center manifold theorem there is a center manifold for the flow of $X = (X_\mu, 0)$ tangent to the eigenspace of $\lambda(0)$ and $\overline{\lambda(0)}$ and to the μ-axis at the point $(0,0,0,0)$. The center manifold may be represented locally as the graph of a function, that is, as $\{(x_1,x_2,f(x_1,x_2,\mu),\mu)$ for (x_1,x_2,μ) in some neighborhood of $(0,0,0)\}$. Also, $f(0,0,0) = df(0,0,0) = 0$ and the projection map $P(x_1,x_2,f(x_1,x_2,\mu),\mu) = (x_1,x_2,\mu)$ is a local chart for the center manifold. In a neighborhood of the origin X is tangent to the center manifold because the center manifold is locally invariant under the flow of X. We consider the push forward of X: $\hat{X}(x_1,x_2,\mu) = TP \circ X(x_1,x_2,f(x_1,x_2,\mu),\mu) = (X_\mu^1(x_1,x_2,f(x_1,x_2,\mu)), X_\mu^2(x_1,x_2,f(x_1,x_2,\mu)),0)$ by linearity of P. If we let $\hat{X}_\mu(x_1,x_2) = (X^1(x_1,x_2,f(x_1,x_2,\mu)), X_\mu^2(x_1,x_2,f(x_1,x_2,\mu)))$, then $\hat{X}_\mu$ is a smooth one-parameter family of vector fields on R^2 such that $\hat{X}_\mu(0,0) = 0$ for all μ. We will show that $\hat{X}_\mu$ satisfies the conditions (except, of course, the stability condition) of the Hopf Bifurcation Theorem. If ϕ_t and $\hat{\phi}_t$ are the flows of X and $\hat{X}$ respectively, then $P \circ \phi_t = \hat{\phi}_t \circ P$ for points on the center manifold. Therefore, if the resultant closed orbits of $\hat{\phi}_t$ are not attracting, those of ϕ_t will not be either. We will also show that if the origin is a vague attractor for $\hat{\phi}_t$ at $\mu = 0$, then the closed orbits of ϕ_t are attracting.

Since the center manifold has the property that it contains all the local recurrence of ϕ_t, the points $(0,0,0,\mu)$ are on it for small μ and so $f(0,0,\mu) = 0$ for small μ. Thus, $\hat{X}_\mu(0,0) = (X^1(0,0,f(0,0,\mu)), X_\mu^2(0,0,f(0,0,\mu))) = (X^1(0,0,0), X^2(0,0,0)) = 0$. $P \circ X = \hat{X} \circ P$ on the center manifold, so $P \circ dX = d\hat{X} \circ P$ for vectors tangent to the center manifold. A typical tangent vector to the center manifold has the form $(u, v, d_1 f(x_1, x_2, \mu)u + d_2 f(x_1, x_2, \mu)v + d_3 f(x_1, x_2, \mu)w, w)$, where $(x_1, x_2, f(x_1, x_2, \mu), \mu)$ is the base point of the vector. Because we wish to calculate $\sigma(d\hat{X}_\mu(0,0))$, we will be interested in the case $w = 0$. Now $P \circ dX(0,0,0,\mu)(u, v, d_1 f(0,0,\mu)u + d_2 f(0,0,\mu)v, 0) = d\hat{X}(0,0,\mu)(u,v,0)$. That is, $dX^i_\mu(0,0,0)(u, v, d_1 f(0,0,\mu)u + d_2 f(0,0,\mu)v) = dX^i_\mu(0,0)(u,v)$ for $i = 1,2$. Let $\lambda \in \sigma(d\hat{X}_\mu(0,0))$. Since $d\hat{X}_\mu(0,0)$ is a two-by-two matrix, λ is an eigenvalue and there is a complex vector (u,v) such that $d\hat{X}_\mu(0,0)(u,v) = (\lambda u, \lambda v)$. We will show that λ is a eigenvalue of $d\hat{X}_\mu(0,0,0)$ and $(u, v, d_1 f(0,0,\mu)u + d_2 f(0,0,\mu)v)$ is an eigenvector. Because X is tangent to the center manifold,

$$X^3(x_1,x_2,f(x_1,x_2,\mu),\mu) = d_1 f(x_1,x_2,\mu) X^1(x_1,x_2,f(x_1,x_2,\mu),\mu) + d_2 f(x_1,x_2,\mu) X^2(x_1,x_2,f(x_1,x_2,\mu),\mu).$$

Therefore,

$$\begin{aligned}
&d_1 X^3(x_1,x_2,f(x_1,x_2,\mu),\mu)u + d_3 X^3(x_1,x_2,\mu),\mu) \circ d_1 f(x_1,x_2,\mu)u \\
&\quad = d_1 d_1 f(x_1,x_2,\mu) \circ X^1(x_1,x_2,f(x_1,x_2,\mu),\mu)u \\
&\quad + d_1 f(x_1,x_2,\mu) \circ d_1 X^1(x_1,x_2,f(x_1,x_2,\mu),\mu)u \\
&\quad + d_1 f(x_1,x_2,\mu) \circ d_3 X^1(x_1,x_2,f(x_1,x_2,\mu),\mu) \circ d_1 f(x_1,x_2,\mu)u \\
&\quad + d_1 d_2 f(x_1,x_2,\mu) X^2(x_1,x_2,f(x_1,x_2,\mu),\mu)u \\
&\quad + d_2 f(x_1,x_2,\mu) \circ d_1 X^2(x_1,x_2,f(x_1,x_2,\mu),\mu)u \\
&\quad + d_2 f(x_1,x_2,\mu) \circ d_3 X^2(x_1,x_2,f(x_1,x_2,\mu),\mu) \circ d_1 f(x_1,x_2,\mu)u
\end{aligned}$$

and

$$d_2X^3(x_1,x_2,f(x_1,x_2,f(x_1,x_2,\mu),\mu)v$$
$$+ d_3X^3(x_1,x_2,f(x_1,x_2,\mu),\mu)\circ d_2f(x_1,x_2,\mu)v$$
$$= d_2d_1f(x_1,x_2,\mu)\circ X^1(x_1,x_2,f(x_1,x_2,\mu),\mu)v$$
$$+ d_1f(x_1,x_2,\mu)\circ d_2X^1(x_1,x_2,f(x_1,x_2,\mu),\mu)v +$$
$$+ d_1f(x_1,x_2,\mu)\circ d_3X^1(x_1,x_2,f(x_1,x_2,\mu),\mu)\circ d_2f(x_1,x_2,\mu)v$$
$$+ d_2d_2f(x_1,x_2,\mu)X^2(x_1,x_2,f(x_1,x_2,\mu),\mu)v$$
$$+ d_2f(x_1,x_2,\mu)\circ d_2X^2(x_1,x_2,f(x_1,x_2,\mu),\mu)v$$
$$+ d_2f(x_1,x_2,\mu)\circ d_3X^2(x_1,x_2,f(x_1,x_2,\mu),\mu)\circ d_2f(x_1,x_2,\mu)v.$$

At the point $(0,0,0,\mu)$, $X^1 = X^2 = 0$ and so we get

$$dX^3(0,0,0,\mu)(u,v,d_1f(0,0,\mu)u + d_2f(0,0,\mu)v)$$
$$= d_1X^3(0,0,0,\mu)u + d_2X^3(0,0,0,\mu)v + d_3X^3(0,0,0,\mu)\circ d_1f(0,0,\mu)u$$
$$+ d_3X^3(0,0,0,\mu)\circ d_2f(0,0,\mu)v$$
$$= d_1f(0,0,\mu)\circ(d_1X^1(0,0,0,\mu)u + d_2X^1(0,0,0,\mu)v$$
$$+ d_3X^1(0,0,0,\mu)\circ d_1f(0,0,\mu)u + d_3X^1(0,0,0,\mu)\circ d_2f(0,0,\mu)v)$$
$$+ d_2f(0,0,\mu)\circ(d_1X^2(0,0,0,\mu)u + d_2X^2(0,0,0,\mu)v$$
$$+ d_3X^2(0,0,0,\mu)\circ d_1f(0,0,\mu)u + d_3X^2(0,0,0,\mu)\circ d_2f(0,0,\mu)v)$$
$$= d_1f(0,0,\mu)\lambda u + d_2f(0,0,\mu)\lambda v$$
$$= \lambda(d_1f(0,0,\mu)u + d_2f(0,0,\mu)v)$$

by the assumption that (u,v) is an eigenvector of $d\hat{X}_\mu(0,0)$ with eigenvalue λ.

When $\mu = 0$, $df = 0$ and we have that

$$d\hat{X}_0(0,0) = \begin{pmatrix} 0 & |\lambda(0)| \\ -|\lambda(0)| & 0 \end{pmatrix}.$$

The eigenvalues of $d\hat{X}_\mu(0,0)$ are continuous in μ because they are roots of a quadratic polynomial. Let these roots be $\alpha_1(\mu)$ and $\alpha_2(\mu)$ so that $\alpha_1(0) = \lambda(0)$ and $\alpha_2(0) = \overline{\lambda(0)}$. Because $\alpha_1(\mu)$ and $\alpha_2(\mu) \in \sigma(dX_\mu(0,0,0))$, if $\alpha_1(\mu) \neq \lambda(\mu)$ and $\alpha_2(\mu) \neq \overline{\lambda(\mu)}$, then $\operatorname{Re} \alpha_i(\mu)$ would be bounded away from zero for small μ. Since this is not true, $\alpha_1(\mu) = \lambda(\mu)$ and $\alpha_2(\mu) = \overline{\lambda(\mu)}$. Furthermore, since $\lambda(\mu)$ and $\overline{\lambda(\mu)}$ are simple eigenvalues of $\sigma(dX_\mu(0,0,0))$, we must have that the center manifold is tangent to the eigenspace of $\{\lambda(\mu),\overline{\lambda(\mu)}\}$ at the point $(0,0,0,\mu)$.

We show now that if $V'''(0) < 0$ for $\hat{X}$, then the closed orbits of ϕ_t are attracting. The map $Q(x_1,x_2,x_3,\mu) = (x_1,x_2,x_3-f(x_1,x_2,\mu),\mu)$ is a diffeomorphism from a neighborhood $\mathcal{U}$ of $(0,0,0,0)$ onto a neighborhood V of $(0,0,0,0)$ where we have chosen $\mathcal{U}$ small enough so that X is tangent to the center manifold M for $(x_1,x_2,f(x_1,x_2,\mu),\mu) \in \mathcal{U}$. Clearly $Q|_M = P$, $\tilde{X}|_{\{x_3=0\}} = \hat{X}$, and $\tilde{\phi}_t|_{\{x_3=0\}} = \hat{\phi}_t$. Therefore, we are immediately reduced to the case of Y_μ a vector field on $R^2 \oplus F$ where R^2 is invariant under Y_μ and Y_μ satisfies the conditions for Hopf Bifurcation with R^2 being the eigenspace of $\lambda(\mu)$ and $\overline{\lambda(\mu)}$ at $(0,0,0,\mu)$. The center manifold for y is $\{(x_1,x_2,0,\mu)\}$. Assume that $V'''(0) < 0$ for $Y = Y_{\{x_3=0\}}$ and let the point $(x_1,0,0,\mu(x_1))$ be on a closed orbit of the flow ϕ_t of Y. Because R^2 is invariant,

$$d\phi_{T(x_1)}(x_1,0,\mu(x_1)) = \begin{pmatrix} a_{11} & a_{12} & a_{13} \\ a_{21} & a_{22} & a_{23} \\ 0 & 0 & d_3\phi^3_{T(x_1)}(x_1,0,0,\mu(x_1)) \end{pmatrix}.$$

By assumption, $(d_3\phi^3_{T(0)}(0,0,0,0)) = \left(e^{T(0)d_3X^3(0,0,0)}\right) = e^{\sigma(T(0)d_3X^3(0,0,0))}$ is inside the unit circle. By continuity, so is $\sigma(d_3\phi^3_{T(x_1)}(x_1,0,0,\mu(x_1)))$. Since $V'''(0) < 0$, the eigenvalues of the Poincaré map in R^2 have absolute value less than 1, so all the eigenvalues of the Poincaré map are inside the unit circle and so the orbit is attracting (see Section 2B).

Summarizing: We have shown that the stability problem for the closed orbits of the flow of X_μ is the same as that for the closed orbits of the flow of $\hat{X}_\mu$, where $\hat{X}_\mu(x_1,x_2) = (X^1_\mu(x_1,x_2,f(x_1,x_2,\mu)),\ X^2_\mu(x_1,x_2,f(x_1,x_2,\mu)))$. Coordinates are chosen so that x_1,x_2 are coordinates in the eigenspace of $dX_0(0,0,0)$ and the third component is in a complementary subspace F. The set $\{(x_1,x_2,f(x_1,x_2,\mu),\mu)$ for (x_1,x_2,μ) in a neighborhood of $(0,0,0)\}$ is the center manifold.

Outline of the Stability Calculation

From the proof of Theorem 4.5, we know that the closed orbits of $\hat{X}_\mu$ will be attracting if $V'''(0) < 0$ (or more generally, see Section 3B, if the first nonzero derivative of V at the origin is negative). The derivatives of V at $(0,0)$ can be computed from those of X_0 at $(0,0,0)$. We do this in two steps. First we compute $V'''(0)$ from the derivatives of $\hat{X}_0$, the vector field pushed to the center manifold, at $(0,0)$ using the equation:

$$V(x_1) = \int_0^{T(x_1)} \hat{X}^1(a_t(x_1,0),b_t(x_1,0))dt \tag{4.2}$$

where (a_t,b_t) is the flow of $\hat{X}$. (Note that in the two-dimensional case, $X = \hat{X}$.) Then we compute the derivatives of $\hat{X}_0$ at $(0,0)$ from those of X_0 at $(0,0,0)$. Since $\hat{X}_\mu(x_1,x_2) = (X_\mu^1(x_1,x_2,f(x_1,x_2,\mu)),\ X_\mu^2(x_1,x_2,f(x_1,x_2,\mu)))$, what we need to know is the derivatives of f at the point $(0,0,0)$. We can find these using the local invariance of the center manifold under the flow of X. We use the equation (see page 107):

$$X^3(x_1,x_2,f(x_1,x_2,\mu)) = d_1f(x_1,x_2,\mu)\circ X^1(x_1,x_2,f(x_1,x_2,\mu)) + d_2f(x_1,x_2,\mu)\circ X^2(x_1,x_2,f(x_1,x_2,\mu)). \tag{4.3}$$

<u>Calculation of $V'''(0)$ in Terms of $\hat{X}$</u>

We now calculate $V'''(0)$ from the derivatives of $\hat{X}_0$ at $(0,0)$ using (4.2). We assume that coordinates have been chosen so that

$$d\hat{X}_0(0,0) = \begin{pmatrix} \frac{\partial \hat{X}_0^1}{\partial x_1}(0,0,0) & \frac{\partial \hat{X}_0^1}{\partial x_2}(0,0,0) \\ \frac{\partial \hat{X}_0^2}{\partial x_1}(0,0,0) & \frac{\partial \hat{X}_0^2}{\partial x_2}(0,0,0) \end{pmatrix} = \begin{pmatrix} 0 & |\lambda(0)| \\ -|\lambda(0)| & 0 \end{pmatrix}. \tag{4.4}$$

This change of variables is not necessary, but it simplifies the computations considerably and, although our method for finding $V'''(0)$ will work if the change of variable has not been made, our formula will not be correct in that case.

From (4.2) we see that

$$V'(x_1) = \int_0^{T(x_1)} \frac{d}{dx_1}[\hat{X}^1(a_t(x_1,0),b_t(x_1,0))]dt$$

$$+ T'(x_1)\hat{X}^1(a_{T(x_1)}(x_1,0),b_{T(x_1)}(x_1,0))$$

$$V''(x_1) = \int_0^{T(x_1)} \frac{d^2}{dx_1^2}[\hat{X}^1(a_t(x_1,0),b_t(x_1,0))]dt$$

$$+ T'(x_1)\frac{d}{dx_1}[\hat{X}^1(a_{T(x_1)}(x_1,0),b_{T(x_1)}(x_1,0))]$$

$$+ T''(x_1)\hat{X}^1(a_{T(x_1)}(x_1,0),b_{T(x_1)}(x_1,0))$$

$$+ T'(x_1)\frac{d}{dx_1}[\hat{X}^1(a_{T(x_1)}(x_1,0),b_{T(x_1)}(x_1,0))].$$

Using the chain rule, we get:

$$V''(x_1) = \int_0^{T(x_1)} \frac{d^2}{dx_1^2}[\hat{X}^1(a_t(x_1,0),b_t(x_1,0))]dt$$

$$+ T'(x_1)\left(\frac{\partial\hat{X}^1}{\partial a}\frac{\partial a_t}{\partial x_1}\bigg|_{(T(x_1),x_1,0)} + \frac{\partial\hat{X}^1}{\partial b}\frac{\partial b_t}{\partial x_1}\bigg|_{(T(x_1),x_1,0)}\right)$$

$$+ T''(x_1)\hat{X}^1(a_{T(x_1)}(x_1,0),b_{T(x_1)}(x_1,0))$$

$$+ T'(x_1)\left(T'(x_1)\frac{\partial\hat{X}_1}{\partial a}\frac{\partial a_t}{\partial t}\bigg|_{(T(x_1),x_1,0)}\right.$$

$$+ \frac{\partial\hat{X}^1}{\partial a}\frac{\partial a_t}{\partial x_1}\bigg|_{(T(x_1),x_1,0)} + T'(x_1)\frac{\partial\hat{X}^1}{\partial b}\frac{\partial b_t}{\partial t}\bigg|_{(T(x_1),x_1,0)}$$

$$\left. + \frac{\partial\hat{X}^1}{\partial b}\frac{\partial b_t}{\partial x_1}\bigg|_{(T(x_1),x_1,0)}\right) + T'(x_1)^2\left(\frac{\partial\hat{X}^1}{\partial a}\frac{\partial a_t}{\partial t}\,(T(x_1),x_1,0)\right.$$

$$\left. + \frac{\partial\hat{X}^1}{\partial b}\frac{\partial b_t}{\partial t}\bigg|_{(T(x_1),x_1,0)}\right) + T''(x_1)\hat{X}^1(a_{T(x_1)}(x_1,0),b_{T(x_1)}(x_1,0)).$$

Differentiating once more,

$$V'''(x_1) = \int_0^{T(x_1)} \frac{d^3}{dx_1^3}[\hat{X}^1(a_t(x_1,0),b_t(x_1,0))]dt$$

$$+ T'(x_1)\frac{d^2}{dx_1^2}[\hat{X}^1(a_{T(x_1)}(x_1,0),b_{T(x_1)}(x_1,0)]$$

$$+ 2T''(x_1)\left(\frac{\partial \hat{X}^1}{\partial a}\frac{\partial a_t}{\partial x_1}\bigg|_{(T(x_1),x_1,0)} + \frac{\partial \hat{X}}{\partial b}\frac{\partial b_t}{\partial x_1}\bigg|_{(T(x_1),x_1,0)}\right)$$

$$+ 2T'(x_1)\Bigg[\frac{\partial^2\hat{X}^1}{\partial a^2}\frac{\partial a_t}{\partial x_1}\left(\frac{\partial a_t}{\partial t}T'(x_1) + \frac{\partial a_t}{\partial x_1}\right)$$

$$+ \frac{\partial^2\hat{X}^1}{\partial a\partial b}\frac{\partial a_t}{\partial x_1}\left(\frac{\partial b_t}{\partial t}T'(x_1) + \frac{\partial b_t}{\partial x_1}\right) + \frac{\partial \hat{X}^1}{\partial a}\left(\frac{\partial^2 a_t}{\partial t\partial x_1}T'(x_1) + \frac{\partial^2 a_t}{\partial x_1^2}\right)$$

$$+ \frac{\partial^2\hat{X}^1}{\partial b^2}\frac{\partial b_t}{\partial x_1}\left(\frac{\partial b_t}{\partial t}T'(x_1) + \frac{\partial b_t}{\partial x_1}\right)$$

$$+ \frac{\partial \hat{X}^1}{\partial b}\left(\frac{\partial^2 b_t}{\partial t\partial x_1}T'(x_1) + \frac{\partial^2 b_t}{\partial x_1^2}\right)\Bigg]\Bigg|_{(T(x_1),x_1,0)}$$

$$+ 2T'(x_1)T''(x_1)\left(\frac{\partial \hat{X}^1}{\partial a}\frac{\partial a_t}{\partial t} + \frac{\partial \hat{X}^1}{\partial b}\frac{\partial b_t}{\partial t}\right)\bigg|_{(T(x_1),x_1,0)}$$

$$+ T'(x_1)^2\Bigg[\frac{\partial^2\hat{X}^1}{\partial a^2}\frac{\partial a_t}{\partial t}\left(\frac{\partial a_t}{\partial t}T'(x_1) + \frac{\partial a_t}{\partial x_1}\right)$$

$$+ \frac{\partial^2\hat{X}^1}{\partial a\partial b}\frac{\partial a_t}{\partial t}\left(\frac{\partial b_t}{\partial t}T'(x_1) + \frac{\partial b_t}{\partial x_1}\right) + \frac{\partial \hat{X}^1}{\partial a}\left(\frac{\partial^2 a_t}{\partial t^2}T'(x_1) + \frac{\partial^2 a_t}{\partial x_1\partial t}\right)$$

$$+ \frac{\partial^2\hat{X}^1}{\partial b^2}\frac{\partial b_t}{\partial t}\left(\frac{\partial b_t}{\partial t}T'(x_1) + \frac{\partial b_t}{\partial x_1}\right)$$

$$+ \frac{\partial \hat{X}^1}{\partial b}\left(\frac{\partial^2 b_t}{\partial t^2}T'(x_1) + \frac{\partial^2 b_t}{\partial x_1\partial t}\right)\Bigg]\Bigg|_{(T(x_1),x_1,0)}$$

$$+ T'''(x_1)\hat{X}^1(a_{T(x_1)}(x_1,0),b_{T(x_1)}(x_1,0))$$

$$+ T''(x_1)\left[\frac{\partial \hat{X}^1}{\partial a}\frac{\partial a_t}{\partial t}T'(x_1) + \frac{\partial \hat{X}^1}{\partial a}\frac{\partial a_t}{\partial x_1} + \frac{\partial \hat{X}^1}{\partial b}\frac{\partial b_t}{\partial t}T'(x_1)\right.$$

$$\left.\left. + \frac{\partial \hat{X}^1}{\partial b}\frac{\partial b_t}{\partial x_1}\right]\right|_{(T(x_1),x_1,0)} .$$

In the case $x_1 = 0$, we can considerably simplify this equation. We know the following about the point $(0,0)$:

$$a_t(0,0) = b_t(0,0) = 0 \quad \text{for all} \quad t \tag{4.5}$$

$$\frac{\partial a_t}{\partial t}(0,0) = \frac{\partial b_t}{\partial t}(0,0) = 0 \tag{4.6}$$

$$d\hat{X}(a_t(0,0),\ b_t(0,0)) = d\hat{X}(0,0) = \begin{pmatrix} 0 & |\lambda(0)| \\ -|\lambda(0)| & 0 \end{pmatrix} \tag{4.7}$$

$$d\hat{\phi}_t(0,0) = e^{td\hat{X}(0,0)} = \begin{pmatrix} \cos|\lambda(0)|t & \sin|\lambda(0)|t \\ -\sin|\lambda(0)|t & \cos|\lambda(0)|t \end{pmatrix} \tag{4.8}$$

$$T(0) = 2\pi/|\lambda(0)| \tag{4.9}$$

and

$$T'(0) = 0. \tag{4.10}$$

<u>Proof of (4.10)</u>. Let $S(x_1) = T(x_1,\mu(x_1))$. Then $S'(0) = 0$ because given small $x > 0$, there is a small $y < 0$ such that $S(x) = S(y)$. Thus, $\frac{S(x) - S(0)}{x}$ and $\frac{S(y) - S(0)}{y}$ have opposite signs. Choosing $x_n \downarrow 0$, we get the result. $S'(0) = \frac{\partial T}{\partial x_1}(0,0) + \mu'(0)\frac{\partial T}{\partial \mu}(0,0)$. But $\mu'(0) = 0$, as was shown in the Proof of Theorem 3.1 (see p. 65).

Therefore, $V'''(0) = \int_0^{2\pi/|\lambda(0)|} \frac{\partial^3 \hat{X}^1}{\partial x_1^3}(a_t(x_1,0),b_t(x_1,0))dt.$

We now evaluate $\left.\frac{d^3 X_1}{dx_1^3}\right|_{a_t(0,0),b_t(0,0)}$ and get:

$$V'''(0) = \int_0^{2\pi/|\lambda(0)|} \left[\frac{\partial^3 \hat{X}}{\partial a^3}(0,0)\cos^3|\lambda(0)|t - \frac{\partial^3 \hat{X}}{\partial b^3}(0,0)\sin^3|\lambda(0)|t \right.$$

$$- 3\frac{\partial^3 \hat{X}^1}{\partial a^2 \partial b}(0,0)\cos^2|\lambda(0)|t\ \sin|\lambda(0)|t$$

$$+ 3\frac{\partial^3 \hat{X}^1}{\partial a \partial b^2}(0,0)\cos|\lambda(0)|t\ \sin^2|\lambda(0)|t$$

$$+ 3\frac{\partial^2 \hat{X}^1}{\partial a^2}(0,0)\ \frac{\partial^2 a_t}{\partial x_1^2}(0,0)\cos|\lambda(0)|t$$

$$- 3\frac{\partial^2 \hat{X}^1}{\partial b^2}(0,0)\ \frac{\partial^2 b_t}{\partial x_1^2}(0,0)\sin|\lambda(0)|t$$

$$+ 3\frac{\partial^2 \hat{X}^1}{\partial a \partial b}(0,0)\left(\frac{\partial^2 b_t}{\partial x_1^2}(0,0)\cos|\lambda(0)|t\right.$$

$$\left.\left. - \frac{\partial^2 a_t}{\partial x_1^2}(0,0)\sin|\lambda(0)|t\right) + |\lambda(0)|\ \frac{\partial^3 b_t}{\partial x_1^3}(0,0)\right]dt$$

$$= \int_0^{2\pi/|\lambda(0)|}\left[3\frac{\partial^2 \hat{X}^1}{\partial a^2}(0,0)\ \frac{\partial^2 a_t}{\partial x_1^2}(0,0)\cos|\lambda(0)|t\right.$$

$$- 3\frac{\partial^2 \hat{X}^1}{\partial b^2}(0,0)\ \frac{\partial^2 b_t}{\partial x_1^2}(0,0)\sin|\lambda(0)|t$$

$$+ 3\frac{\partial^2 \hat{X}^1}{\partial a \partial b}(0,0)\left(\frac{\partial^2 b_t}{\partial x_1^2}(0,0)\cos|\lambda(0)|t - \frac{\partial^2 a_t}{\partial x_1^2}\sin|\lambda(0)|t\right)$$

$$\left. + |\lambda(0)|\ \frac{\partial^3 b_t}{\partial x_1^3}(0,0)\right]dt.$$

In order to get a formula for $V'''(0)$ depending only on the derivatives of $\hat{X}$ at the origin, we must evaluate the derivatives of the flow (e.g., $\frac{\partial^3 b_t}{\partial x_1^3}(0,0)$) from those of $\hat{X}$

at $(0,0)$. This can be done because the origin is a fixed point of the flow of $\hat{X}$. Because this idea is important, we state it in a more general case.

(4.1) Theorem. *Let* X *be a* C^k *vector field on* R^n *such that* $X(0) = 0$ (*or* $X(p) = 0$). *Let* Φ_t *be the time* t *map of the flow of* X. *The first three* (*or, the first* j) *derivatives of* Φ_t *at* 0 *can be calculated from the first three* (*or, the first* j) *derivatives of* X *at* 0.

Proof. Consider $\dfrac{\partial \Phi_t^i}{\partial x_j}(0)$:

$$\frac{\partial}{\partial t}\frac{\partial \Phi_t^i}{\partial x_j}(0) = \frac{\partial}{\partial x_j}\frac{\partial \Phi_t^i}{\partial t}(0) = \frac{\partial}{\partial x_j} X^i \circ \Phi_t(0) = \frac{\partial X^i}{\partial x_k} \circ \Phi_t(0)\,\frac{\partial \Phi_t^k}{\partial x_j}(0)$$

$$= \frac{\partial X^i}{\partial x_k}(0)\,\frac{\partial \Phi_t^k}{\partial x_j}(0)$$

because $\Phi_t(0) = 0$. Furthermore, $\dfrac{\partial \Phi_0^i}{\partial x_j}(0) = \delta_{ij}$ because $\Phi_0(x) = x$ for all x. So $d\Phi_t(0)$ satisfies the differential equation $\frac{\partial}{\partial t}(d\Phi_t(0)) = dX(0)\cdot d\Phi_t(0)$ and $d\Phi_0(0) = I$. Thus, $d\Phi_t(0) = e^{t dX(0)}$.

Consider $\dfrac{\partial^2 \Phi_t^i}{\partial x_j \partial x_k}(0)$:

$$\frac{\partial}{\partial t}\frac{\partial^2 \Phi_t^i}{\partial x_j \partial x_k}(0) = \frac{\partial^2}{\partial x_j \partial x_k}\frac{\partial \Phi_t^i}{\partial t}(0) = \frac{\partial^2}{\partial x_j \partial x_k} X^i \circ \Phi_t(0)$$

$$= \frac{\partial}{\partial x_j}\left(\frac{\partial X^i}{\partial x_\ell} \circ \Phi_t\,\frac{\partial \Phi_t^\ell}{\partial x_k}\right)(0) = \frac{\partial^2 X^i}{\partial x_p \partial x_\ell} \circ \Phi_t\,\frac{\partial \Phi_t^p}{\partial x_j}\,\frac{\partial \Phi_t^\ell}{\partial x_k}$$

$$+ \frac{\partial X^i}{\partial x_\ell} \circ \Phi_t\,\frac{\partial^2 \Phi_t^\ell}{\partial x_j \partial x_k}(0)$$

$$= \frac{\partial^2 X^i}{\partial x_p \partial x_\ell}(0)\,\frac{\partial \Phi_t^p}{\partial x_j}(0)\,\frac{\partial \Phi_t^\ell}{\partial x_k}(0) + \frac{\partial X^i}{\partial x_\ell}(0)\,\frac{\partial^2 \Phi_t^\ell}{\partial x_j \partial x_k}(0).$$

Furthermore, $\dfrac{\partial^2 \Phi_0^i}{\partial x_j \partial x_k}(0) = 0$. We get the differential equation:

$$\frac{\partial^2 \Phi_t^i}{\partial x_j \partial x_k}(0) = d^2X(0)\left(\frac{\partial \Phi_t}{\partial x_j}(0), \frac{\partial \Phi_t}{\partial x_k}(0)\right) + dX(0) \cdot \frac{\partial^2 \Phi_t}{\partial x_j \partial x_k} .$$

The solution is:

$$\frac{\partial^2 \Phi_t}{\partial x_j \partial x_k}(0) = e^{tdX(0)} \int_0^t e^{-sdX(0)} d^2X(0)\left(\frac{\partial \Phi_s}{\partial x_j}(0), \frac{\partial \Phi_s}{\partial x_k}(0)\right) ds$$

$$+ e^{tdX(0)} \frac{\partial^2 \Phi_0}{\partial x_j \partial x_k}(0).$$

$$\frac{\partial^2 \Phi_t}{\partial x_j \partial x_k}(0) = e^{tdX(0)} \int_0^t e^{-sdX(0)} d^2X(0)\left(\frac{\partial \Phi_s}{\partial x_j}(0), \frac{\partial \Phi_s}{\partial x_k}(0)\right) ds.$$

Finally consider $\dfrac{\partial^3 \Phi_t^i}{\partial x_j \partial x_k \partial x_h}(0)$:

$$\frac{\partial}{\partial t}\left(\frac{\partial^3 \Phi_t^i}{\partial x_j \partial x_k \partial x_h}\right)(0) = \frac{\partial^3}{\partial x_j \partial x_k \partial x_h}\left(\frac{\partial \Phi_t^i}{\partial t}\right)(0) = \frac{\partial^3}{\partial x_j \partial x_k \partial x_h} X^i \circ \Phi_t(0)$$

$$= \frac{\partial}{\partial x_h}\left(\frac{\partial^2 X^i}{\partial x_p \partial x_\ell} \circ \Phi_t \frac{\partial \Phi_t^p}{\partial x_j} \frac{\partial \Phi_t^\ell}{\partial x_k} + \frac{\partial X^i}{\partial x_\ell} \circ \Phi_t \frac{\partial^2 \Phi_t^\ell}{\partial x_j \partial x_k}\right)(0)$$

$$= \left[\frac{\partial^3 X^i}{\partial x_p \partial x_\ell \partial x_q} \circ \Phi_t \frac{\partial \Phi_t^q}{\partial x_h} \frac{\partial \Phi_t^p}{\partial x_j} \frac{\partial \Phi_t^\ell}{\partial x_k}\right.$$

$$+ \frac{\partial^2 X^i}{\partial x_p \partial x_\ell} \circ \Phi_t\left(\frac{\partial^2 \Phi_t^p}{\partial x_j \partial x_h} \frac{\partial \Phi_t^\ell}{\partial x_k} + \frac{\partial \Phi_t^p}{\partial x_j} \frac{\partial^2 \Phi_t^\ell}{\partial x_k \partial x_h}\right)$$

$$\left. + \frac{\partial^2 X^i}{\partial x_p \partial x_\ell} \circ \Phi_t \frac{\partial \Phi_t^p}{\partial x_h} \frac{\partial^2 \Phi_t^\ell}{\partial x_j \partial x_k} + \frac{\partial X^i}{\partial x_\ell} \circ \Phi_t \frac{\partial^3 \Phi_t^\ell}{\partial x_j \partial x_k \partial x_h}\right](0)$$

$$= \frac{\partial^3 X^i}{\partial x_p \partial x_\ell \partial x_q}(0) \frac{\partial \Phi_t^q}{\partial x_h}(0) \frac{\partial \Phi_t^p}{\partial x_j}(0) \frac{\partial \Phi_t^\ell}{\partial x_k}(0)$$

$$+ \frac{\partial^2 X^i}{\partial x_p \partial x_\ell}(0) \left(\frac{\partial^2 \Phi_t^p}{\partial x_j \partial x_h}(0) \frac{\partial \Phi_t^\ell}{\partial x_k}(0) + \frac{\partial \Phi_t^p}{\partial x_j}(0) \frac{\partial^2 \Phi_t^\ell}{\partial x_k \partial x_h}(0) \right.$$

$$\left. + \frac{\partial^2 \Phi_t^\ell}{\partial x_j \partial x_k}(0) \frac{\partial \Phi_t^p}{\partial x_h}(0) \right) + \frac{\partial X^i}{\partial x_\ell}(0) \frac{\partial^3 \Phi_t^\ell}{\partial x_j \partial x_k \partial x_h}(0).$$

Furthermore, $\frac{\partial^3 \Phi_0^i}{\partial x_j \partial x_k \partial x_h}(0) = 0$. We get the differential equation:

$$\frac{\partial}{\partial t}\left(\frac{\partial^3 \Phi_t^i}{\partial x_j \partial x_k \partial x_h}(0) \right) = d^3X(0)\left(\frac{\partial \Phi_t}{\partial x_j}(0), \frac{\partial \Phi_t}{\partial x_k}(0), \frac{\partial \Phi_t}{\partial x_h}(0) \right)$$

$$+ d^2X(0)\left(\frac{\partial \Phi_t}{\partial x_k}(0), \frac{\partial^2 \Phi_t}{\partial x_j \partial x_h}(0) \right) + d^2X(0)\left(\frac{\partial \Phi_t}{\partial x_j}(0), \frac{\partial^2 \Phi_t}{\partial x_k \partial x_h}(0) \right)$$

$$+ d^2X(0)\left(\frac{\partial \Phi_t}{\partial x_h}(0), \frac{\partial^2 \Phi_t}{\partial x_j \partial x_k}(0) \right) + dX(0)\left(\frac{\partial^3 \Phi_t}{\partial x_j \partial x_k \partial x_h}(0) \right)$$

and

$$\frac{\partial^3 \Phi_0}{\partial x_j \partial x_k \partial x_h}(0) = 0.$$

The solution is:

$$\frac{\partial^3 \Phi_t}{\partial x_j \partial x_k \partial x_h}(0)$$

$$= e^{tdX(0)} \int_0^t e^{-sdX(0)} \left\{ d^3X(0)\left(\frac{\partial \Phi_t}{\partial x_j}(0), \frac{\partial \Phi_t}{\partial x_k}(0), \frac{\partial \Phi_t}{\partial x_h}(0) \right) \right.$$

$$+ d^2X(0)\left(\frac{\partial \Phi_t}{\partial x_k}(0), \frac{\partial^2 \Phi_t}{\partial x_j \partial x_h}(0) \right) + d^2X(0)\left(\frac{\partial \Phi_t}{\partial x_j}(0), \frac{\partial^2 \Phi_t}{\partial x_k \partial x_h}(0) \right)$$

$$+ \; d^2X(0)\left(\frac{\partial\Phi_t}{\partial x_h}(0), \frac{\partial^2\Phi_t}{\partial x_j\, x_k}(0)\right)\Bigg\}>ds.$$

In the case we are considering, $d\hat{X}(0,0) = \begin{pmatrix} 0 & |\lambda(0)| \\ -|\lambda(0)| & 0 \end{pmatrix}$

and so $d\hat{\phi}_t(0,0) = \begin{pmatrix} \cos|\lambda(0)|t & \sin|\lambda(0)|t \\ -\sin|\lambda(0)|t & \cos|\lambda(0)|t \end{pmatrix}$. We wish to

calculate $\dfrac{\partial^2\hat{\phi}_t}{\partial x_1^2}(0,0) = \left(\dfrac{\partial^2 a_t}{\partial x_1^2}(0,0), \dfrac{\partial^2 b_t}{\partial x_1^2}(0,0)\right)$ and

$\dfrac{\partial^3 b_t}{\partial x_1^3}(0,0)$. First note that $e^{-sdX(0)} = e^{-s\begin{pmatrix} 0 & |\lambda(0)| \\ -|\lambda(0)| & 0 \end{pmatrix}}$

$= \begin{pmatrix} \cos|\lambda(0)|s & -\sin|\lambda(0)|s \\ \sin|\lambda(0)|s & \cos|\lambda(0)|s \end{pmatrix}$. Also,

$$d^2\hat{X}(0)\left(\frac{\partial\hat{\phi}_s}{\partial x_1}(0), \frac{\partial\hat{\phi}_s}{\partial x_1}(0)\right) = \Bigg(\frac{\partial^2\hat{X}^1}{\partial x_1^2}(0,0)\cos^2|\lambda(0)|s$$

$$- \; 2\frac{\partial^2\hat{X}^1}{\partial x_1\partial x_2}(0,0)\cos|\lambda(0)|s\,\sin|\lambda(0)|s$$

$$+ \; \frac{\partial^2\hat{X}^1}{\partial x_2^2}(0,0)\sin^2|\lambda(0)|s, \; \frac{\partial^2\hat{X}^2}{\partial x_1^2}(0,0)\cos^2|\lambda(0)|s$$

$$- \; 2\frac{\partial^2\hat{X}^2}{\partial x_1\partial x_2}(0,0)\cos|\lambda(0)|s\,\sin|\lambda(0)|s$$

$$+ \; \frac{\partial^2\hat{X}^2}{\partial x_2^2}(0,0)\sin^2|\lambda(0)|s\Bigg).$$

Consequently,

$$e^{-sd\hat{X}(0)}\, d^2\hat{X}(0)\left(\frac{\partial\hat{\phi}_s}{\partial x_1}(0), \frac{\partial\hat{\phi}_s}{\partial x_1}(0)\right) =$$

$$= \left[\frac{\partial^2 \hat{X}^1}{\partial x_1^2}(0)\cos^3|\lambda(0)|s - \frac{\partial^2 \hat{X}^2}{\partial x_2^2}(0)\sin^3|\lambda(0)|s\right.$$

$$+ \left(-2\frac{\partial^2 \hat{X}^1}{\partial x_1 \partial x_2}(0) - \frac{\partial^2 \hat{X}^2}{\partial x_1^2}(0)\right)\sin|\lambda(0)|s\, \cos^2|\lambda(0)|s$$

$$+ \left(2\frac{\partial^2 \hat{X}^2}{\partial x_1 \partial x_2}(0) + \frac{\partial^2 \hat{X}^1}{\partial x_2^2}(0)\right)\cos|\lambda(0)|s\, \sin^2|\lambda(0)|s,\ \frac{\partial^2 \hat{X}^2}{\partial x_1^2}(0)\cos^3|\lambda(0)|s$$

$$+ \frac{\partial^2 \hat{X}^1}{\partial x_2^2}(0)\ \sin^3|\lambda(0)|s + \left(-2\frac{\partial^2 \hat{X}^2}{\partial x_1 \partial x_2}(0) + \frac{\partial^2 \hat{X}^1}{\partial x_1^2}(0)\right)\sin|\lambda(0)|s\, \cos^2|\lambda(0)|s$$

$$\left. + \left(-2\frac{\partial^2 \hat{X}^1}{\partial x_1 \partial x_2}(0) + \frac{\partial^2 \hat{X}^2}{\partial x_2^2}(0)\right)\cos|\lambda(0)|s\, \sin^2|\lambda(0)|s\right],$$

and thus

$$\int_0^t e^{-sd\hat{X}(0)}\ d^2\hat{X}(0)\left(\frac{\partial \hat{\phi}_s}{\partial x_1}(0),\ \frac{\partial \hat{\phi}_s}{\partial x_1}(0)\right)ds$$

$$= \frac{1}{3|\lambda(0)|}\left[\left(-2\frac{\partial^2 \hat{X}^2}{\partial x_2^2} - 2\frac{\partial^2 \hat{X}^1}{\partial x_1 \partial x_2} - \frac{\partial^2 \hat{X}^2}{\partial x_1^2}\right) + 3\frac{\partial^2 \hat{X}^1}{\partial x_1^2}\sin|\lambda(0)|t\right.$$

$$+ 3\frac{\partial^2 \hat{X}^2}{\partial x_2^2}\cos|\lambda(0)|t + \left(-\frac{\partial^2 \hat{X}^1}{\partial x_1^2} + 2\frac{\partial^2 \hat{X}^2}{\partial x_1 \partial x_2} + \frac{\partial^2 \hat{X}^1}{\partial x_2^2}\right)\sin^3|\lambda(0)|t$$

$$+ \left(-\frac{\partial^2 \hat{X}^2}{\partial x_2^2} + 2\frac{\partial^2 \hat{X}^1}{\partial x_1 \partial x_2} + \frac{\partial^2 \hat{X}^2}{\partial x_1^2}\right)\cos^3|\lambda(0)|t,$$

$$\left(2\frac{\partial^2 \hat{X}^1}{\partial x_2^2} - 2\frac{\partial^2 \hat{X}^2}{\partial x_1 \partial x_2} + \frac{\partial^2 \hat{X}^1}{\partial x_1^2}\right)$$

$$+ 3 \frac{\partial^2 \hat{X}^1}{\partial x_1^2} \sin|\lambda(0)|t - 3 \frac{\partial^2 \hat{X}^1}{\partial x_2^2} \cos|\lambda(0)|t$$

$$+ \left(- \frac{\partial^2 \hat{X}^2}{\partial x_1^2} - 2 \frac{\partial^2 \hat{X}^1}{\partial x_1 \partial x_2} + \frac{\partial^2 \hat{X}^2}{\partial x_2^2} \right) \sin^3|\lambda(0)|t$$

$$\left. + \left(\frac{\partial^2 \hat{X}^1}{\partial x_2^2} + 2 \frac{\partial^2 \hat{X}^2}{\partial x_1 \partial x_2} - \frac{\partial^2 \hat{X}^1}{\partial x_1^2} \right) \cos^3|\lambda(0)|t \right]$$

(where all derivatives are evaluated at the origin).

Putting this all together,

$$e^{td\hat{X}(0)} \int_0^t e^{-sd\hat{X}(0)} d^2\hat{X}(0) \left(\frac{\partial \hat{\phi}_s}{\partial x_1}(0), \frac{\partial \hat{\phi}_s}{\partial x_1}(0) \right) ds = \left(\frac{\partial^2 a_t}{\partial x_1^2}(0), \frac{\partial^2 b_t}{\partial x_1^2}(0) \right)$$

$$= \frac{1}{3|\lambda(0)|} \left[\left(- 2 \frac{\partial^2 \hat{X}^2}{\partial x_2^2} - 2 \frac{\partial^2 \hat{X}^1}{\partial x_1 \partial x_2} - \frac{\partial^2 \hat{X}^2}{\partial x_1^2} \right) \cos|\lambda(0)|t \right.$$

$$+ \left(2 \frac{\partial^2 \hat{X}^1}{x_2^2} - 2 \frac{\partial^2 \hat{X}^2}{x_1 \, x_2} + \frac{\partial^2 \hat{X}^1}{x_1^2} \right) \sin|\lambda(0)|t$$

$$+ \left(3 \frac{\partial^2 \hat{X}^1}{\partial x_1^2} - 3 \frac{\partial^2 \hat{X}^1}{\partial x_2^2} \right) \sin|\lambda(0)|t \cos|\lambda(0)|t$$

$$+ 3 \frac{\partial^2 \hat{X}^2}{\partial x_2^2} \cos^2|\lambda(0)|t + 3 \frac{\partial^2 \hat{X}^2}{\partial x_1^2} \sin^2|\lambda(0)|t$$

$$+ \left(- \frac{\partial^2 \hat{X}^1}{\partial x_1^2} + 2 \frac{\partial^2 \hat{X}^2}{\partial x_1 \partial x_2} + \frac{\partial^2 \hat{X}^1}{\partial x_2^2} \right) \cos|\lambda(0)|t \sin^3|\lambda(0)|t$$

$$+ \left(\frac{\partial^2 \hat{X}^1}{\partial x_2^2} + 2 \frac{\partial^2 \hat{X}^2}{\partial x_1 \partial x_2} - \frac{\partial^2 \hat{X}^1}{\partial x_1^2} \right) \sin|\lambda(0)|t \cos^3|\lambda(0)|t$$

$$+ \left(-\frac{\partial^2\hat{X}^2}{\partial x_2^2} + 2\frac{\partial^2\hat{X}^1}{\partial x_1\partial x_2} + \frac{\partial^2\hat{X}^2}{\partial x_1^2}\right)\cos^4|\lambda(0)|t$$

$$+ \left(-\frac{\partial^2\hat{X}^2}{\partial x_1^2} - 2\frac{\partial^2\hat{X}^1}{\partial x_1\partial x_2} + \frac{\partial^2\hat{X}^2}{\partial x_2^2}\right)\sin^4|\lambda(0)|t,$$

$$\left(2\frac{\partial^2\hat{X}^1}{\partial x_2^2} - 2\frac{\partial^2\hat{X}^2}{\partial x_1\partial x_2} + \frac{\partial^2\hat{X}^1}{\partial x_1^2}\right)\cos|\lambda(0)|t$$

$$+ \left(2\frac{\partial^2\hat{X}^2}{\partial x_2^2} + 2\frac{\partial^2\hat{X}^1}{\partial x_1\partial x_2} + \frac{\partial^2\hat{X}^2}{\partial x_1^2}\right)\sin|\lambda(0)|t$$

$$+ \left(3\frac{\partial^2\hat{X}^2}{\partial x_1^2} - 3\frac{\partial^2\hat{X}^2}{\partial x_2^2}\right)\sin|\lambda(0)|t\,\cos|\lambda(0)|t$$

$$- 3\frac{\partial^2\hat{X}^1}{\partial x_1^2}\sin^2|\lambda(0)|t - 3\frac{\partial^2\hat{X}^1}{\partial x_2^2}\cos^2|\lambda(0)|t$$

$$+ \left(\frac{\partial^2\hat{X}^2}{\partial x_2^2} - 2\frac{\partial^2\hat{X}^1}{\partial x_1\partial x_2} - \frac{\partial^2\hat{X}^2}{\partial x_1^2}\right)\sin|\lambda(0)|t\,\cos^3|\lambda(0)|t$$

$$+ \left(-\frac{\partial^2\hat{X}^2}{\partial x_1^2} - 2\frac{\partial^2\hat{X}^1}{\partial x_1\partial x_2} + \frac{\partial^2\hat{X}^2}{\partial x_2^2}\right)\cos|\lambda(0)|t\,\sin^3|\lambda(0)|t$$

$$+ \left(\frac{\partial^2\hat{X}^1}{\partial x_1^2} - 2\frac{\partial^2\hat{X}^2}{\partial x_1\partial x_2} - \frac{\partial^2\hat{X}^1}{\partial x_2^2}\right)\sin^4|\lambda(0)|t$$

$$+ \left.\left(\frac{\partial^2\hat{X}^1}{\partial x_2^2} + 2\frac{\partial^2\hat{X}^2}{\partial x_1\partial x_2} - \frac{\partial^2\hat{X}^1}{\partial x_1^2}\right)\cos^4|\lambda(0)|t\right].$$

Before computing $\frac{\partial^3 b_t}{\partial x_1^3}(0)$, which is a lengthy calculation, we will use the information above to simplify our expression for $V'''(0)$. To do this, we must compute

$$\int_0^{2\pi/|\lambda(0)|} \frac{\partial^2 a_t}{\partial x_1^2}(0,0)\cos|\lambda(0)|t\,dt,$$

$$\int_0^{2\pi/|\lambda(0)|} \frac{\partial^2 a_t}{\partial x_1^2}(0,0)\sin|\lambda(0)|t\,dt,$$

$$\int_0^{2\pi/|\lambda(0)|} \frac{\partial^2 b_t}{\partial x_1^2}(0,0)\cos|\lambda(0)|t\,dt,$$

and

$$\int_0^{2\pi/|\lambda(0)|} \frac{\partial^2 b_t}{\partial x_1^2}(0,0)\sin|\lambda(0)|t\,dt.$$

The results are:

$$\int_0^{2\pi/|\lambda(0)|} \frac{\partial^2 a_t}{\partial x_1^2}(0,0)dt = \frac{\pi}{|\lambda(0)|^2}\left[\frac{\partial^2\hat{X}^2}{\partial a^2} + \frac{\partial^2\hat{X}^2}{\partial b^2}\right]$$

$$\int_0^{2\pi/|\lambda(0)|} \frac{\partial^2 a_t}{\partial x_1^2}(0,0)\cos|\lambda(0)|t\,dt = \frac{\pi}{3|\lambda(0)|^2}\left(-2\frac{\partial^2\hat{X}^2}{\partial b^2} - 2\frac{\partial^2\hat{X}^1}{\partial a\partial b} - \frac{\partial^2\hat{X}^2}{\partial a^2}\right)$$

$$\int_0^{2\pi/|\lambda(0)|} \frac{\partial^2 a_t}{\partial x_1^2}(0,0)\sin|\lambda(0)|t\,dt = \frac{\pi}{3|\lambda(0)|^2}\left(2\frac{\partial^2\hat{X}^1}{\partial b^2} - 2\frac{\partial^2\hat{X}^2}{\partial a\partial b} + \frac{\partial^2\hat{X}^1}{\partial a^2}\right)$$

$$\int_0^{2\pi/|\lambda(0)|} \frac{\partial^2 b_t}{\partial x_1^2}(0,0)\cos|\lambda(0)|t\,dt = \frac{\pi}{3|\lambda(0)|^2}\left(2\frac{\partial^2\hat{X}^1}{\partial b^2} - 2\frac{\partial^2\hat{X}^2}{\partial a\partial b} + \frac{\partial^2\hat{X}^1}{\partial a^2}\right)$$

$$\int_0^{2\pi/|\lambda(0)|} \frac{\partial^2 b_t}{\partial x_1^2}(0,0)\sin|\lambda(0)|t\,dt = \frac{\pi}{3|\lambda(0)|^2}\left(2\frac{\partial^2\hat{X}^2}{\partial b^2} + 2\frac{\partial^2\hat{X}^1}{\partial a\partial b} + \frac{\partial^2\hat{X}^2}{\partial a^2}\right)$$

Therefore,

$$V'''(0) = |\lambda(0)|\int_0^{2\pi/|\lambda(0)|} \frac{\partial^2 b_t}{\partial x_1^3}(0,0)dt$$

$$+ 3\int_0^{2\pi/|\lambda(0)|}\left[\frac{\partial^2\hat{X}^1}{\partial a^2}(0,0)\,\frac{\partial^2 a_t}{\partial x_1^2}(0,0)\cos|\lambda(0)|t\right.$$

$$- \frac{\partial^2 \hat{X}^1}{\partial b^2}(0,0) \frac{\partial^2 b_t}{\partial x_1^2}(0,0)\sin|\lambda(0)|t$$

$$+ \frac{\partial^2 \hat{X}^1}{\partial a \partial b}(0,0)\left[\frac{\partial^2 b_t}{\partial x_1^2}(0,0)\cos|\lambda(0)|t\right.$$

$$\left.\left. - \frac{\partial^2 a_t}{\partial x_1^2}(0,0)\sin|\lambda(0)|t\right]\right]dt$$

i.e.

$$V'''(0) = |\lambda(0)| \int_0^{2\pi/|\lambda(0)|} \frac{\partial^3 b_t}{\partial x_1^3}(0,0)dt$$

$$+ \frac{\pi}{|\lambda(0)|^2} \frac{\partial^2 \hat{X}^1}{\partial a^2}\left(-2\frac{\partial^2 \hat{X}^2}{\partial b^2} - 2\frac{\partial^2 \hat{X}^1}{\partial a \partial b} - \frac{\partial^2 \hat{X}^2}{\partial a^2}\right)$$

$$+ \frac{\pi}{|\lambda(0)|^2} \frac{\partial^2 \hat{X}^1}{\partial b^2}\left(-2\frac{\partial^2 \hat{X}^2}{\partial b^2} - 2\frac{\partial^2 \hat{X}^1}{\partial a \partial b} - \frac{\partial^2 \hat{X}^2}{\partial a^2}\right)$$

$$= |\lambda(0)| \int_0^{2\pi/|\lambda(0)|} \frac{\partial^3 b_t}{\partial x_1^3}(0,0)dt$$

$$- \frac{\pi}{|\lambda(0)|^2}\left(2\frac{\partial^2 \hat{X}^1}{\partial a^2}\frac{\partial^2 \hat{X}^2}{\partial b^2} + \frac{\partial^2 \hat{X}^1}{\partial b^2}\frac{\partial^2 \hat{X}^2}{\partial a^2} + 2\frac{\partial^2 \hat{X}^1}{\partial a^2}\frac{\partial^2 \hat{X}^1}{\partial a \partial b} + 2\frac{\partial^2 \hat{X}^1}{\partial b^2}\frac{\partial^2 \hat{X}^1}{\partial a \partial b}\right.$$

$$\left. + \frac{\partial^2 \hat{X}^1}{\partial a^2}\frac{\partial^2 \hat{X}^2}{\partial a^2} + 2\frac{\partial^2 \hat{X}^1}{\partial b^2}\frac{\partial^2 \hat{X}^2}{\partial b^2}\right)$$

(where all derivatives are taken at the origin).

To compute $\frac{\partial^3 b_t}{\partial x_1^3}(0,0)$, we use the equation:

$$\left(\frac{\partial^3 a_t}{\partial x_1^3}(0,0), \frac{\partial^3 b_t}{\partial x_1^3}(0,0)\right)$$

$$= e^{td\hat{X}(0)} \int_0^t e^{-sd\hat{X}(0)} \left\{ d^3\hat{X}(0)\left(\frac{\partial \hat{\phi}_s}{\partial x_1}(0,0), \frac{\partial \hat{\phi}_s}{\partial x_1}(0,0), \frac{\partial \hat{\phi}_s}{\partial x_1}(0,0)\right)\right.$$

$$\left. + 3d^2\hat{X}(0)\left(\frac{\partial \hat{\phi}_s}{\partial x_1}(0,0), \frac{\partial^2 \hat{\phi}_s}{\partial x_1^2}(0,0)\right)\right\}ds.$$

The calculation involved is quite long,* but is straightforward, so we will merely indicate how it is done and then state the results. The lengthy computation alluded to is:

$$\int_0^{2\pi/|\lambda(0)|} e^{td\hat{X}(0)} \int_0^t e^{-sdX(0)} d^2\hat{X}(0)\left(\frac{\partial \Phi_s}{\partial x_1}(0,0),\ \frac{\partial^2 \Phi_s}{\partial x_1^2}(0,0)\right) ds\, dt$$

$$= \Bigg(\text{something},\ \frac{2\pi}{|\lambda(0)|^3}\frac{\partial^2 \hat{X}_1}{\partial x_1^2}\frac{\partial^2 \hat{X}_2}{\partial x_2^2} + \frac{5\pi}{4|\lambda(0)|^3}\frac{\partial^2 \hat{X}_1}{\partial x_1^2}\frac{\partial^2 \hat{X}_1}{\partial x_1 \partial x_2}$$

$$+ \frac{7\pi}{4|\lambda(0)|^3}\frac{\partial^2 \hat{X}_1}{\partial x_1^2}\frac{\partial^2 \hat{X}_2}{\partial x_1^2} + \frac{5\pi}{4|\lambda(0)|}\frac{\partial^2 \hat{X}_1}{\partial x_1 \partial x_2}\frac{\partial^2 \hat{X}_1}{\partial x_2^2}$$

$$+ \frac{3\pi}{4|\lambda(0)|^3}\frac{\partial^2 \hat{X}_2}{\partial x_1 \partial x_2}\frac{\partial^2 \hat{X}_2}{\partial x_1^2} + \frac{5\pi}{4|\lambda(0)|^3}\frac{\partial^2 \hat{X}_1}{\partial x_2^2}\frac{\partial^2 \hat{X}_2}{\partial x_2^2}$$

$$+ \frac{\pi}{|\lambda(0)|^3}\frac{\partial^2 \hat{X}_1}{\partial x_2^2}\frac{\partial^2 \hat{X}_2}{\partial x_1^2} + \frac{3\pi}{4|\lambda(0)|^3}\frac{\partial^2 \hat{X}_2}{\partial x_1 \partial x_2}\frac{\partial^2 \hat{X}_2}{\partial x_2^2}\Bigg); \quad (\text{at } (0,0))$$

and one easily sees that

$$\int_0^{2\pi/|\lambda(0)|} e^{td\hat{X}(0)} \int_0^t e^{-sd\hat{X}(0)} d^3\hat{X}(0)\left(\frac{\partial \Phi_s}{\partial x_1}(0,0),\ \frac{\partial \Phi_s}{\partial x_1}(0,0),\ \frac{\partial \Phi_s}{\partial x_1}(0,0)\right) ds\, dt$$

$$= \frac{3\pi}{4|\lambda(0)|^2}\left(\frac{\partial^3 \hat{X}_1}{\partial x_1^3} + \frac{\partial^3 \hat{X}_1}{\partial x_1 \partial x_2^2} + \frac{\partial^3 \hat{X}_1}{\partial x_1^2 \partial x_2} + \frac{\partial^3 \hat{X}_2}{\partial x_2^2}\right).$$

The final result of the computations is, therefore,

---

*We are not joking! One has to be prepared to shack up with the previous calculations for several days. Details will be sent only on serious request.

$$\int_0^{2\pi/|\lambda(0)|} \frac{\partial^3 b_t}{\partial x_1^3}(0,0)dt$$

$$= \frac{3\pi}{4|\lambda(0)|^2}\left[\frac{\partial^3 \hat{X}^1}{\partial a^3}(0,0) + \frac{\partial^3 \hat{X}^1}{\partial a \partial b^2}(0,0) + \frac{\partial^3 \hat{X}^2}{\partial a^2 \partial b}(0,0) + \frac{\partial^3 \hat{X}^2}{\partial b^3}(0,0)\right]$$

$$+ \frac{\pi}{|\lambda(0)|^3}\left[2\frac{\partial^2 \hat{X}^1}{\partial a^2}(0,0)\frac{\partial^2 \hat{X}^2}{\partial b^2}(0,0) + \frac{\partial^2 \hat{X}^1}{\partial b^2}(0,0)\frac{\partial^2 \hat{X}^2}{\partial a^2}(0,0)\right.$$

$$+ \frac{5}{4}\frac{\partial^2 \hat{X}^1}{\partial a^2}(0,0)\frac{\partial^2 \hat{X}^1}{\partial a \partial b}(0,0) + \frac{3}{4}\frac{\partial^2 \hat{X}^2}{\partial b^2}(0,0)\frac{\partial^2 \hat{X}^2}{\partial a \partial b}(0,0)$$

$$+ \frac{7}{4}\frac{\partial^2 \hat{X}^1}{\partial a^2}(0,0)\frac{\partial^2 \hat{X}^2}{\partial a^2}(0,0) + \frac{5}{4}\frac{\partial^2 \hat{X}^2}{\partial b^2}(0,0)\frac{\partial^2 \hat{X}^1}{\partial b^2}(0,0)$$

$$\left. + \frac{5}{4}\frac{\partial^2 \hat{X}^1}{\partial a \partial b}(0,0)\frac{\partial^2 \hat{X}^1}{\partial b^2}(0,0) + \frac{3}{4}\frac{\partial^2 \hat{X}^2}{\partial a \partial b}(0,0)\frac{\partial^2 \hat{X}^2}{\partial a^2}(0,0)\right].$$

Thus we get our formula:

(4.2) Formula

$$V'''(0) = \frac{3\pi}{4|\lambda(0)|}\left[\frac{\partial^3 \hat{X}^1}{\partial a^3}(0,0) + \frac{\partial^3 \hat{X}^1}{\partial a \partial b^2}(0,0)\right.$$

$$\left. + \frac{\partial^3 \hat{X}^2}{\partial a^2 \partial b}(0,0) + \frac{\partial^3 \hat{X}^2}{\partial b^3}(0,0)\right]$$

$$+ \frac{3\pi}{4|\lambda(0)|^2}\left[-\frac{\partial^2 \hat{X}^1}{\partial a^2}(0,0)\frac{\partial^2 \hat{X}^1}{\partial a \partial b}(0,0)\right.$$

$$+ \frac{\partial^2 \hat{X}^2}{\partial b^2}(0,0)\frac{\partial^2 \hat{X}^2}{\partial a \partial b}(0,0)$$

$$+ \frac{\partial^2 \hat{X}^2}{\partial a^2}(0,0)\frac{\partial^2 \hat{X}^2}{\partial a \partial b}(0,0) - \frac{\partial^2 \hat{X}^1}{\partial b^2}(0,0)\frac{\partial^2 \hat{X}^1}{\partial a \partial b}(0,0)$$

$$\left. + \frac{\partial^2 \hat{X}^1}{\partial a^2}(0,0)\frac{\partial^2 \hat{X}^2}{\partial a^2}(0,0) - \frac{\partial^2 \hat{X}^1}{\partial b^2}(0,0)\frac{\partial^2 \hat{X}^2}{\partial b^2}(0,0)\right].$$

In two dimensions, where $X = \hat{X}$, this can be used directly to test stability if $d\hat{X}_0(0,0)$ is in the form on p. 108.

Expressing the Stability Condition in Terms of X.

We now use the equation

$$X_\mu^3(x_1,x_2,f(x_1,x_2,\mu)) = d_1f(x_1,x_2,\mu) \circ X^1(x_1,x_2,f(x_1,x_2,\mu)) + d_2f(x_1,x_2,\mu) \circ X^2(x_1,x_2,f(x_1,x_2,\mu))$$

near $(0,0,0,0)$

to compute the derivatives of $\hat{X}_0$ at $(0,0)$ in terms of those of X_0 at $(0,0,0)$. Since no differentiation with respect to μ occurs, we drop all reference to μ. Upon differentiating this with respect to x_1 and to x_2, we get:

$$\begin{aligned}
&d_1X^3(x_1,x_2,f(x_1,x_2)) + d_3X^3(x_1,x_2,f(x_1,x_2)) \circ d_1f(x_1,x_2) \\
&\quad = d_1d_1f(x_1,x_2) \circ X^1(x_1,x_2,f(x_1,x_2)) \\
&\quad + d_1f(x_1,x_2) \circ d_1X^1(x_1,x_2,f(x_1,x_2)) \\
&\quad + d_1f(x_1,x_2) \circ d_3X^1(x_1,x_2,f(x_1,x_2)) \circ d_1f(x_1,x_2) \\
&\quad + d_1d_2f(x_1,x_2) \circ X^2(x_1,x_2,f(x_1,x_2)) \\
&\quad + d_2f(x_1,x_2) \circ d_1X^2(x_1,x_2,f(x_1,x_2)) \\
&\quad + d_2f(x_1,x_2) \circ d_3X^2(x_1,x_2,f(x_1,x_2)) \circ d_1f(x_1,x_2)
\end{aligned}$$

and

$$\begin{aligned}
&d_2X^3(x_1,x_2,f(x_1,x_2)) + d_3X^3(x_1,x_2,f(x_1,x_2)) \circ d_2f(x_1,x_2) \\
&\quad = d_2d_2f(x_1,x_2) \circ X^2(x_1,x_2,f(x_1,x_2))
\end{aligned}$$

$$+ d_2 f(x_1,x_2) \circ d_2 X^2(x_1,x_2,f(x_1,x_2))$$

$$+ d_2 f(x_1,x_2) \circ d_3 X^2(x_1,x_2,f(x_1,x_2)) \circ d_2 f(x_1,x_2)$$

$$+ d_1 d_2 f(x_1,x_2) \circ X^1(x_1,x_2,f(x_1,x_2))$$

$$+ d_1 f(x_1,x_2) \circ d_2 X^1(x_1,x_2,f(x_1,x_2))$$

$$+ d_1 f(x_1,x_2) \circ d_3 X^1(x_1,x_2,f(x_1,x_2)) \circ d_2 f(x_1,x_2).$$

We now differentiate the first equation with respect to x_1 and x_2 and the second with respect to x_2 only. We get three expressions. These expressions are easy to write down and to evaluate at the point $x_1 = x_2 = 0$, where

$$f = df = 0 \quad \text{and} \quad dX = \begin{pmatrix} 0 & |\lambda(0)| & 0 \\ -|\lambda(0)| & 0 & 0 \\ 0 & 0 & d_3 X^3 \end{pmatrix}.$$

This procedure yields

$$d_1 d_1 X^3(0,0,0) + d_3 X^3(0,0,0) \circ d_1 d_1 f(0,0) = -2|\lambda(0)| d_1 d_2 f(0,0)$$

$$d_1 d_2 X^3(0,0,0) + d_3 X^3(0,0,0) \circ d_1 d_2 f(0,0)$$
$$= |\lambda(0)| d_1 d_1 f(0,0) - |\lambda(0)| d_2 d_2 f(0,0)$$

$$d_2 d_2 X^3(0,0,0) + d_3 X^3(0,0,0) \circ d_2 d_2 f(0,0) = 2|\lambda(0)| d_1 d_2 f(0,0)$$

i.e.

$$\begin{bmatrix} d_3 X^3(0,0,0) & 2|\lambda(0)| & 0 \\ -|\lambda(0)| & d_3 X^3(0,0,0) & |\lambda(0)| \\ 0 & -2|\lambda(0)| & d_3 X^3(0,0,0) \end{bmatrix} \begin{pmatrix} d_1 d_1 f(0,0) \\ d_1 d_2 f(0,0) \\ d_2 d_2 f(0,0) \end{pmatrix}$$

$$= \begin{pmatrix} -d_1 d_1 X^3(0,0,0) \\ -d_1 d_2 X^3(0,0,0) \\ -d_2 d_2 X^3(0,0,0) \end{pmatrix}$$

See formula (4A.6) on p. 134 for the expression for $d_i d_j f$ obtained by inverting the 3×3 matrix on the left. Note that since the determinant is $d_3 X^3(0,0,0)\,(d_3 X^3(0,0,0)^2 + 4|\lambda(0)|^2)$, and since $\sigma(d_3 X^3(0,0,0)) \subset \sigma(dX(0,0,0)) \subset \{z \mid \operatorname{Re} z < 0\} \cup \{\lambda(0), \overline{\lambda(0)}\}$ implies both $d_3 X^3(0,0,0)$ and $d_3 X^3(0,0,0)^2 + 4|\lambda(0)|^2 = (d_3 X^3(0,0,0) + 2|\lambda(0)|i)(d_3 X^3(0,0,0) - 2|\lambda(0)|i)$ are invertible, the matrix is invertible.

Finally we must compute the first three derivatives of $\hat{X}_0$ at $(0,0)$ in terms of those of X_0 at $(0,0,0)$. Remember that

$$\hat{X}^i(x_1,x_2) = X^i(x_1,x_2,f(x_1,x_2)) \quad \text{for} \quad i = 1,2.$$

Therefore,

$$\begin{aligned} d_j \hat{X}^i(x_1,x_2) &= d_j X^i(x_1,x_2,f(x_1,x_2)) \\ &\quad + d_3 X^i(x_1,x_2,f(x_1,x_2)) \circ d_j f(x_1,x_2) \\ &\quad \text{for} \quad i,j = 1,2. \end{aligned}$$

So

$$d\hat{X}(0,0) = \begin{pmatrix} 0 & |\lambda(0)| \\ -|\lambda(0)| & 0 \end{pmatrix}.$$

Differentiating again, we get:

$$
\begin{aligned}
d_k d_j \hat{X}^i(x_1,x_2) &= d_k d_j X^i(x_1,x_2,f(x_1,x_2)) \\
&+ d_3 d_j X^i(x_1,x_2,f(x_1,x_2)) \circ d_k f(x_1,x_2) \\
&+ d_k d_3 X^i(x_1,x_2,f(x_1,x_2)) \circ d_j f(x_1,x_2)) \\
&+ d_3 d_3 X^i(x_1,x_2,f(x_1,x_2)) \circ d_k f(x_1,x_2) \circ d_j f(x_1,x_2) \\
&+ d_3 X^i(x_1,x_2,f(x_1,x_2)) \circ d_k d_j f(x_1,x_2), \quad i,j,k = 1,2.
\end{aligned}
$$

Evaluating at $t = 0$, we get:

$$
d_k d_j \hat{X}^i(0,0) = d_k d_j X^i(0,0,0), \qquad i,j,k = 1,2.
$$

We differentiate once more at evaluate at 0:

$$
\begin{aligned}
d_\ell d_k d_j \hat{X}^i(0,0) &= d_\ell d_k d_j X^i(0,0,0) + d_3 d_j X^i(0,0,0) \circ d_\ell d_k f(0,0) \\
&+ d_k d_3 X^i(0,0,0) \circ d_\ell d_j f(0,0) \\
&+ d_\ell d_3 X^i(0,0,0) \circ d_k d_j f(0,0), \quad i,j,k, = 1,2.
\end{aligned}
$$

This can be inserted into the previous results to give an explicit expression for $V'''(0)$ on p.

Below in Section 4A, we shall summarize the results algorithmically so that this proof need not be traced through, and in Section 4B examples and exercises illustrating the method are given.

(4.3) Exercise. In Exercise 1.16 make a stability analysis for the pair of bifurcated fixed points. In that proof, write $f(\alpha,0) = \alpha + A\alpha^3 + \cdots$ and show that we have stability if $A < 0$ and supercritical bifurcation and instability with subcritical bifurcation if $A > 0$. Develop an explicit formula for A and apply it to the ball in the hoop example. (Reference: Ruelle-Takens [1], p. 189-191.)

SECTION 4A

HOW TO USE THE STABILITY FORMULA; AN ALGORITHM

The above calculations are admittedly a little long, but they are not difficult. Here we shall summarize the results of the calculation in the form of a specific algorithm that can be followed for any given vector field. In the two dimensional case the algorithm ends rather quickly. In general, it is much longer. Examples will be given in Section 4B following.

Stability is determined by the sign of $V'''(0)$, so our object is to calculate this number. We assume there is no difficulty in calculating the spectrum of the linearized problem.

Before stating the procedure for calculation of $V'''(0)$, let us recall the set up and the overall operation.

Let $X_\mu: E \to E$ be a C^k $(k \geq 5)$ vector field on a Banach space E (if X_μ is a vector field on a manifold, one must work in a chart to compute the stability condition). Assume $X_\mu(a(\mu)) = 0$ for all μ and let the spectrum of

$dX_\mu(a(\mu))$ satisfy:

For $\mu < \mu_0$, $\sigma(dX_\mu(a(\mu)) \subset \{z \mid \text{Re } z < 0\}$. $dX_\mu(a(\mu))$ has two complex conjugate, simple eigenvalues $\lambda(\mu)$ and $\overline{\lambda(\mu)}$. At $\mu = \mu_0$, $\lambda(\mu)$ and $\lambda(\mu)$ cross the imaginary axis with non-zero speed and $\lambda(\mu_0) \neq 0$. The rest of $\sigma(dX_\mu(a(\mu))$ remains in the left half plane bounded away from the imaginary axis.

I. Under these circumstances

(A) Bifurcation to periodic orbits takes place, as described in Theorem 3.1.

Choose coordinates so that $X_{\mu_0} = (X^1_{\mu_0}, X^2_{\mu_0}, X^3_{\mu_0})$ where $X^1_{\mu_0}$ and $X^2_{\mu_0}$ are coordinates in the eigenspace to $\lambda(\mu_0)$ and $\overline{\lambda(\mu_0)}$ and $X^3_{\mu_0}$ is the coordinate in some complementary subspace. Choose the coordinates so that*

$$dX_{\mu_0}(a(\mu_0)) = \begin{bmatrix} 0 & |\lambda(\mu_0)| & 0 \\ -|\lambda(\mu_0)| & 0 & 0 \\ 0 & 0 & d_3X^3_{\mu_0}(a(\mu_0)) \end{bmatrix}. \quad (4.1)$$

(B) If the coefficient $V'''(0)$ computed in (II) below is negative, the periodic orbits occur for $\mu > \mu_0$ and are attracting. If $V'''(0) > 0$, the orbits occur for $\mu < \mu_0$, are repelling on the center manifold, and so are unstable in general.

(C) If $V'''(0) = 0$, the test yields no information

---

*See Examples 4B.2 and 4B.8. Computer programs are available for this step. See for instance, "A Program to Compute the Real Schur Form of a Real Square Matrix" by B.N. Parlett and R. Feldman; ERL Memorandum M 526 (1975), Univ. of Calif., Berkeley.

and the procedures outlined in Section 4 must be used to compute $V^{(5)}(0)$. Good luck.

II. Write out the expression

$$
\begin{aligned}
V'''(0) = \frac{3\pi}{4|\lambda(\mu_0)|}\Bigg(& \frac{\partial^3 \hat{x}^1_{\mu_0}}{\partial x_1^3}(a_1(\mu_0),a_2(\mu_0)) \\
& + \frac{\partial^3 \hat{x}^1_{\mu_0}}{\partial x_1 \partial x_2^2}(a_1(\mu_0),a_2(\mu_0)) + \frac{\partial^3 \hat{x}^2_{\mu_0}}{\partial x_1^2 \partial x_2}(a_1(\mu_0),a_2(\mu_0)) \\
& + \frac{\partial^3 \hat{x}^2_{\mu_0}}{\partial x_2^3}(a_1(\mu_0),a_2(\mu_0))\Bigg) \\
+ \frac{3\pi}{4|\lambda(\mu_0)|^2}\Bigg(& -\frac{\partial^2 x^1_{\mu_0}}{\partial x_1^2}(a_1(\mu_0),a_2(\mu_0))\,\frac{\partial^2 x^1_{\mu_0}}{\partial x_1 \partial x_2}(a_1(\mu_0),a_2(\mu_0)) \\
& + \frac{\partial^2 x^2_{\mu_0}}{\partial x_2^2}(a_1(\mu_0),a_2(\mu_0))\,\frac{\partial^2 x^2_{\mu_0}}{\partial x_1 \partial x_2}(a_1(\mu_0),a_2(\mu_0)) \\
& + \frac{\partial^2 x^2_{\mu_0}}{\partial x_1^2}(a_1(\mu_0),a_2(\mu_0))\,\frac{\partial^2 x^2_{\mu_0}}{\partial x_1 \partial x_2}(a_1,(\mu_0),a_2(\mu_0)) \\
& - \frac{\partial^2 x^1_{\mu_0}}{\partial x_2^2}(a_1(\mu_0),a_2(\mu_0))\,\frac{\partial^2 x^1_{\mu_0}}{\partial x_1 \partial x_2}(a_1(\mu_0),a_2(\mu_0)) \\
& + \frac{\partial^2 x^1_{\mu_0}}{\partial x_1^2}(a_1(\mu_0),a_2(\mu_0))\,\frac{\partial^2 x^2_{\mu_0}}{\partial x_1^2}(a_1(\mu_0),a_2(\mu_0)) \\
& - \frac{\partial^2 x^1_{\mu_0}}{\partial x_2^2}(a_1(\mu_0),a_2(\mu_0))\,\frac{\partial^2 x^2_{\mu_0}}{\partial x_2^2}(a_1(\mu_0),a_2(\mu_0))\Bigg).
\end{aligned}
\tag{4A.2}
$$

(A) If your space is two dimensional, let $\hat{X}^1 = X^1$, $\hat{X}^2 = X^2$. Take off the hats; you are done with the computation of $V'''(0)$ and the results may be read off from I.

Otherwise, go to Step B.

(B) In expression (4A.2) fill in

$$d_j\hat{X}^i_{\mu_0}(a_1(\mu_0),a_2(\mu_0)) = d_jX^i_{\mu_0}(a(\mu_0)) \quad \text{for } i,\ j = 1,2. \tag{4A.3}$$

$$d_kd_j\hat{X}^i_{\mu_0}(a_1(\mu_0),a_2(\mu_0)) = d_kd_jX^i_{\mu_0}(a(\mu_0)), \quad \text{for } i,j,k = 1,2. \tag{4A.4}$$

and

$$\begin{aligned} d_\ell d_k d_j\hat{X}^i_{\mu_0}(a_1(\mu_0),a_2(\mu_0)) &= d_\ell d_k d_j X^i_{\mu_0}(a(\mu_0)) \\ &+ d_3d_jX^i_{\mu_0}(a(\mu_0)) \circ d_\ell d_k f(a_1(\mu_0),a_2(\mu_0)) \\ &+ d_3d_kX^i_{\mu_0}(a(\mu_0)) \circ d_\ell d_j f(a_1(\mu_0),a_2(\mu_0)) \\ &+ d_3d_\ell X^i_{\mu_0}(a(\mu_0)) \circ d_k d_j f(a_1(\mu_0),a_2(\mu_0)) \quad \text{for } i,j,k,\ell = 1,2. \end{aligned} \tag{4A.5}$$

(C) In this expression you now have, fill in d_id_jf given by:

$$\begin{pmatrix} d_1d_1f(a_1(\mu_0),a_2(\mu_0)) \\ d_1d_2f(a_1(\mu_0),a_2(\mu_0)) \\ d_2d_2f(a_1(\mu_0),a_2(\mu_0)) \end{pmatrix} =$$

$$\Delta^{-1}\begin{bmatrix} 2|\lambda(\mu_0)|^2+(d_3X^3_{\mu_0}(a(\mu_0)))^2 & -2|\lambda(\mu_0)|d_3X^3_{\mu_0}(a(\mu_0)) & 2|\lambda(\mu_0)|^2 \\ |\lambda(\mu_0)|d_3X^3_{\mu_0}(a(\mu_0)) & (d_3X^3_{\mu_0}(a(\mu_0)))^2 & -|\lambda(\mu_0)|d_3X^3_{\mu_0}(a(\mu_0)) \\ 2|\lambda(\mu_0)|^2 & 2|\lambda(\mu_0)|d_3X^3_{\mu_0}(a(\mu_0)) & 2|\lambda(\mu_0)|^2+(d_3X^3_{\mu_0}(a(\mu_0)))^2 \end{bmatrix} \begin{pmatrix} -d_1d_1X^3_{\mu_0}(a(\mu_0)) \\ -d_1d_2X^3_{\mu_0}(a(\mu_0)) \\ -d_2d_2X^3_{\mu_0}(a(\mu_0)) \end{pmatrix} \tag{4A.6}$$

where $\Delta = (d_3 X^3_{\mu_0}(a(\mu_0)))((d_3 X^3_{\mu_0}(a(\mu_0)))^2 + 4|\lambda(0)|^2)$.

(Note: if X^3 is linear, all derivatives of f are zero.)

(D) If you have done it correctly your expression for $V'''(0)$ is now entirely in terms of known quantities and in an explicit example, is a known real number and you may go to Step I to read off the results.

Remarks. S. Wan has recently obtained a proof of the stability formula using complex notation, which is somewhat simpler. It also yields information on the period (it is closely related to the expression $\beta_0 + i a_0 b_0$ from Section 3C; stability being the real part, the period being the imaginary part). The formulas have been programmed and interesting numerical work is being done by B. Hassard (SUNY at Buffalo).

SECTION 4B
EXAMPLES

We now consider a few examples to illustrate how the above procedure works. The first few examples are all simple, designed to illustrate basic points. We finish in Example 4B.8 with a fairly intricate example from fluid mechanics (the Lorenz equations).

(4B.1) Example (see Hirsch-Smale [1], Chapter 10 and Zeeman [2] for motivation). Consider the differential equation $\frac{d^2x}{dt^2} + \left(\frac{dx}{dt}\right)^3 - a\frac{dx}{dt} + x = 0$, a special case of Lienard's equation. Before applying the Hopf Bifurcation theorem we make this into a first order differential equation on R^2. Let $y = \frac{dx}{dt}$. Then we get the system

$$\frac{dx}{dt} = y, \quad \frac{dy}{dt} = -y^3 + ay - x.$$

Let $X_a(x,y) = (y, -y^3+ay-x)$. Now $X_a(0,0) = 0$ for all a and

$$dX_a(0,0) = \begin{pmatrix} 0 & 1 \\ -1 & +a \end{pmatrix}.$$

The eigenvalues are $\frac{a \pm \sqrt{a^2-4}}{2}$. Consider a such that $|a| < 2$. In this case $\operatorname{Im} \lambda(a) \neq 0$, where $\lambda(a) = \frac{a + \sqrt{a^2-4}}{2} = \frac{a}{2} + i \frac{\sqrt{4-a^2}}{2}$. Furthermore, for $-2 < a < 0$, $\operatorname{Re} \lambda(a) < 0$ and for $a = 0$, $\operatorname{Re} \lambda(a) = 0$ and for $2 > a > 0$, $\operatorname{Re} \lambda(a) > 0$ and $\left.\frac{d(\operatorname{Re} \lambda(a))}{da}\right|_{a=0} = \frac{1}{2}$. Therefore, the Hopf Bifurcation theorem applies and we conclude that there is a one parameter family of closed orbits of $X = (X_a, 0)$ in a neighborhood of $(0,0,0)$. To find out if these orbits are stable and if they occur for $a > 0$, we look at $X_0(x,y) = (y, -y^3-x)$. $dX_0(0,0) = \begin{pmatrix} 0 & 1 \\ -1 & 0 \end{pmatrix}$ and $\lambda(0) = i$. Recall that to use the stability formula developed in the above section we must choose coordinates so that

$$dX_0(0,0) = \begin{pmatrix} 0 & \operatorname{Im} \lambda(0) \\ -\operatorname{Im} \lambda(0) & 0 \end{pmatrix} = \begin{pmatrix} 0 & 1 \\ -1 & 0 \end{pmatrix}.$$

Thus, the original coordinates are appropriate to the calculation; an example where this is not true will be given below. We calculate the partials of X_0 at $(0,0)$ up to order three:

$$\frac{\partial^n X_1}{\partial x^j \partial y^{n-j}} (0,0) = 0 \quad \text{for all } n > 1$$

since $X_1(x,y) = y$.

$$\frac{\partial^2 X_2}{\partial x_1^2} (0,0) = 0, \quad \frac{\partial^2 X_2}{\partial x_1 \partial x_2} (0,0) = 0, \quad \frac{\partial^2 X_2}{\partial x_2^2} (0,0) = 0,$$

$$\frac{\partial^3 X_2}{\partial x_1^3} (0,0) = 0, \quad \frac{\partial^3 X_2}{\partial x_1^2 \partial x_2} (0,0) = 0, \quad \frac{\partial^3 X_2}{\partial x_1 \partial x_2^2} (0,0) = 0, \quad \frac{\partial^3 X_2}{\partial x_2^3} (0,0) = -6.$$

Thus, $V'''(0) = \frac{3\pi}{4}(-6) < 0$, so the periodic orbits are attracting and bifurcation takes place above criticality.

(4B.2) Example. On R^2 consider the vector field

$$X_\mu(x,y) = (x+y,-x^3-x^2y+(\mu-2)x+(\mu-1)y).$$

$$X_\mu(0,0) = 0$$

and

$$dX_\mu(0,0) = \begin{pmatrix} 1 & 1 \\ \mu-2 & \mu-1 \end{pmatrix}.$$

The eigenvalues are $\frac{\mu \pm \sqrt{\mu^2-4}}{2} = \frac{\mu \pm i\sqrt{4-\mu^2}}{2}$. Let $-1 < \mu < 1$, then the conditions for Hopf Bifurcation to occur at $\mu = 0$ are fulfilled with $\lambda(0) = 1$. Consider $X_0(x,y) = (x+y,-x^3-x^2y-2x-y)$. $dX_0(0,0) = \begin{pmatrix} 1 & 1 \\ -2 & -1 \end{pmatrix}$, which is not in the required form. We must make a change of coordinates so that $dX_0(0,0)$ becomes $\begin{pmatrix} 0 & 1 \\ -1 & 0 \end{pmatrix}$. That is, we must find vectors $\hat{e}_1$ and $\hat{e}_2$ so that $dX_0(0,0)\hat{e}_1 = -\hat{e}_2$ and $dX_0(0,0)\hat{e}_2 = \hat{e}_1$. The vectors $\hat{e}_1 = (1,-1)$ and $\hat{e}_2 = (0,1)$ will do. (A procedure for finding $\hat{e}_1$ and $\hat{e}_2$ is to find α and $\bar{\alpha}$ the complex eigenvectors; we may then take $\hat{e}_1 = \alpha + \bar{\alpha}$ and $\bar{e}_2 = i(\alpha-\bar{\alpha})$. See Section 4, Step 1 for details.) $X_0(x\hat{e}_1+y\hat{e}_2) = X_0(x,y-x) = (y,-x^3-x^2(y-x)-2x-(y-x)) = (y,-x^2y-x-y) = y\hat{e}_1 + (-x^2y-x)\hat{e}_2$. Therefore in the new coordinate system, $X_0(x,y) = (y,-x^2y-x)$.

$$\frac{\partial^n X_1}{\partial x^j \partial y^{n-j}}(0,0) = 0 \quad \text{for all} \quad n > 1.$$

$$\frac{\partial^2 X_2}{\partial y^2}(0,0) = 0, \quad \frac{\partial^2 X_2}{\partial x \partial y}(0,0) = 0, \quad \frac{\partial^2 X_2}{\partial y^2}(0,0) = 0,$$

$$\frac{\partial^3 X_2}{\partial x^3}(0,0) = 0, \quad \frac{\partial^3 X_2}{\partial x^2 \partial y}(0,0) = -2, \quad \frac{\partial^3 X_2}{\partial x \partial y^2}(0,0) = 0, \quad \frac{\partial^3 X_2}{\partial y^3}(0,0) = 0.$$

Therefore, $V'''(0) = \frac{3\pi}{4|\lambda(0)|}(-2) < 0$. The orbits are stable and bifurcation takes place above criticality.

(4B.3) Example. The van der Pol Equation

The van der Pol equation $\frac{d^2x}{dt^2} + \mu(x^2-1)\frac{dx}{dt} + x = 0$ is important in the theory of the vacuum tube. (See Minorsky [1], LaSalle-Lefschetz [1] for details.) As is well known, for all $\mu > 0$, there is a stable oscillation for the solution of this equation. It is easy to check that the eigenvalue conditions for the Hopf Bifurcation theorem are met so that bifurcation occurs at the right for $\mu = 0$. However, if $\mu = 0$ the equation is a linear rotation, so $V^{(n)}(0) = 0$ for all n. For $\mu = 0$, all circles centered at the origin are closed orbits of the flow. By uniqueness, these are the closed orbits given by the Hopf Theorem. Thus, we cannot use the Hopf Theorem on the problem as stated here to get the existence of stable oscillations for $\mu > 0$. In fact, the closed orbits bifurcate off the circle of radius two (see LaSalle-Lefschetz [1], p. 190 for a picture). In order to obtain them from the Hopf Theorem one needs to make a change of coordinates bringing the circle of radius 2 into the origin. In fact the general van der Pol equation $u'' + f(u)u' + g(u) = 0$ can be transformed into the general Lienard equation $x' = y - F(x)$, $y' = -g(x)$ by means of $x = u$, $y = u' + F(u)$. This change of coordinates reduces the present example to 4B.1. See Brauer-Nohel [1, p. 219 ff.] for general information on these matters.)

(4B.4) Example. On R^3, let $X_\mu(x,y,z) = (\mu x+y+6x^2, -x+\mu y+yz, (\mu^2-1)y-x-z+x^2)$. Then $X_\mu(0,0,0) = 0$ and

$dX_\mu(0,0,0) = \begin{pmatrix} \mu & 1 & 0 \\ -1 & \mu & 0 \\ -1 & \mu^2-1 & -1 \end{pmatrix}$ which has eigenvalues -1 and $\mu \pm i$. For $\mu = 0$, the eigenspace of $\pm i$ for $dX_0(0,0,0)$ is spanned by $\{(1,0,-1),(0,1,0)\}$. The complementary subspace is spanned by $(0,0,1)$. With respect to this basis $X_\mu(x,y,z) = (\mu x+y+6x^2, -x+\mu y+yz, \mu x+\mu^2 y-z+x^2)$. We now compute the stability condition. $|\lambda(0)| = 1$ and $d_3 X_0^3(0,0,0) = -1$.

$$\begin{pmatrix} d_1 d_1 f(0,0) \\ d_1 d_2 f(0,0) \\ d_2 d_2 f(0,0) \end{pmatrix} = \begin{pmatrix} 3 & 2 & 2 \\ -1 & -1 & 1 \\ 2 & -2 & 3 \end{pmatrix} \frac{1}{5} \begin{pmatrix} 2 \\ 0 \\ 0 \end{pmatrix} = \begin{pmatrix} 6/5 \\ -2/5 \\ 4/5 \end{pmatrix}$$

$$d_\ell d_j d_k \hat{X}_0^1(0,0) = 0$$

$$d_1 d_1 d_1 \hat{X}_0^2(0,0) = 0$$

$$d_1 d_1 d_2 \hat{X}_0^2(0,0) = 1\cdot(6/5) = 6/5$$

$$d_1 d_2 d_2 \hat{X}_0^2(0,0) = 3\cdot 1\cdot 4/5 = 12/5.$$

Therefore, $V'''(0) = \frac{3\pi}{4}(6/5 + 12/5) > 0$, so the orbits are unstable.

The next two exercises discuss some easy two dimensional examples.

(4B.5) Exercise. Let $X(x,y) = A_\mu \binom{x}{y} + B(x,y)$ where $B(x,y) = (ax^2+cy^2, dx^2+fy^2)$ and $A = \begin{pmatrix} \mu & 1 \\ -1 & \mu \end{pmatrix}$. Show that $\mu_0 = 0$ is a bifurcation point and that a stable periodic orbit develops for $\mu > 0$ provided that $cf > ad$. (This example is a two dimensional prototype of the Navier-Stokes equations; note that X is linear plus quadratic.)

(4B.6) Exercise (see Arnold [2]). Let $\dot{z} = z(iw+\mu+cz\bar{z})$ in R^2 using complex notation. Show that bifurcation to periodic orbits takes place at $z = \mu = 0$. Show that these orbits are stable if $c < 0$.

For some additional easy two dimensional examples, see Minorsky [1], p. 173-177. These include an oscillator instability of an amplifier in electric circuit theory and oscillations of ships. The reader can also study the example $\ddot{x} + \sin x + \varepsilon\dot{x} = M$ for a pendulum with small friction and being acted on by a torque M. See Arnold [1], p. 94 and Andronov-Chailkin [1].

The following is a fairly simple three-dimensional exercise to warm the reader up for the following example.

(4B.7) Exercise. Let $X_\mu(x,y,z,w) = (\mu x+y+z-w,-x+\mu y,$ $-z,-w+y^3)$. Show that bifurcation to attracting closed orbits takes place at $(x,y,z,w) = (0,0,0,0)$ and $\mu = 0$. (Answer: $V'''(0) = -9\pi/4$).

The following example, the most intricate one we shall discuss, has many interesting features. For example, change in the physical parameters can alter the bifurcation from sub to supercritical; in the first case, complicated "Lorenz attractors" (see Section 12) appear in plcace of closed orbits.

(4B.8) Example (suggested by J.A. Yorke and D. Ruelle). The Lorenz Equations (see Lorenz [1]).

The Lorenz equations are an idealization of the equations of motion of the fluid in a layer of uniform depth when the temperature difference between the top and the bottom is maintained at a constant value. The equations are

$$\frac{dx}{dt} = -\sigma x + \sigma y$$

$$\frac{dy}{dt} = -xz + rx - y$$

$$\frac{dz}{dt} = xy - bz.$$

Lorenz [1] says that "... x is proportional to the intensity of the convective motion, while y is proportional to the temperature difference between the ascending and descending currents, similar signs of x and y denoting that warm fluid is rising and cold fluid is descending. The variable z is proportional to the distortion of the vertical temperature profile from linearity, a positive value indicating that the strongest gradients occur near the boundaries". $\sigma = K^{-1}\nu$ is the Prandtl number, where K is the coefficient of thermal expansion and ν is the viscosity; r, the Rayleigh number, is the bifurcation parameter.

For $r > 1$, the system has a pair of fixed points at $x = y = \pm\sqrt{b(r-1)}$, $z = r - 1$. The linearization of the vector field at the fixed point $x = y = +\sqrt{b(r-1)}$, $z = r - 1$ is

$$M \equiv \begin{bmatrix} -\sigma & \sigma & 0 \\ 1 & -1 & -\sqrt{b(r-1)} \\ \sqrt{b(r-1)} & \sqrt{b(r-1)} & -b \end{bmatrix} = dX_r(\sqrt{b(r-1)}, \sqrt{b(r-1)}, r-1).$$

The characteristic polynomial of this matrix is

$$x^3 + (\sigma+b+1)x^2 + (r+\sigma)bx + 2\sigma b(r-1) = 0,$$

which has one negative and two complex conjugate roots. For $\sigma > b + 1$, a Hopf bifurcation occurs at $r = \frac{\sigma(\sigma+b+3)}{(\sigma-b-1)}$. We shall now prove this and determine the stability.

Let the characteristic polynomial be written

$$(x-\lambda)(x-\bar{\lambda})(x-\alpha) = 0, \text{ where } \lambda = \lambda_1 + i\lambda_2,$$

i.e. $x^3 - (2\lambda_1+\alpha)x^2 + (|\lambda|^2+2\lambda_1\alpha)x - |\lambda|^2\alpha = 0.$

Clearly, this has two pure imaginary roots iff the product of the coefficients of x^2 and x equals the constant term. That is, iff $(\sigma+b+1)(r_0+\sigma) = 2\sigma b(r_0-1)$ or $r_0 = \frac{\sigma(\sigma+b+3)}{(\sigma-b-1)}$. Thus, we have the bifurcation value. We now wish to calculate $\lambda_1'(r_0)$. Equating coefficients of like powers of x, we get

$$(\sigma+b+1) = 2\lambda_1 + \alpha$$

$$(r+\sigma)b = |\lambda|^2 + 2\lambda_1\alpha$$

and

$$-2\sigma b(r-1) = |\lambda|^2\alpha.$$

Thus, $\alpha = -(\sigma+b+1+2\lambda_1)$ and $(r+\sigma)b\alpha = 2\lambda_1\alpha^2 - 2\lambda b(r-1)$, so that $-(\sigma+b+1+2\lambda_1)(r+\sigma)b = -2\sigma b(r-1) + 2\lambda_1(\sigma+b+1+2\lambda_1)^2$. Differentiating with respect to r, setting $r = r_0$, and recalling that $\lambda_1(r_0) = 0$, we obtain

$$\lambda_1'(r_0) = \frac{b(\sigma-b-1)}{2[b(r_0+\sigma) + (\sigma+b+1)^2]} > 0 \quad \text{for} \quad \sigma > b+1.$$

Thus, the eigenvalues cross the imaginary axis with non-zero speed, so a Hopf bifurcation occurs at $r_0 = \frac{\sigma(\sigma+b+3)}{\sigma-b-1}$. We will compute $V'''(0)$ for arbitrary σ, b and will evaluate it at the physically significant values $\sigma = 10$, $b = 8/3$. At r_0, α is minus the coefficient of x^2, so $\alpha = -(\sigma+b+1)$; and at r_0, $|\lambda|^2$ is the coefficient of x, so $|\lambda|^2 = \frac{2\sigma b(\sigma+1)}{\sigma-b-1}$.

Following I(A) of Section 4A, must compute a basis for R^3 in which

$$dX_{r_0}(\sqrt{b(r_0-1)},\sqrt{b(r_0-1)},r_0-1) =$$

$$M = \begin{bmatrix} -\sigma & \sigma & 0 \\ 1 & -1 & -\sqrt{b(r_0-1)} \\ \sqrt{b(r_0-1)} & \sqrt{b(r_0-1)} & -b \end{bmatrix}$$

becomes

$$\begin{bmatrix} 0 & \sqrt{\dfrac{2\sigma b(\sigma+1)}{\sigma-b-1}} & 0 \\ -\sqrt{\dfrac{2\sigma b(\sigma+1)}{\sigma-b-1}} & & 0 \\ 0 & 0 & -(\sigma+b+1) \end{bmatrix}.$$

The basis vectors will be u,v,w where $Mu = -|\lambda|v$, $Mv = -|\lambda|u$, $Mw = \alpha w$. An eigenvector of M with eigenvalue α is $\left(-\sigma, b+1, \sqrt{\dfrac{(\sigma+b+1)(\sigma-b-1)b}{\sigma+1}}\right)$. The eigenspace of M corresponding to the eigenvalues $\lambda,\bar{\lambda}$ is the orthogonal complement of the eigenvector of M^t corresponding to the eigenvalue α. This eigenvector is $\left(\sigma+b-1, -(\sigma-b-1), -\sqrt{\dfrac{b(\sigma+b+1)(\sigma-b-1)}{\sigma+1}}\right)$. We will choose $u = (-(\sigma-b-1), -(\sigma+b-1), 0)$. Because $M^2 = \begin{pmatrix} -|\lambda|^2 & 0 \\ 0 & -|\lambda|^2 \end{pmatrix}$ on the eigenspace of $\lambda,\bar{\lambda}$, we may choose $v = -\dfrac{1}{|\lambda|}Mu =$ $\left(\sqrt{\dfrac{2\sigma b(\sigma-b-1)}{\sigma+1}}, -\sqrt{\dfrac{2b(\sigma-b-1)}{\sigma(\sigma+1)}}, (\sigma-1)\sqrt{\dfrac{2(\sigma+b+1)}{\sigma}}\right)$. We now have our new basis and, as in Example 4B.4, after writing the differential equations in this basis to get $\hat{X}^1,\hat{X}^2,\hat{X}^3$, we are ready to compute $V'''(0)$. This is a very lengthy computation so we give only the results.* Following (II) of Section 4A,

*Details will be sent on request.

the second derivative terms of $V'''(0)$ are

$$\frac{3\pi}{4|\lambda|^2}\left(-\frac{\partial^2 X^1}{\partial x_1^2}\frac{\partial^2 X^1}{\partial x_1\partial x_2}+\frac{\partial^2 X^2}{\partial x_2^2}\frac{\partial^2 X^2}{\partial x_1\partial x_2}+\frac{\partial^2 X^2}{\partial x_1^2}\frac{\partial^2 X^2}{\partial x_1\partial x_2}\right.$$

$$\left.-\frac{\partial^2 X^1}{\partial x_1^2}\frac{\partial^2 X^1}{\partial x_1\partial x_2}+\frac{\partial^2 X^1}{\partial x_1^2}\frac{\partial^2 X^1}{\partial x_2^2}\frac{\partial^2 X^2}{\partial x_2^2}\right)=$$

$$\frac{3\pi(\sigma-b-1)^2}{4\sigma b(\sigma+1)^3\omega^2}\sqrt{\frac{2b(\sigma-b-1)}{\sigma(\sigma+1)}}\Big\{[2\sigma^2 b^2(\sigma+b-1)-2\sigma b^2(\sigma-b-1)$$

$$+2\sigma b(\sigma-1)(\sigma-b-1)(\sigma+b+1)+2\sigma(\sigma-1)^2(\sigma+1)(\sigma+b+1)][b(\sigma+1)(\sigma+b-1)(\sigma-b-1)$$

$$-2b^2(\sigma-b-1)+2b(\sigma-1)(\sigma-b-1)(\sigma+b+1)+2(\sigma-1)^2(\sigma+1)(\sigma+b+1)]$$

$$+[b(\sigma+1)(\sigma-b-1)^2+(\sigma^2-1)(\sigma-b-1)(\sigma+b+1)-\sigma(\sigma^2-1)(\sigma+b-1)(\sigma+b+1)$$

$$-\sigma b(\sigma+1)(\sigma+b-1)(\sigma-b-1)-(\sigma^2-1)(\sigma+b+1)(\sigma-b-1)^2][(\sigma^2-1)(\sigma+b-1)(\sigma+b+1)$$

$$+b(\sigma+1)(\sigma+b-1)(\sigma-b-1)-2b(\sigma-1)(\sigma+b+1)-2B^2(\sigma-b-1)$$

$$+2B(\sigma-1)(\sigma-b-1)(\sigma+b+1)]+2b\sigma(\sigma-1)(\sigma+1)^2(\sigma-b-1)(\sigma+b+1)(\sigma+b-1)^2$$

$$+2b^2\sigma(\sigma+1)^2(\sigma-b-1)^2(\sigma+b-1)^2-[8b^2\sigma(\sigma-1)(\sigma-b-1)(\sigma+b+1)$$

$$+8b\sigma(\sigma-1)^2(\sigma+1)(\sigma+b+1)-8b^3\sigma(\sigma-b-1)][(\sigma-1)(\sigma-b-1)(\sigma+b+1)$$

$$-b(\sigma-b-1)-(\sigma-1)(\sigma+b+1)]\Big\}\equiv A_1\xi.$$

The third derivative terms are

$$\frac{3\pi}{4|\lambda|}\left(\frac{\partial^3\hat{X}^1}{\partial x_1^3}+\frac{\partial^3\hat{X}^1}{\partial x_1\partial x_2^2}+\frac{\partial^3\hat{X}^2}{\partial x_1^2\partial x_2}+\frac{\partial^3\hat{X}^2}{\partial x_2^3}\right)$$

$$=\frac{3\pi}{4|\lambda|}\left\{\frac{\partial^2 f}{\partial x_1^2}\left(3\frac{\partial^2 X^1}{\partial x_1\partial x_2}+\frac{\partial^2 X^2}{\partial x_2\partial x_3}\right)\right.$$

$$\left.+\frac{\partial^2 f}{\partial x_2^2}\left(3\frac{\partial^2 X^2}{\partial x_2\partial x_3}+\frac{\partial^2 X^1}{\partial x_1\partial x_3}\right)+2\frac{\partial^2 f}{\partial x_1\partial x_2}\left(\frac{\partial^2 X^1}{\partial x_1\partial x_2}+\frac{\partial^2 X^2}{\partial x_1\partial x_3}\right)\right\}$$

$$\frac{3\pi(\sigma-1)(\sigma-b-1)^2}{2\sigma b(\sigma+1)^3\omega^2}\sqrt{\frac{2b(\sigma-b-1)}{\sigma(\sigma+1)}}\frac{1}{[(\sigma+b+1)^2(\sigma-b-1)+8\sigma b(\sigma+1)]}$$

$$\Big\{-4\sigma^2b^2(\sigma^2-1)(\sigma-b+1) + \sigma b(\sigma+b-1)(\sigma-b-1)(\sigma+b+1)^2$$

$$+ 2\sigma b(\sigma+1)(\sigma-b-1)(\sigma+b+1)(\sigma^2+b-1)$$

$$- 2\sigma^2b^2(\sigma+1)(\sigma+b-1)(\sigma+b+1)][3\sigma b(\sigma+1)(\sigma+b-1) - b(\sigma+1)(2b+1)(\sigma-b-1)$$

$$- 2b(\sigma-b-1)^2(\sigma+b+1) - 4(\sigma^2-1)(\sigma-b-1)(\sigma+b+1) + (\sigma^2-1)(b+2)(\sigma+b+1)]$$

$$+ [4\sigma^2b^2(\sigma+1)^2(\sigma+b-1) - 2\sigma b(\sigma+1)(\sigma+b+1)(\sigma-b-1)(\sigma^2+b-1)$$

$$+ 2\sigma^2b^2(\sigma+1)(\sigma+b+1)(\sigma+b-1) - 8\sigma^2b^2(\sigma+1)(\sigma^2+b-1)$$

$$- 2\sigma b^2(\sigma+b+1)^2(\sigma-b-1)(\sigma^2+b-1)][2b(\sigma+b+1)(\sigma-b-1)^2$$

$$- 4(\sigma^2-1)(\sigma-b-1)\sigma+b+1) + \sigma b(\sigma+1)(\sigma+b-1)$$

$$+ b(\sigma+1)(2b+5)(\sigma-b-1) + 3(b+2)(\sigma^2-1)(\sigma+b+1)]$$

$$- [2\sigma b^2(\sigma+1)(\sigma+b-1)(\sigma+b+1) + 4\sigma b(\sigma+b+1)(\sigma^2+b-1)$$

$$- (\sigma+b+1)^2(\sigma-b-1)(\sigma^2+b-1) + \sigma b(\sigma+b+1)^2(\sigma+b-1)]$$

$$[2\sigma b^2(\sigma+1)(b+2)(\sigma-b-1) + 2\sigma b^2(\sigma-b-1)^2(\sigma+b+1)$$

$$- \sigma(\sigma-1)(\sigma+1)^2(\sigma+b+1)(\sigma-b-1) + \sigma b(\sigma+1)^2(\sigma-b-1)(\sigma+b-1)$$

$$- (\sigma-1)(\sigma+1)^2(b+1)(\sigma-b-1)(\sigma+b+1) - b(b+1)(\sigma+1)^2(\sigma-b-1)^2$$

$$- b(\sigma+1)(\sigma+b+1)(\sigma-b-1)^3]\Big\} \equiv A_2\xi$$

where

$$\omega = \frac{(\sigma-1)}{(\sigma+1)}\sqrt{\frac{2(\sigma+b+1)}{\sigma}}\,[(b+1)(\sigma+1)(\sigma-b-1) + \sigma(\sigma+1)(\sigma+b-1) + b(\sigma+b+1)(\sigma-b-1)]$$

and

$$\xi = \frac{3\pi(\sigma-b-1)^2}{2\sigma b(\sigma+1)^3\omega^2}\sqrt{\frac{2b(\sigma-b-1)}{\sigma(\sigma+1)}}\,.$$

Since $V'''(0) = (A_1+A_2)\xi$ and $\xi > 0$, the periodic orbits resulting from the Hopf bifurcation are stable if $A_1 + A_2 < 0$, and unstable if $A_1 + A_2 > 0$. For $\sigma = 10$, $b = 8/3$, $A_1 \cong 1.63 \times 10^9$, $A_2 \cong 0.361 \times 10^9$, $A_1 + A_2 \cong 1.99 \times 10^9$;

therefore, the orbits are unstable, i.e. the bifurcation is subcritical.

Our calculations thus prove the conjecture of Lorenz [1], who believed the orbits to be unstable because of numerical work he had done. For different σ or b however, the sign may change, so one cannot conclude that the closed orbits are always unstable. A simple computer program determines the regions of stability and instability in the b-σ plane. See Figure 4B.1

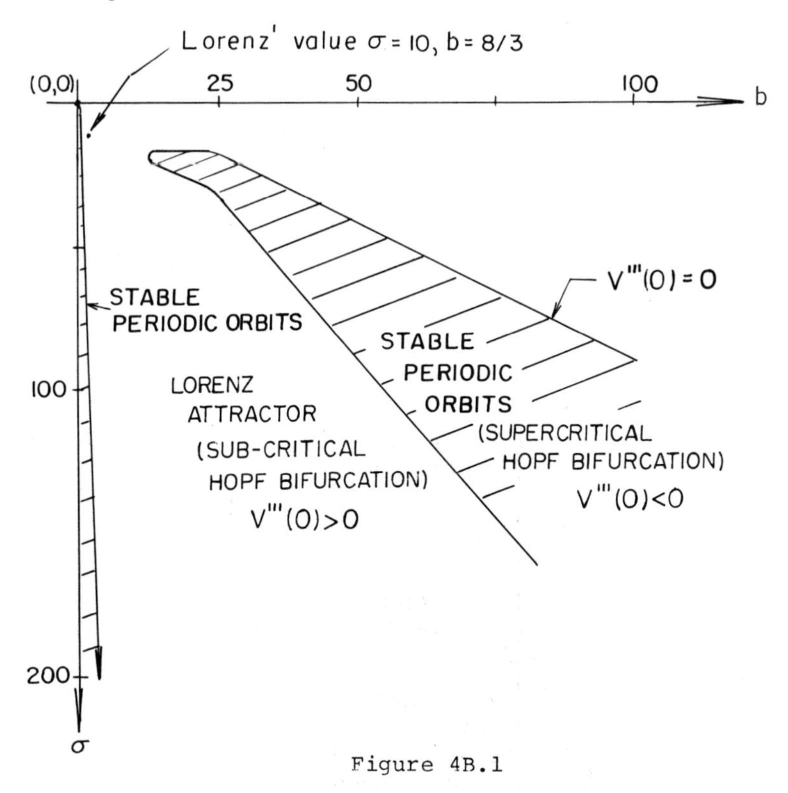

Figure 4B.1

We also investigate the behavior of the system for fixed $b > 0$ as $\sigma \to \infty$. This has also been done by Martin and McLaughlin [1]. Our result agrees with theirs. We proceed as follows:

$$\{(\sigma+b+1)^2(\sigma-b-1) + 8\sigma b(\sigma+1)\}A_1 + A_2 = p(b,\sigma)$$

is a polynomial of degree 11 in σ. For b fixed, the highest order term is $(8b^2+12b)\sigma^{11}$. If $b > 0$ this coefficient is positive, so for large positive σ (with b fixed), $V'''(0) > 0$ and the bifurcation is subcritical.

This example may be important for understanding eventual theorems of turbulence (see discussion in Section 9). The idea is shown in Figure 4B.2. For further information on the behavior of solutions above criticality, see Lorenz [1] and Section 12. (L. Howard has built a device to simulate the dynamics of these equations.)

(4B.9) Exercise. Analyze the behavior of $V'''(0)$ as $b \to \infty$ for fixed σ and as $b \to \infty$ for $\sigma = \beta b$, for various $\beta > 0$.

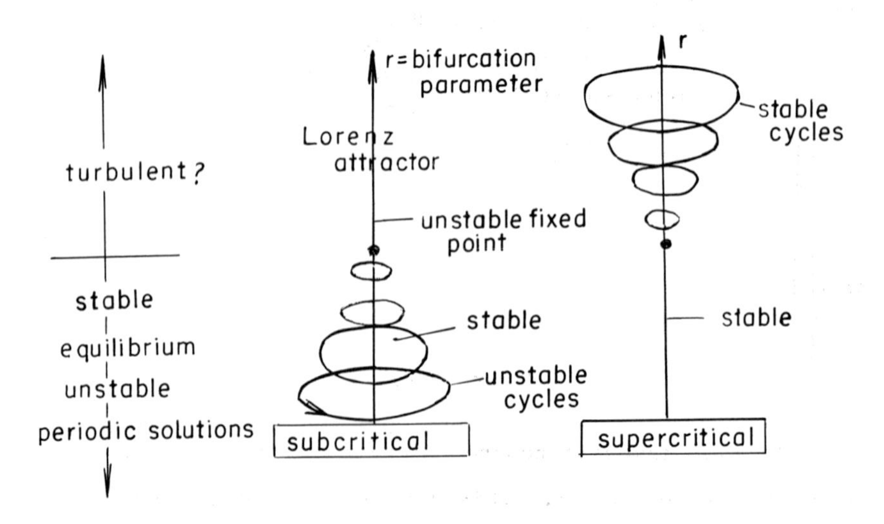

Figure 4B.2

(4B.10) Exercise. By a change of variable, analyze the stability of the fixed point $x = y = -b(r-1)$, $z = r - 1$. Show that Hopf bifurcation occurs at $r_0 = \frac{\sigma(\sigma+b+3)}{\sigma-b-1}$ and that the orbits obtained are attracting iff those obtained from the analysis above are.

(4B.11) Exercise. Prove that for $r > 1$, the matrix

$$\begin{bmatrix} -\sigma & \sigma & 0 \\ 1 & -1 & \sqrt{b(r-1)} \\ \sqrt{b(r-1)} & \sqrt{b(r-1)} & -1 \end{bmatrix}$$

has one negative and two complex conjugate roots.

(4B.12) Exercise. Let F denote the vector field on R^3 defined by the right hand side of the Lorentz equations.

(a) Note that $\operatorname{div} F = -\sigma - 1 - b$, a constant. Use this to estimate the order of magnitude of the contractions in the principal directions at the fixed points.

(b) If $V = (x,y,z)$, show that the inner product $\langle F,V\rangle$ is a quadratic function of x,y,z. By considering $\frac{d}{dt}\langle V,V\rangle$, show that this implies that solutions of the Lorenz equations are globally defined in t. [Note that most quadratic equations, eq: $\dot{x} = x^2$ do not have global t solutions.]

(4B.13) Exercise. The following equations arise in the oscillatory Zhabotinskii reaction (cf. Hastings-Murray [1]):

$$\dot{x} = s(y-xy+x-qx^2)$$

$$\dot{y} = \frac{1}{s}(fz-y-xy)$$

$$\dot{z} = w(x-y)$$

(compare the Lorenz equations!). Let f be the bifurcation parameter and let, eg: $s = 7.7 \times 10$, $q = 8.4 \times 10^{-6}$, $w = 1.61 \times 10^{-1}$. Show that a Hopf bifurcation occurs at $f = f_c$ where

$$2q(2+3f_c) = (2f_c+q-1)[(1-f_c-q) + \{(1-f_c-q)^2 + 4q(1+f_c)\}^{1/2}].$$

Show that for the above values, the bifurcation is subcritical. S. Hastings informs us that the bifurcation picture looks like that in Figure 4B.3 (the existence of stable closed orbits for supercritical values is proven in Hastings-Murray [1]).

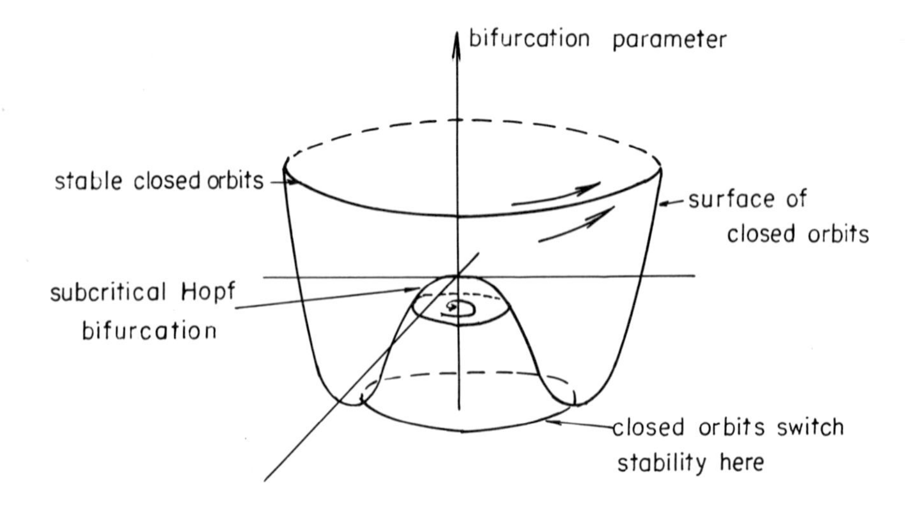

Figure 4B.3

SECTION 4C
HOPF BIFURCATION AND THE METHOD OF AVERAGING

by

S. Chow and J. Mallet-Paret

The method of averaging* provides an algorithm for preparing a bifurcation problem, that is, putting it into a normal form. Once this is done, one may more readily determine certain qualitative features of the bifurcation, by means of the implicit function theorem (or contraction mapping principle) and the center manifold theorem.

Consider first the Hopf bifurcation problem

$$\dot{z} = f(z,\alpha) \tag{4C.1}$$

about the equilibrium $z = 0$. Thus assume $z \in R^n$, $\alpha \in (-\alpha_0,\alpha_0)$, f takes values in R^n and $f(0,\alpha) = 0$. For simplicity, an ordinary differential equation is considered although we could just as well consider a partial differential equation. In

*The method has been used by a number of authors, such as Halanay, Hale, Meyer, Diliberto, etc. cf. Kurzweil [1] and articles in Lefschetz [1].

this case z and $\dot{z}$ belong to (generally different) Banach spaces X_1 and X_2 and f is smooth from $X_1 \times (-\alpha_0, \alpha_0)$ to X_2. We may write (1) in the form

$$\begin{aligned} \dot{z} &= A(\alpha)z + g(z,\alpha) \\ |g(z,\alpha)| &= O(|z|^2). \end{aligned} \tag{4C.2}$$

The standard hypotheses (see Sections 1,3) on the spectrum of $A(\alpha)$ hold, namely that it posses a pair $\lambda(\alpha), \overline{\lambda(\alpha)}$ of complex conjugate eigenvalues of the form

$$\lambda(\alpha) = \gamma(\alpha) + i\omega(\alpha)$$

where

$$\gamma(0) = 0, \ \nu \overset{\text{def}}{=} \gamma'(0) \neq 0$$

and

$$\omega \overset{\text{def}}{=} \omega(0) \neq 0$$

and that the remainder of the spectrum of $A(\alpha)$ stays uniform positive distance away from the imaginary axis. Decompose $z \in R^n$ as

$$z = (x,y) \quad P \oplus Q = R^2 \oplus R^{n-2}$$

according to the spectrum of $A(0)$, so that P is the eigenspace corresponding to $\lambda(0), \overline{\lambda(0)}$ and Q is its complement. With this decomposition we may assume

$$A(\alpha) = \begin{pmatrix} A_P(\alpha) & O(\alpha) \\ O(\alpha) & A_Q(\alpha) \end{pmatrix}$$

where

$$A_P(\alpha) = \begin{pmatrix} \gamma(\alpha) & -\omega(\alpha) \\ \omega(\alpha) & \gamma(\alpha) \end{pmatrix}$$

and the spectrum of $A_Q(\alpha)$ stays uniformly away from the imaginary axis. We may also represent x in polar coordinates

$$x = (r \cos \theta, r \sin \theta).$$

Consider a periodic solution bifurcating from $(x,y,\alpha) = (0,0,0)$. The differential equation for r is

$$\dot{r} = \gamma(\alpha)r + O(r^2) = \alpha\nu r + O(\alpha^2 r) + O(r^2),$$

and it follows that when r attains its maximum on the solution, $\dot{r} = 0$, and hence

$$\alpha = O(r). \tag{4C.3}$$

Now the periodic solution also lies on the center manifold Σ, described by

$$\Sigma:\ y = \phi(r,\theta,\alpha). \tag{4C.4}$$

The fixed point $(r,y) = (0,0)$ lies on Σ for all α; moreover, Σ is tangent to $Px(-\alpha_0,\alpha_0)$ at $(r,\alpha) = (0,0)$ which implies (4C.4) has the form

$$\Sigma:\ y = r\psi(r,\theta,\alpha)$$
$$\psi(0,\theta,0) = 0.$$

Thus, we have on the periodic solution

$$y = O(r^2) + O(r\alpha) = O(r^2). \tag{4C.5}$$

Choosing $\varepsilon > 0$ of the same order as the amplitude of the solution, we may scale the equation by making the replacements

$$r \to \varepsilon r, \quad \alpha \to \varepsilon\alpha, \quad y \to \varepsilon y.$$

The estimates (4C.3), (4C.5) imply then

$$\left.\begin{aligned} r &= O(1) \\ \dot{x} &= O(1) \\ y &= O(\varepsilon) \end{aligned}\right\} \text{ in scaled coordinates.} \tag{4C.6}$$

The precise relation between ε and α will be determined later when α will be chosen as a particular function of ε. We will in fact show then that

$$\alpha = 0 \quad \text{in scaled coordinates.}$$

Expand the differential equation (4C.2) in a Taylor series, in scaled coordinates. It is not difficult to show the estimates (4C.6) imply the equation takes the form

$$\begin{aligned} \dot{x} &= A_P(\varepsilon\alpha) + B_2x^2 + \varepsilon^3 B_3 x^3 + Gxy + O(\varepsilon^2\alpha) + O(\varepsilon^3) \\ \dot{y} &= A_Q y + \varepsilon J x^2 + O(\varepsilon\alpha) + O(\varepsilon^2) \end{aligned} \tag{4C.7}$$

where

$$\begin{aligned} B_j &= (B_j^1, B_j^2) = \text{homogeneous polynomial of degree } j \text{ in} \\ &\qquad x \in R^2, \text{ taking values in } R^2 \\ G&: R^2 \times R^{n-2} \to R^2 \quad \text{bilinear} \\ A_Q &= A_Q(0) \\ J &= R^2 \times R^2 \to R^2 \quad \text{symmetric, bilinear.} \end{aligned}$$

In polar coordinates (4C.7) becomes

$$\left.\begin{aligned} \dot{r} &= \varepsilon\alpha\nu r + \varepsilon r^2 C_3(\theta) + \varepsilon^2 r^3 C_4(\theta) + \varepsilon r G_2(\theta) y \\ &\qquad + O(\varepsilon^2\alpha) + O(\varepsilon^3) \\ \dot{\theta} &= \omega + \varepsilon r D_3(\theta) + O(\varepsilon\alpha) + O(\varepsilon^2) \\ \dot{y} &= A_Q y + \varepsilon r^2 J(\cos\theta, \sin\theta)^2 + O(\varepsilon\alpha) + O(\varepsilon^2) \end{aligned}\right\} \tag{4C.8}$$

where

$$C_j(\theta) = (\cos\theta)B^1_{j-1}(\cos\theta, \sin\theta) + (\sin\theta)B^2_{j-1}(\cos\theta, \sin\theta)$$

$$D_j(\theta) = (\cos\theta)B^2_{j-1}(\cos\theta, \sin\theta) - (\sin\theta)B^1_{j-1}(\cos\theta, \sin\theta)$$

= homogeneous trigonometric polynomials of degree j

$G_2(\theta)$ = homogeneous trigonometric polynomial of degree 2 taking values in Q^* the dual of Q.

The goal of the method of averaging is to "average out" the dependence of $\dot{r}$ on θ and y, that is, to find a new radial coordinate $\bar{r}$ in which the equation for $\dot{\bar{r}}$ is

$$\dot{\bar{r}} = F(\bar{r},\varepsilon).$$

If this were done, then all periodic solutions would simply be circles $\bar{r} = \bar{r}(\varepsilon)$ satisfying $F(\bar{r}(\varepsilon),\varepsilon) = 0$. Actually it is not necessary to entirely eliminate dependence on ε and y; generally all that is required is the absence of ε and y from a finite number of terms in the Taylor series expression in ε and y. For example, in (4C.8), generically it is enough to average the $\varepsilon, \varepsilon y$ and ε^2 terms.

More precisely, consider any differential equation

$$\dot{r} = \sum_{j=1}^{\infty} \sum_{k=0}^{\infty} \varepsilon^j R_{jk}(r,\theta,\alpha)y^k$$

$$\dot{\theta} = \omega + O(\varepsilon)$$

$$\dot{y} = A_Q y + O(\varepsilon).$$

The series for $\dot{r}$ may be a finite Taylor series with remainder. In order to average a given term, say R_{pq}, define a new coordinate $\bar{r}$ by

$$\bar{r} = r + \varepsilon^p u(r,\theta,\alpha)y^q.$$

In the new coordinates, the coefficient of $\varepsilon^p y^q$ becomes $\bar{R}_{pq}$

where

$$\bar{R}_{pq}(r,\theta,\alpha) = \frac{\partial u}{\partial \theta}\omega + quy^q A_Q + R_{pq}(r,\theta,\alpha).$$

Two cases are considered.

<u>Case I</u>, $q = 0$. In this case we choose u to be

$$u(r,\theta,\alpha) = -\frac{1}{\omega}\int_0^\theta R_{p0}(r,\xi,\alpha)d\xi + \frac{\theta}{2\pi\omega}\int_0^{2\pi} R_{p0}(r,\xi,\alpha)d\xi.$$

Observe that u is 2π-periodic in θ and $\overline{R_{p0}}$ is independent of θ and, in fact, is the mean value

$$\bar{R}_{p0}(r,\alpha) = \frac{1}{2\pi}\int_0^{2\pi} R_{p0}(r,\xi,\alpha)d\xi.$$

Therefore, we have averaged the coefficient of $\varepsilon^p y^0$.

<u>Case II</u>, $q > 0$. Here we wish to choose u so that $\overline{R_{pq}}$ is identically zero thus eliminating the $\varepsilon^p y^q$ term. Therefore, we seek a 2π-periodic function $u(r,\theta,\alpha)$ satisfying

$$\frac{\partial u}{\partial \theta}\omega + quA_Q + R_{pq}(r,\theta,\alpha) = 0. \tag{4C.9}$$

By considering R_{pq} as a forcing term in (4C.9), it follows that such a unique u always exists if and only if the homogeneous equation

$$\frac{\partial u}{\partial \theta}\omega + quA_Q = 0$$

has no nontrivial solution 2π-periodic solution. It can be shown that this is the case provided that

$$\frac{1}{\omega_i}\sum_{j=1}^{n-2} n_j\lambda_j \neq \text{integer}$$

$$\text{for all integers} \quad n_j \geq 0, \quad \sum_{j=1}^{n-2} n_j = q$$

$$\lambda_1, \ldots, \lambda_{n-2} = \text{eigenvalues of } A_Q.$$

In particular this can always be done if either $q = 1$, or A_Q is stable.

We return to the bifurcation problem (4C.8) and now average the $\varepsilon, \varepsilon y$ and ε^2 in the above fashion by means of the transformation

$$\bar{r} = r + \varepsilon u(r,\theta,\alpha) + \varepsilon w(r,\theta,\alpha)y + \varepsilon^2 v(r,\theta,\alpha).$$

In fact, the transformation has the form

$$\bar{r} = r + \varepsilon r^2 u(\theta) + \varepsilon r w(\theta) y + \varepsilon^2 r^3 v(\theta),$$

and this yields the equation for $\bar{r}$

$$\left.\begin{aligned}
\dot{\bar{r}} &= \varepsilon[\alpha\nu\bar{r} + \bar{r}^2 C_3(\theta) + \bar{r}^2 u'(\theta)\omega] \\
&+ \varepsilon\bar{r}[G_2(\theta) + w(\theta)A_Q + w'(\theta)\omega]y \\
&+ \varepsilon^2\bar{r}^3[C_4(\theta) + u'(\theta)D_3(\theta) + w(\theta)J(\cos\theta, \sin\theta)^2 \\
&- 2u(\theta)u'(\theta)\omega + v'(\theta)\omega] \\
&+ O(\varepsilon^2\alpha) + O(\varepsilon^3).
\end{aligned}\right\} \quad (4C.10)$$

Since C_3 has mean value zero, we choose

$$u(\theta) = -\frac{1}{\omega}\int_0^\theta C_3(\xi)d\xi$$

so that the coefficient of ε in (10) is $\alpha\nu\bar{r}$. Set $w(\theta)$ equal to the unique solution of

$$G_2(\theta) + w(\theta)A_Q + w'(\theta)\omega = 0$$

$$w(\theta) \quad \text{of period} \quad 2\pi.$$

so the εy term vanishes. Finally, we may choose $v(\theta)$ to make the coefficient of $\varepsilon^2\bar{r}^3$ the constant

$$K = \text{mean}[C_4(\theta) + u'(\theta)D_3(\theta) + w(\theta)J(\cos\theta, \sin\theta) - 2u(\theta)u'(\theta)\omega]$$

$$= \frac{1}{2\pi}\int_0^{2\pi} C_4(\theta) - \frac{1}{\omega}C_3(\theta)D_3(\theta) + w(\theta)J(\cos\theta, \sin\theta)^2 .$$

Thus in the new coordinates (r,θ,y), (4C.8) becomes

$$\left.\begin{aligned}\dot{\bar{r}} &= \varepsilon\alpha\nu\bar{r} + \varepsilon^2\bar{r}^3K + O(\varepsilon^2\alpha) + O(\varepsilon^3)\\ \dot{\theta} &= \omega + O(\varepsilon)\\ \dot{y} &= A_Q y + O(\varepsilon).\end{aligned}\right\} \tag{4C.11}$$

The equation for $\dot{y}$ may be neglected now as we restrict to the center manifold $y = r\psi(\varepsilon r,\theta,\varepsilon\alpha)$. Moreover, it is not difficult to show that the unique branch of periodic solutions bifurcating from the origin has the form

$$\left.\begin{aligned}\bar{r} &= \left|\frac{\nu}{K}\right|^{1/2} + O(\varepsilon)\\ \alpha &= -\varepsilon\ \text{sgn}(\nu K)\end{aligned}\right\} \text{ in scaled, averaged coordinates}$$

that is,

$$\left.\begin{aligned}r &= \left|\frac{\nu}{K}\right|^{1/2}\varepsilon + O(\varepsilon^2)\\ y &= O(\varepsilon^2)\\ \alpha &= -\varepsilon^2\text{sgn}(\nu K)\end{aligned}\right\} \text{ in original coordinates.} \tag{4C.12}$$

The amplitude of the bifurcating solution is therefore approximately $\left(-\frac{\alpha\nu}{K}\right)^{1/2}$ and one sees the period is near $\frac{2\pi}{\omega}$.

In case $K = 0$, one simply averages higher order terms

in ε and y in the same manner. The possible normal forms one arrives at in this case are

$$\dot{\bar{r}} = \varepsilon\alpha\nu\bar{r} + \varepsilon^{2p}\bar{r}^{2p+1}K' + O(\varepsilon^2\alpha) + O(\varepsilon^{2p+1})$$

$$\dot{\theta} = \omega + O(\varepsilon)$$

for integers $p \geq 2$ and $K' \neq 0$. The bifurcating solution in this case has the form

$$\left.\begin{aligned} r &= \left|\frac{\nu}{K'}\right|^{1/2p}\varepsilon + O(\varepsilon^2) \\ y &= 0(\varepsilon^2) \\ \alpha &= -\varepsilon^{2p}\operatorname{sgn}(\nu K') \end{aligned}\right\} \text{ in original coordinates}$$

so has amplituded near $\left(-\frac{\alpha\nu}{K'}\right)^{1/2p}$ and period near $\frac{2\pi}{\omega}$. Observe that in all these cases bifurcation takes place only on one side of $\alpha = 0$. For cases in which all bifurcating solutions occur at $\alpha = 0$ (for example in the proof of the Lyapunov center theorem - see Section 3C), the method of averaging gives no information.

For more details of the above method, as well as several applications, see Chow and Mallet-Paret [1]. We mention here two examples treated in this paper using averaging.

(1) Delay Differential Equations (Wright's Equation). The equation

$$\dot{z}(t) = -az(t-1)[1+z(t)]$$

arises in such diverse areas as population models and number theory, and is one of the most deeply studied delay equations.

For $a > \frac{\pi}{2}$, topological fixed point techniques prove the existence of a periodic solution. Using averaging techniques, one can analyze the behavior of this solution near $a = \frac{\pi}{2}$. In particular, for $\frac{\pi}{2} < a < \frac{\pi}{2} + \varepsilon$ the solution bifurcates from $z = 0$, is stable, and has the asymptotic form

$$z(t) = K(a - \tfrac{\pi}{2})^{1/2}\cos(\tfrac{\pi}{2} t) + O(a - \tfrac{\pi}{2})$$

$$K = \left(\frac{40}{3\pi-2}\right)^{1/2} \approx 2.3210701.$$

(2) Diffusion Equations. Linear equations with non-linear boundary conditions, such as

$$u_t = u_{xx} \qquad t \geq 0 \qquad 0 < x < 1$$

$$u_x(0,t) = 0 \qquad u_x(1,t) = ag(u(0,t),\ u(1,t))$$

occur in various problems in biology and chemical reactions. (See, for example, Aronson [2].) Here we take

$$g(u,v) = \alpha u + \beta v + O(u^2+v^2),$$

so the linearized equation around $u = 0$ has the boundary conditions

$$u_x(0,t) = 0 \qquad u_x(1,t) = a[\alpha u(0,t) + \beta u(1,t)].$$

For appropriate parameter values (α,β) a pair of eigenvalues of this problem crosses the imaginary axis with non-zero speed as a passes the critical value a_0. The stability of the resulting Hopf bifurcation can be determined by averaging.

The power of the averaging method is that it can handle a rather wide variety of bifurcation problems. We mention two here.

(3) Almost Periodic Equations. Consider

$$\dot{z}(t) = A(\alpha)z + g(z,t,\alpha) \tag{4C.13}$$

where $A(\alpha)$ is as before, g is almost periodic in t uniformly for (z,α) in compact sets and $g = O(|z|^2)$. Suppose in addition that the periods $\frac{2\pi}{N\omega}$ for $N = 1,2,3,4$ are bounded away from the fundamental periods for g. Then an averaging procedure similar to that described above yields a normal form in scaled coordinates given by (4C.11). Here however the higher order terms (but not the constant K) are almost periodic in t. Thus this manifold can be thought of as a cylinder in the $(x,t) \in R^n \times R^1$ space, where each section $t = \text{const.}$ is a circle near $x = 0$. The cylinder is almost periodic in t with the same fundamental periods as those in g.

(4) A special case of (3) occurs in studying the bifurcation of an invariant torus from a periodic orbit of an autonomous equation. In an appropriate local coordinate system around the orbit, the autonomous equation takes the form (4C.13) where t represents the (periodic) coordinate around the orbit and z the normal to the orbit. The condition on the fundamental periods of g reduces to the

standard condition that the periodic orbit have no characteristic multipliers which are N^{th} roots of unity, for $N = 1,2,3,4$. The invariant cylinder that is obtained if $K \neq 0$ is periodic in t and thus is actually a two dimensional torus around the periodic orbit.

SECTION 5

A TRANSLATION OF HOPF'S ORIGINAL PAPER

BY L. N. HOWARD AND N. KOPELL

"Abzweigung einer periodischen Lösung von einer stationären Lösung eines Differentialsystems" Berichten der Mathematisch-Physischen Klasse der Sächsischen Akademie der Wissenschaften zu Leipzig. XCIV. Band Sitzung vom 19. Januar 1942.

Bifurcation of a Periodic Solution from a Stationary Solution of a System of Differential Equations

by

Eberhard Hopf

Dedicated to Paul Koebe on his 60th birthday

1. Introduction

Let

$$\dot{x}_i = F_i(x_1,\ldots,x_n,\mu) \quad (i = 1,\ldots,n)$$

or, in vector notation,

$$\dot{\underline{x}} = \underline{F}(\underline{x},\mu) \tag{1.1}$$

be a real system of differential equations with real parameter μ, where $\underline{F}$ is analytic in $\underline{x}$ and μ for $\underline{x}$ in a domain G and $|\mu| < c$. For $|\mu| < c$ let (1.1) possess an analytic family of stationary solutions $\underline{x} = \tilde{\underline{x}}(\mu)$ lying in G:

$$\underline{F}(\tilde{\underline{x}}(\mu),\mu) = 0.$$

As is well known, the characteristic exponents of the stationary solution are the eigenvalues of the eigenvalue problem

$$\lambda\underline{a} = \underline{L}_{\mu}\underline{a}$$

where $\underline{L}_{\mu}$ stands for the linear operator, depending only on μ, which arises after neglect of the nonlinear terms in the series expansion of F about $\underline{x} = \tilde{\underline{x}}$. The exponents are either real or pairwise complex conjugate and depend on μ.

Suppose one assumes simply that there is a stationary solution $\underline{x}_0$ in G for the special value $\mu = 0$ and that none of the characteristic exponents is 0; then, as is well known, it automatically follows that there is a unique stationary solution $\tilde{\underline{x}}(\mu)$ in a suitable neighborhood of $\underline{x} = \underline{x}_0$ for every sufficiently small $|\mu|$, and $\tilde{\underline{x}}(\mu)$ is analytic at $\mu = 0$.

On passing through $\mu = 0$ let us now assume that none of the characteristic exponents vanishes, but a conjugate pair crosses the imaginary axis. This situation commonly occurs in nonconservative mechanical systems, for example, in hydrodynamics. The following theorem asserts, that with this hypothesis, there is always a periodic solution of equation (1.1) in the neighborhood of the values $\underline{x} = \underline{x}_0$ and $\mu = 0$.

Theorem. *For* $\mu = 0$, *let exactly two characteristic exponents be pure imaginary. Their continuous extensions* $\alpha(\mu)$, $\overline{\alpha}(\mu)$ *shall satisfy the conditions*

$$\alpha(0) = -\overline{\alpha}(0) \neq 0, \quad \mathrm{Re}(\alpha'(0)) \neq 0. \tag{1.2}$$

Then, there exists a family of real periodic solutions $\underline{x} = \underline{x}(t,\varepsilon)$, $\mu = \mu(\varepsilon)$ *which has the properties* $\mu(0) = 0$ *and* $\underline{x}(t,0) = \tilde{\underline{x}}(0)$, *but* $\underline{x}(t,\varepsilon) \neq \tilde{\underline{x}}(\mu(\varepsilon))$, *for all sufficiently small* $\varepsilon \neq 0$. $\varepsilon(\mu)$ *and* $\underline{x}(t,\varepsilon)$ *are analytic at the point* $\varepsilon = 0$ *and correspondingly at each point* $(t,0)$. *The same holds for the period* $T(\varepsilon)$ *and*

$$T(0) = 2\pi/|\alpha(0)|.$$

For arbitrarily large L *there are two positive numbers* a *and* b *such that for* $|\mu| < b$, *there exist no periodic solutions besides the stationary solution and the solutions of the semi-family* $\varepsilon > 0$ *whose period is smaller than* L *and which lie entirely in* $|\underline{x}-\tilde{\underline{x}}(\mu)| < a$.*
For sufficiently small μ, the periodic solutions generally exist only for $\mu > 0$ or only for $\mu < 0$; it is also possible that they exist only for $\mu = 0$.

As is well known, the characteristic exponents of the periodic solution $\underline{x}(t,\varepsilon)$ are the eigenvalues of the eigenvalue problem

$$\dot{\underline{v}} + \lambda \underline{v} = \underline{L}_{t,\varepsilon}(\underline{v}) \tag{1.3}$$

where $\underline{v}(t)$ has the same period $T = T(\varepsilon)$ as the solution.

---

*The other half-family must represent the same solution curves.

$\underline{L}$ is the linear operator obtained by linearizing around the periodic solution. It depends periodically on t with the period T and at $\varepsilon = 0$ is analytic in t and ε. The characteristic exponents are only determined $\mathrm{mod}(2\pi i/T)$ and depend continuously on ε. One of them, of course, is zero; for $\underline{F}$ does not depend explicitly on t, so

$$\lambda = 0, \quad \underline{v} = \dot{\underline{x}}(t,\varepsilon)$$

is a solution of the eigenvalue problem. For $\varepsilon \to 0$ the exponents, $\mathrm{mod}(2\pi i/T_0)$, go continuously into those of the stationary solution $\underline{x}(0)$ of (1.1) with $\mu = 0$. By assumption then exactly two exponents approach the imaginary axis. One of them is identically zero. The other $\beta = \beta(\varepsilon)$ must be real and analytic at $\varepsilon = 0$, $\beta(0) = 0$. It follows directly from the above theorem that the coefficients μ_1 and β_1 in the power series expansion

$$\mu = \mu_1\varepsilon + \mu_2\varepsilon^2 + \ldots$$

$$\beta = \beta_1\varepsilon + \beta_2\varepsilon^2 + \ldots$$

satisfy $\mu_1 = \beta_1 = 0$. In addition to that it will be shown below that the simple relationship

$$\beta_2 = -2\mu_2 \mathrm{Re}(\alpha'(0)) \tag{1.4}$$

holds; I have not run across it before.

In the general case $\mu_2 \neq 0$, this relationship gives information about the stability conditions. If, for example, for $\mu < 0$ all the characteristic exponents of the stationary solution $\underline{x} = \tilde{\underline{x}}(\mu)$ have a negative real part (stability, a small neighborhood of $\tilde{\underline{x}}$ collapses onto $\tilde{\underline{x}}$ as $t \to \infty$), then there are the following alternatives. Either the periodic solutions branch off after the destabilization of the stationary solution ($\mu > 0$); in this case all characteristic exponents of the periodic solution have negative real part (stability; a thin tube around the periodic solutions collapses onto these as $t \to \infty$). Alternatively, the family exists before, that is for $\mu < 0$; then the periodic solutions are unstable.*

Since in nature only stable solutions can be observed for a sufficiently long time of observation, the bifurcation of a periodic solution from a stationary solution is observable only through the latter becoming unstable. Such observations are well known in hydromechanics. For example, in the flow around a solid body; the motion is stationary if the velocity of the oncoming stream is low enough; yet if the latter is sufficiently large it can become periodic (periodic vortex shedding). Here we are talking about examples of nonconservative systems (viscosity of the fluid).+ In conservative systems, of course, the hypothesis (1.2) is never fulfilled; if λ is a characteristic exponent, $-\lambda$ always is as well.

In the literature, I have not come across the bifurcation problem considered on the basis of the hypothesis

---

*In $n = 2$ dimensions, this is immediately clear.

+I do not know of a hydrodynamical example of the second case. One could conclude the existence of the unstable solutions if, with the most careful experimenting, (very slow variation of the parameters) one always observes a sudden breaking off of the stationary motion at exactly the same point.

(1.2). However, I scarcely think that there is anything essentially new in the above theorem. The methods have been developed by Poincaré perhaps 50 years ago,* and belong today to the classical conceptual structure of the theory of periodic solutions in the small. Since, however, the theorem is of interest in non-conservative mechanics it seems to me that a thorough presentation is not without value. In order to facilitate the extension to systems with infinitely many degrees of freedom, for example the fundamental equations of motion of a viscous fluid, I have given preference to the more general methods of linear algebra rather than special techniques (e.g. choice of a special coordinate system).

Of course, it can equally well happen that at $\mu = 0$ a real characteristic exponent $\alpha(\mu)$ of the stationary solution $\underline{\tilde{x}}(\mu)$ crosses the imaginary axis, i.e.,

$$\alpha(0) = 0, \quad \alpha'(0) \neq 0$$

---

*Les méthodes nouvelles de la mécanique céleste. The above periodic solutions represent the simplest limiting case of Poincaré's periodic solutions of the second type ("genre"). Compare Vol. III, chapter 28, 30-31. Poincaré, having applications to celestial mechanics in mind, has only thoroughly investigated these solutions (with the help of integral invariants) in the case of canonical systems of differential equations, where the situation is more difficult than above. Poincaré uses the auxiliary parameter ε in Chap. 30 in the calculation of coefficients (the calculation in our §4 is essentially the same), but not in the proof of existence which thereby becomes simpler.

In a short note in Vol. I, p. 156, Painlevé is touched upon: Les petits mouvements périodiques des systèmes, Comptes Rendus Paris XXIV (1897), p. 1222. The general theorem stated there refers to the case $\mu = 0$ in our system (.), but it cannot be generally correct. For the validity of this statement $\underline{F}$ must satisfy special conditions.

while the others remain away from it. In this case it is not periodic but other stationary solutions which branch off.*
We content ourselves with the statement of the theorems in this simpler case. There is an analytic family, $\underline{x} = \underline{x}^*(\varepsilon)$, $\mu = \mu^*(\varepsilon)$ of stationary solutions, different from $\tilde{x}$, with $\mu(0) = 0$, $x^*(0) = \tilde{x}(0)$. If $\mu_1 \neq 0$ (the general case) then the solutions exist for $\mu > 0$ and for $\mu < 0$. For the characteristic exponent $\beta(\varepsilon)$ which goes through zero, the analog of (1.4) holds:

$$\beta_1 = -\mu_1 \alpha'(0).$$

If $\underline{\tilde{x}}$ is stable for $\mu < 0$ and unstable for $\mu > 0$ then just the opposite holds for $\underline{x}^*$. (If one observes $\tilde{x}$ for $\mu < 0$, than one will observe x^* for $\mu > 0$.) In the exceptional case $\mu_1 = 0$, the situation is different. If $\mu_2 \neq 0$, then the new solutions exist only for $\mu > 0$ or only for $\mu < 0$. There are then two solutions for fixed μ, (one with ε positive, one with ε negative). Here we have

$$\beta_2 = -2\mu_2 \alpha'(0).$$

From this one can obtain statements about stability analogous to those above. In this case either both solutions x^* are stable or both are unstable.

2. The Existence of the Periodic Solutions.

Without restriction of generality one can assume that the stationary solution lies at the origin, i.e.,

---

*An example from hydrodynamics is the fluid motion between two concentric cylinders (G. I. Taylor).

$$\underline{F}(0,\mu) = 0.$$

Let the development of $\underline{F}$ in powers of the x_i be

$$\underline{F}(\underline{x},\mu) = \underline{L}_\mu(\underline{x}) + \underline{Q}_\mu(\underline{x},\underline{x}) + \underline{K}_\mu(\underline{x},\underline{x},\underline{x}) + \cdots, \tag{2.1}$$

where the vector functions

$$\underline{L}_\mu(\underline{x}), \quad \underline{Q}_\mu(\underline{x},\underline{y}), \quad \underline{K}_\mu(\underline{x},\underline{y},\underline{z}), \ \ldots$$

are linear functions of each argument and also symmetric in these vectors.

The substitution

$$\underline{x} = \varepsilon\underline{y} \tag{2.2}$$

carries (1.1) into

$$\dot{\underline{y}} = \underline{L}_\mu(\underline{y}) + \varepsilon\underline{Q}_\mu(\underline{y},\underline{y}) + \varepsilon^2\underline{K}_\mu(\underline{y},\underline{y},\underline{y}) + \cdots \tag{2.3}$$

The right hand side is analytic in ε, μ, $\underline{y}$ at the point $\varepsilon = \mu = 0$, $\underline{y} = \underline{y}^0$ ($\underline{y}^0$ arbitrary). We consider the case $\varepsilon = 0$ in (2.3), that is, the homogeneous linear differential equation

$$\dot{\underline{z}} = \underline{L}_\mu(\underline{z}). \tag{2.4}$$

For the question of existence, this has the deciding significance.

The complex conjugate characteristic exponents $\alpha(\mu)$, $\overline{\alpha}(\mu)$, which were referred to in the hypothesis, are simple for all small $|\mu|$. In the associated solutions

$$e^{\alpha t}\underline{a}, \quad e^{\overline{\alpha} t}\underline{\overline{a}} \tag{2.5}$$

of (2.4), the complex vector $\underline{a}$ is consequently determined up to a complex scalar factor; $\overline{a}$ is the conjugate vector.

Furthermore, there are no solutions of the form

$$e^{\alpha t}(t\underline{b} + \underline{c}), \quad \underline{b} \neq 0. \tag{2.6}$$

$\alpha(\mu)$ is analytic at $\mu = 0$. One can choose a fixed real vector $\underline{e} \neq 0$ so that for all small $|\mu|$, $\underline{a} \cdot \underline{e} \neq 0$ for $\underline{a} \neq 0$. $\underline{a} = \underline{a}(\mu)$ is then uniquely determined by the condition

$$\underline{a}(\mu)\cdot\underline{e} = \frac{1}{\alpha(\mu)-\bar{\alpha}(\mu)} \qquad (\bar{\underline{e}} = \underline{e} \neq 0). \tag{2.7}$$

By hypothesis,

$$\bar{\alpha}(0) = -\alpha(0) \neq 0. \tag{2.8}$$

$\underline{a}(\mu)$ is analytic at $\mu = 0$.

The real solutions of (2.4), which are linear combinations of (2.5), have the form

$$\underline{z} = ce^{\alpha t}\underline{a} + \bar{c}e^{\bar{\alpha}t}\bar{\underline{a}} \tag{2.9}$$

with complex scalar c. They form a family depending on two real parameters; one of these parameters is a proportionality factor, while the other represents an additive constant in t (the solutions form only a one parameter family of curves). Because $\bar{\underline{e}} = \underline{e}$, we have

$$\left.\begin{aligned} \underline{z}\cdot\underline{e} &= c\ \underline{a}\cdot\underline{e} + \bar{c}\ \overline{\underline{a}\cdot\underline{e}} \\ \dot{\underline{z}}\cdot\underline{e} &= c\ \alpha\ \underline{a}\cdot\underline{e} + \bar{c}\ \bar{\alpha}\ \overline{\underline{a}\cdot\underline{e}} \end{aligned}\right\} \quad \text{at} \quad t = 0.$$

For $c = 1$, (2.9) is

$$\underline{z} = e^{\alpha t}\underline{a} + e^{\bar{\alpha}t}\bar{\underline{a}} = \underline{z}(t,\mu); \tag{2.10}$$

because of (2.7), this $\underline{z}$ satisfies the conditions:

$$t = 0: \quad \underline{z}\cdot\underline{e} = 0, \quad \frac{d}{dt}(\underline{z}\cdot\underline{e}) = 1. \tag{2.11}$$

This is the unique solution of the form (2.9) satisfying these conditions; for from

$$t = 0: \quad \underline{z}\cdot\underline{e} = \dot{\underline{z}}\cdot\underline{e} = 0$$

and from (2.9), (2.7) and (2.8) it follows that $c = 0$; thus $\underline{z} = 0$.

By hypothesis, for $\mu = 0$, α, $\overline{\alpha}$ are the only ones among the characteristic exponents which are pure imaginary. Hence, for $\mu = 0$, (2.9) gives all the real and periodic solutions of (2.4). Their period is

$$T_0 = \frac{2\pi}{|\alpha(0)|} \quad . \tag{2.12}$$

In particular, for $\mu = 0$, (2.10) is the only real and periodic solution with the properties (2.11).

For later use we also notice that, for $\mu = 0$, (2.4) can have no solutions of the form

$$t\,\underline{p}(t) + \underline{q}(t)$$

where $\underline{p}$ and $\underline{q}$ have a common period and $\underline{p}$ is not identically zero. Otherwise (2.4) would break up into the two equations

$$\dot{\underline{p}} = \underline{L}_0(\underline{p}), \quad \underline{p} + \dot{\underline{q}} = \underline{L}_0(\underline{q})$$

and $\underline{p}$ would be a nontrivial linear combination of the solutions (2.5). The Fourier expansion of $\underline{q}(t)$ would then lead to a solution of the form (2.6).

By differentiation of (2.4) with respect to μ at $\mu = 0$ one obtains the non-homogeneous differential equation

$$\dot{\underline{z}}' = \underline{L}_0(\underline{z}') + \underline{L}_0'(\underline{z}); \quad \underline{L}_0' = \frac{d}{d\mu}\underline{L}_\mu, \quad \mu = 0, \tag{2.13}$$

for the μ-derivative of (2.10):

$$\underline{z}' = t(\alpha' e^{\alpha t}\underline{a} + \bar{\alpha}' e^{\bar{\alpha} t}\bar{\underline{a}}) + (e^{\alpha t}\underline{a}' + e^{\bar{\alpha} t}\bar{\underline{a}}') \quad (\mu = 0).$$

The factor of t is a solution of (2.4). If one expresses it linearly in terms of the solution (2.10) and $\dot{\underline{z}}$, it follows from (2.8) that

$$\underline{z}' = t(\mathrm{Re}(\alpha')\underline{z} + \frac{\mathrm{Im}(\alpha')}{\alpha}\dot{\underline{z}}) + \underline{h}(t) \tag{2.14}$$

with

$$\underline{h}(t + T_0) = \underline{h}(t). \tag{2.15}$$

Now let

$$\underline{y} = \underline{y}(t,\mu,\varepsilon,\underline{y}^0)$$

be the solution of (2.3), which satisfies the initial condition $y = y^0$ for $t = 0$. According to well known theorems it depends analytically on all its arguments at each point $(t,0,0,\underline{y}^0)$. It is periodic with the period T if and only if the equation

$$\underline{y}(T,\mu,\varepsilon,\underline{y}^0) - \underline{y}^0 = 0 \tag{2.16}$$

is satisfied. If one denotes by $\underline{z}^0$ the fixed initial value of the fixed solution (2.10) of (2.4), $\mu = 0$, then (2.16) is satisfied by the values

$$T = T_0,\ \mu = \varepsilon = 0,\ \underline{y}^0 = \underline{z}^0. \tag{2.17}$$

The problem is: for given ε, solve equation (2.16) for T, μ and $\underline{y}^0$. These are n equations with $n + 2$ unknowns. In order to make the solution unique, we add the two equations

$$\underline{y}^0 \cdot \underline{e} = 0, \quad \dot{\underline{y}}^0 \cdot \underline{e} = 1 \tag{2.18}$$

where $\underline{e}$ is the real vector introduced above and where

$\dot{\underline{y}}^0 = \dot{\underline{y}}$ for $t = 0$. The introduction of these conditions implies no restriction on the totality of solutions in the small, as will be demonstrated in the next section. For the initial values $\mu = \varepsilon = 0$, $\underline{y}^0 = \underline{z}^0$, it follows from (2.11) that these equations are satisfied by the solution (2.10).

Now for all sufficiently small $|\varepsilon|$, (2.16) and (2.18) have exactly one solution

$$T = T(\varepsilon), \quad \mu = \mu(\varepsilon), \quad \underline{y}^0 = \underline{y}^0(\varepsilon) \tag{2.19}$$

in a suitable neighborhood of the system of values

$$T = T_0, \quad \mu = 0, \quad \underline{y}^0 = \underline{z}^0, \tag{2.20}$$

if the following is the case: the system of linear equations formed by taking the differential (at the place (2.17)) with respect to the variables T, μ, ε, $\underline{y}^0$ is uniquely solvable for given $d\varepsilon$. Equivalently, there are such functions (2.19) if these linear equations for $d\varepsilon = 0$ have only the zero solution $dT = d\mu = d\underline{y}^0 = 0$. This is the case, as will now be shown.

We have

$$\dot{\underline{y}} = \underline{L}_\mu(\underline{y}), \quad \underline{y} = \underline{y}(t,\mu,0,\underline{y}^0). \tag{2.21}$$

In particular

$$\underline{y}(t,\mu,0,\underline{z}^0) = \underline{z}(t,\mu) \tag{2.22}$$

is the solution to (2.10). The differential $d\underline{y}(t,\mu,0,\underline{y}^0)$ is the sum of the differentials with respect to the separate arguments when the others are all fixed. If we introduce for the differentials

$$dt, \quad d\mu, \quad d\underline{y}^0$$

as independent constants or vectors the notations:

$$\rho, \quad \sigma, \quad \underline{u}^0$$

then, the differential referred to becomes

$$\rho\dot{\underline{y}} + \sigma\underline{y}' + \underline{u}.$$

Here $\dot{\underline{y}}$ and $\underline{y}' = \partial\underline{y}/\partial\mu$ are taken at $T = T_0$, $\mu = 0$, $\underline{y}^0 = \underline{z}^0$ and $\underline{u}$ is the solution of

$$\dot{\underline{u}} = \underline{L}_0(\underline{u})$$

with the initial value $\underline{u}^0$ for $t = 0$. According to (2.22), $\dot{\underline{y}} = \dot{\underline{z}}(t,0)$. If one sets $\underline{y}' = \underline{v}$, then $\underline{v}(t)$ is the solution of

$$\dot{\underline{v}} = \underline{L}_0(\underline{v}) + \underline{L}_0'(\underline{z}), \quad \underline{v}(0) = 0. \tag{2.23}$$

The linear vector equation arising from (2.16) is then

$$\rho\dot{\underline{z}}(T_0) + \sigma\underline{v}(T_0) + \underline{u}(T_0) - \underline{u}(0) = 0, \tag{2.24}$$

where $\underline{z}(t)$ denotes the solution (2.10) of

$$\dot{\underline{z}} = \underline{L}_0(\underline{z}). \tag{2.25}$$

$\underline{u}(t)$ is any solution of this homogeneous linear differential equation with constant $\underline{L}_0$, and $\underline{v}(t)$ is the solution of (2.23). We show now that (2.24) is possible only if $\rho = \sigma = 0$ and $\underline{u}(t) = 0$.

Now for all t

$$\rho\dot{\underline{z}}(t) + \sigma[\underline{v}(t+T_0) - \underline{v}(t)] + \underline{u}(t+T_0) - \underline{u}(t) = 0. \tag{2.26}$$

This is true because $\underline{z}(t)$ has period T_0, so it follows from (2.23) that the square bracket is a solution of (2.25). $\dot{z}$ is also a solution of (2.25)[+], so the whole left

---

[+] In the original, this number is (2.23).

side of (2.26) is a solution of (2.25). By (2.24) and the fact that $\underline{v}(0) = 0$, the initial value of this solution is zero, and thus it is identically zero. Now from (2.13) and (2.23) it follows that

$$\underline{v}(t) = \underline{z}'(t) + \underline{g}(t), \quad \dot{\underline{g}} = \underline{L}_0(\underline{g}).$$

Thus, by (2.14) and (2.15), the square bracket in (2.26) has the value

$$T_0[\mathrm{Re}(\alpha')\underline{z}(\underline{t}) + \frac{\mathrm{Im}(\alpha')}{\alpha}\dot{\underline{z}}(t)] + [\underline{g}(t+T_0) - \underline{g}(t)].$$

If one sets $\underline{u} + \sigma\underline{g} = \underline{w}$ and

$$\sigma T_0 \mathrm{Re}(\alpha')\underline{z}(t) + [\rho+\sigma T_0 \frac{\mathrm{Im}(\alpha')}{\alpha}]\dot{\underline{z}}(t) = \tilde{\underline{z}}(t), \tag{2.27}$$

it follows that

$$\tilde{\underline{z}}(t) + \underline{w}(t+T_0) - \underline{w}(t) = 0,$$

where $\underline{w}(t)$ is a solution and $\tilde{\underline{z}}(t)$ a periodic solution of (2.25). This means that

$$\underline{w}(t) = -\frac{t}{T_0}\tilde{\underline{z}}(t) + q(t)$$

with periodic q. However, as we stated before, such solutions cannot exist unless $\tilde{\underline{z}} = 0$.

Since $\underline{z}, \dot{\underline{z}}$ are linearly independent, it follows from (2.27) and from the hypothesis (1.2) that $\sigma = 0$ and $\rho = 0$. Thus, by (2.24), $\underline{u}(t)$ has period T_0.

Finally, since $d\underline{y}^0 = \underline{u}^0$, and $d\dot{\underline{y}}^0 = \dot{\underline{u}}$, at $t = 0$, it follows from the equations (2.18) that

$$\underline{u} \cdot \underline{e} = \dot{\underline{u}} \cdot \underline{e} = 0, \quad \text{at} \quad t = 0.$$

A periodic solution of $\dot{\underline{u}} = \underline{L}_0(\underline{u})$ with these properties must vanish, as we have stated above. With this the proof of the existence of a periodic family is concluded.†

The solutions (2.19) are analytic at $\varepsilon = 0$

$$T = T_0(1 + \tau_1\varepsilon + \tau_2\varepsilon^2 + \ldots),$$
$$\mu = \mu_1\varepsilon + \mu_2\varepsilon^2 + \ldots \quad . \tag{2.28}$$

The periodic solutions $\underline{y}(t,\varepsilon)$ of (2.3), and the periodic family of solutions

$$\underline{x}(t,\varepsilon) = \varepsilon\underline{y}(t,\varepsilon) \tag{2.29}$$

of (1.1)+, are analytic at every point $(t,0)$.

One obtains exactly the same periodic solutions if one begins with a multiple mT_0 of the period instead of T_0, that is if one operates in a neighborhood of the system of values

$$T = mT_0, \ \mu = 0, \ y^0 = z^0 \tag{2.30}$$

instead of (2.20). Nothing essential is altered in the proof.

3. Completion of the Proof of the Theorem.

For arbitrarily large $L > T_0$ there are two positive numbers a and b with the following property. Every periodic solution $\underline{x}(t) \neq 0$ of (1.1)+, whose period is smaller than L, which belongs to a μ with $|\mu| < b$ and which lies in $|\underline{x}| < a$, belongs to the family (2.29), (2.28), $\varepsilon > 0$ if a suitable choice is made for the origin of t.

If this were not the case, there would be a sequence

---

† See editorial comments in §5A below.

+ In the original paper, this number is (1).

of periodic solutions $\underline{x}_k(t) \neq 0$[++] having bounded periods $T_k < L$, and of corresponding μ-values, with

$$\kappa_k = \operatorname*{Max}_t |\underline{x}_k(t)| \to 0, \quad \mu_k \to 0 \tag{3.1}$$

and such that no pair $\underline{x}_k(t)$, μ_k belongs to the above family. We let

$$\underline{y}_k(t) = \frac{1}{\kappa_k} \underline{x}_k(t).$$

$\underline{y}_k$ is a solution of (2.3), with κ_k instead of ε, and $\underline{y}_k$ satisfies

$$\operatorname*{Max}_t |\underline{y}_k(t)| = 1.$$

One considers first a subsequence for which the initial values converge, $\underline{y}_k^0 \to \underline{z}^0$. Then, uniformly for $|t| < L$, we have $\underline{y}_k(t) \to \underline{z}(t)$, where $\dot{\underline{z}} = \underline{L}_0(\underline{z})$ and $\underline{z}(0) = \underline{z}^0$. Since the maximum of $|\underline{z}| = 1$, $\underline{z}$ is not identically zero. $\underline{z}$ is of the form (2.9)[+], $c \neq 0$, and it has the fundamental period T_0. If one shifts the origin of t in $\underline{z}(t)$ to the place where $\underline{z} \cdot \underline{e} = 0$, one finds that $\dot{\underline{z}} \cdot \underline{e} \neq 0$ there. This quantity can be taken to be positive; otherwise, since

$$\underline{z}(t + \frac{1}{2} T_0) = -\underline{z}(t),$$

one could achieve this by shifting from $t = 0$ by $\frac{1}{2} T_0$. Consequently,

$$\underline{z}^0 \cdot \underline{e} = 0, \quad \dot{\underline{z}}^0 \cdot \underline{e} > 0.$$

From this it follows that in the neighborhood of $\underline{z}^0$, and for

---

[++] In the original paper, the sequences x_k, T_k, etc. are not indexed.

[+] In the original, this number is (2.8).

small κ and $|\mu|$, all solutions of the differential equation (2.3) (κ instead of ε) cut the hyperplane $\underline{y} \cdot \underline{e} = 0$ once. In this intersection let $t = 0$. Then, for the sequence $\underline{y}_k(t)$, κ_k, μ_k under consideration, with this choice of origin, we always have $\underline{y}_k{}^0 \to \underline{z}^0$. Also

$$\dot{\underline{y}}_k^0 \cdot \underline{e} = 0, \ \rho_k = \dot{\underline{y}}_k^0 \cdot \underline{e} \to \dot{\underline{z}}^0 \cdot \underline{e} = \rho > 0$$

and $\kappa_k \to 0$, $\mu_k \to 0$. If one now sets

$$\tilde{\underline{y}}_k(t) = \frac{1}{\rho_k} \underline{y}_k(t) = \frac{1}{\rho_k \kappa_k} \underline{x}_k(t), \quad \rho_k \kappa_k = \varepsilon_k$$

then $\tilde{\underline{y}}_k$ is a solution of (2.3)[+], for the parameter values $\varepsilon_k > 0$ and μ_k. For it, we have

$$\tilde{\underline{y}}_k \cdot \underline{e} = 0, \ \dot{\tilde{\underline{y}}}_k \cdot \underline{e} = 1, \ \text{at} \ t = 0. \tag{2.18}$$

The periods in the sequence of solutions must converge to a multiple of T_0, mT_0. Furthermore $\varepsilon_k \to 0$. However, this implies that from some point on in the sequence one enters the neighborhood mentioned above of (2.20) or (2.30) in which, for all sufficiently small ε, there is only one solution of the system of equations under consideration. The solutions of our sequence must then belong to the above family and in fact with $\varepsilon > 0$, which conflicts with the assumption. Thus the assertion is proved.[††]

From the fact we have just proved it now follows that if $\mu(\varepsilon) \not\equiv 0$, then the first coefficient which is different from 0 in $\mu = \mu_1\varepsilon + \mu_2\varepsilon^2 + \ldots$ is of even order; the same

---

[+] In the original, this number is (3).

[††] See editorial comments in §5A below.

holds for the expansion $T = T_0 \ (1 + \tau_1\varepsilon + \tau_2\varepsilon^2 + \ldots)$. For the solutions of the family corresponding to $\varepsilon < 0$, and the associated μ and T-values, must already be present among those for $\varepsilon > 0$.[++] In particular we have

$$\mu_1 = \tau_1 = 0. \tag{3.2}$$

The periodic solutions exist, for sufficiently small $|\mu|$ and $|\underline{x}|$, only for $\mu > 0$, or only for $\mu < 0$, or only for $\mu = 0$.

4. Determination of the Coefficients.

We shall need the following result, which gives a criterion for the solvability of the inhomogeneous system of differential equations

$$\dot{\underline{w}} = \underline{L}(\underline{w}) + \underline{q}, \quad (\underline{L} = \underline{L}_0) \tag{4.1}$$

where $\underline{q}(t)$ has a period T_0. Let

$$\dot{\underline{z}}^* = -\underline{L}^*(\underline{z}^*) \tag{4.2}$$

be the differential equation which is adjoint to the homogeneous one; L^* is the adjoint operator to L (transposed matrix), defined by

$$\underline{L}(\underline{u}) \cdot \underline{v} = \underline{u} \cdot \underline{L}^*(\underline{v}).^{+}$$

Then (4.1) has a periodic solution $\underline{w}$ with the period T_0, if and only if

---

[++] And indeed with a shift of the t-origin of approximately $T_0/2$.

[+] In the following, the inner product of two complex vectors a, b is defined by $\sum \bar{a}_i b_i$.

$$\int_0^{T_0} \underline{q}\cdot\underline{z}^* dt = 0 \tag{4.3}$$

for all solutions of (4.2) which have the period T_0.

This result follows from the known criterion for the solvability of an ordinary system of linear equations. The necessity follows directly from (4.1) and (4.2). That the condition is sufficient can be shown in the following way: The adjoint equation has the same characteristic exponents and therefore it also has two solutions of the form

$$e^{\alpha t}a^*, \quad e^{-\alpha t}\bar{a}^*, \quad \alpha = \alpha(0) = -\bar{\alpha}(0), \tag{4.4}$$

from which all periodic solutions can be formed by linear combinations. Furthermore, the development of $\underline{q}(t)$ in Fourier series shows that it suffices to consider the case

$$\underline{q} = e^{-\alpha t}\underline{b}$$

and the analogous case with α instead of $-\alpha$. In (4.1) let us insert

$$\underline{w} = e^{-\alpha t}\underline{c}.$$

(4.1) then becomes

$$(\alpha I + \underline{L})\underline{c} = \underline{b}.$$

(4.4) and (4.2) imply

$$(\alpha I + \underline{L})^*\underline{a}^* = 0$$

while (4.3) says $\underline{b}\cdot\underline{a}^* = 0$. From this everything follows with the help of the theorem referred to.

Secondly, we shall need the following fact. For any solution $\underline{z} \neq 0$ of $\dot{\underline{z}} = \underline{L}(\underline{z})$ having period T_0, there is always a solution $\underline{z}^*$ of the adjoint equation, with the same period, such that

$$\int_0^{T_0} \underline{z} \cdot \underline{z}^* dt \neq 0.^*$$

Otherwise, the equation $\dot{\underline{w}} = \underline{L}(\underline{w}) + \underline{z}$ would have a solution $\underline{w}$, and $\underline{w} - t\underline{z}$[+] would be a solution of the homogeneous differential equation, which contradicts the simplicity of the characteristic exponent α.

Let $\underline{z}_1^*$ and $\underline{z}_2^*$ be two linearly independent solutions of (4.2) with the period T_0. Let

$$[\underline{q}]_i = \int_0^{T_0} \underline{q} \cdot \underline{z}_i^* \, dt \qquad (i = 1,2).$$

Then the criterion for solvability of (4.1) under the given conditions is

$$[\underline{q}]_1 = [\underline{q}]_2 = 0. \tag{4.5}$$

We also note that z_1^*, z_2^* can be chosen in such a way that

$$[\underline{z}]_1 = [\dot{\underline{z}}]_2 = 1, \qquad [\underline{z}]_2 = [\dot{\underline{z}}]_1 = 0 \tag{4.6}$$

where $\underline{z}$ is the solution (2.10) of (2.4) with $\mu = 0$ (biorthogonalization).

The problem of the determination of the coefficients for the power series representation of the periodic family can now be solved in a general way. If one defines the new independent variable s by

$$t = s(1 + \tau_2 \varepsilon^2 + \tau_3 \varepsilon^3 + \dots) \tag{4.7}$$

then according to (2.28) the period in the family of solutions

---

[*] Also, the integrand is always constant.

[+] In the original, the statement reads "w+tz", which is incorrect.

$\underline{y} = \underline{y}(s,\varepsilon)$ is constantly equal to T_0. $\underline{y}$, as a function of s (or t) and ε, is analytic at every point $(s,0)$. One has

$$\underline{y} = \underline{y}_0(s) + \varepsilon \underline{y}_1(s) + \varepsilon^2 \underline{y}_2(s) + \cdots, \tag{4.8}$$

where all the $\underline{y}_i$ have the period T_0. The derivative with respect to s will again be denoted by a dot. We write for simplicity

$$\underline{L}_0 = \underline{L},\ \underline{L}_0' = \underline{L}',\ \underline{Q}_0 = \underline{Q},\ \underline{K}_0 = \underline{K},\ \ldots$$

Then, using (3.2), and inserting (4.7) and (4.8) in (2.3), one obtains the recursive equations

$$\dot{\underline{y}}_0 = \underline{L}(\underline{y}_0) \qquad (\underline{y}_0 = \underline{z}) \tag{4.9}$$

$$\dot{\underline{y}}_1 = \underline{L}(\underline{y}_1) + \underline{Q}(\underline{y}_0,\underline{y}_0) \tag{4.10}$$

$$\begin{aligned} -\tau_2\dot{\underline{y}}_0 + \dot{\underline{y}}_2 = \underline{L}(\underline{y}_2) + \mu_2\underline{L}'(\underline{y}_0) &+ 2\underline{Q}(\underline{y}_0,\underline{y}_1) \\ &+ \underline{K}(\underline{y}_0,\underline{y}_0,\underline{y}_0) \end{aligned} \tag{4.11}$$

....

from which the $\underline{y}_i$, μ_i, τ_i are to be determined. In addition to these, we have the conditions following from (2.18)

$$\underline{y}_k \cdot \underline{e} = \dot{\underline{y}}_k \cdot \underline{e} = 0, \quad \text{at} \quad s = 0 \tag{4.12}$$

for $k = 1,2,\ldots$. In the equations, we again write t instead of s. By (4.10) and (4.12), y_1 is uniquely determined as a periodic function with period T_0. From (4.11) $\underline{L}'$ must first be eliminated with the help of (2.13). Since the parenthesis in the first summand of (2.14) is a solution of $\dot{\underline{z}} = \underline{L}(\underline{z})$, (2.13) can be written

$$\mathrm{Re}(\alpha')\underline{z} + \frac{\mathrm{Im}(\alpha')}{\alpha}\dot{\underline{z}} + \dot{\underline{h}} = \underline{L}(\underline{h}) + \underline{L}'(\underline{z}) \tag{4.13}$$

Let

$$\underline{y}_2 - \mu_2\underline{h} = \underline{v}, \tag{4.14}$$

which, according to (2.15), has the period T_0. Since $\underline{z} = \underline{y}_0$, it follows that

$$-\mu_2\mathrm{Re}(\alpha')\underline{y}_0 - (\tau_2 + \mu_2\frac{\mathrm{Im}(\alpha')}{\alpha})\dot{\underline{y}}_0 + \dot{\underline{v}}$$
$$= \underline{L}(\underline{v}) + 2\underline{Q}(\underline{y}_0,\underline{y}_1) + \underline{K}(\underline{y}_0,\underline{y}_0,\underline{y}_0). \tag{4.15}$$

Thus, by (4.6)

$$\mu_2\mathrm{Re}(\alpha') = -[2\underline{Q}(\underline{y}_0,\underline{y}_1) + \underline{K}(\underline{y}_0,\underline{y}_0,\underline{y}_0)]_1,$$
$$\tau_2 + \mu_2\frac{\mathrm{Im}(\alpha')}{\alpha} = -[2Q(\underline{y}_0,\underline{y}_1) + K(\underline{y}_0,\underline{y}_0,\underline{y}_0)]_2. \tag{4.16}$$

By hypothesis (1.2), μ_2 and τ_2 are determined from this. One then solves (4.15) for $\underline{v}$ and obtains $\underline{y}_2$ from (4.14) and (4.12), $k = 2$, in a unique way.

In an analogous fashion all the higher coefficients are obtained from the subsequent recursion formulas. In general $\mu_2 \neq 0$. If μ_2 is positive then the periodic solutions exist only for $\mu > 0$; the corresponding statement holds for $\mu_2 < 0$.[†††]

5. The Characteristic Exponents of the Periodic Solution.

In the following we shall sometimes make use of determinants; this, however, can be avoided. In the linearization about the periodic solutions of (2.3),

$$\dot{\underline{u}} = \underline{L}_{t,\varepsilon}(\underline{u}) \tag{5.1}$$

[†††] See editorial comments in §5A below.

we have, by (2.3),

$$\underline{L}_{t,\varepsilon}(\underline{u}) = \underline{L}_{\mu}(\underline{u}) + 2\varepsilon\underline{Q}_{\mu}(\underline{y},\underline{u}) + 3\varepsilon^2\underline{K}_{\mu}(\underline{y},\underline{y},\underline{u})+\dots \qquad (5.2)$$

A fundamental system $\underline{u}_i(t,\varepsilon)$ formed with fixed initial conditions depends analytically on (t,ε). The coefficients in $\underline{u}_i(T,\varepsilon) = \sum a_{i\nu}(\varepsilon)\underline{u}_{\nu}(0)$ are analytic at $\varepsilon = 0$. The determinantal equation

$$||a_{ik}(\varepsilon) - \zeta\delta_{ik}|| = 0, \quad \zeta = e^{\lambda T(\varepsilon)}, \qquad (5.3)$$

determines the characteristic exponents λ_k and the solutions $\underline{v}$, of (1.3), where

$$\underline{u} = e^{\lambda t}\underline{v} \quad .$$

Since (5.1) is solved by $\underline{u} = \dot{\underline{y}}$, $\zeta = 1$ is a root of (5.3). The exponent β, which was spoken of in the introduction, corresponds to a simple root of the equation obtained by dividing out $\zeta - 1$. $\beta(\varepsilon)$ is thus real and analytic at $\varepsilon = 0$, $\beta = \beta_2\varepsilon^2 + \dots$ (β_1 is also equal to zero for the same reasons as μ_1 and τ_1). Now if β is not $\equiv 0$, then there is some minor of order $n - 1$ in the determinant (5.3) (with the corresponding ζ) which is not 0. From this it follows that (1.3), $\lambda = \beta$, has a solution $\underline{v} \not\equiv 0$ which is analytic at $\varepsilon = 0$. Even if $\beta \equiv 0$, there is a minor of order $n - 2$ which is not zero. As we know, in this case, there is a solution of (5.1), analytic at $\varepsilon = 0$, of the form $\underline{u} = t\underline{v} + \underline{w}$ with periodic $\underline{v}$, $\underline{w}$, where either $\underline{v} \not\equiv 0$, or $\underline{v} = 0$ and $\underline{w}$ is linearly independent of the solution $\underline{u} = \dot{\underline{y}}$.* That $t\underline{v} + \underline{w}$ is a solution implies that

---

* Cf. e.g. F. R. Moulton, Periodic Orbits, Washington, 1920, p. 26.

$$\dot{\underline{v}} = \underline{L}_{t,\varepsilon}(\underline{v}), \quad \underline{v} + \dot{\underline{w}} = \underline{L}_{t,\varepsilon}(\underline{w}). \tag{5.4}$$

After these preliminary observations we shall calculate β_2. We assume here that $\mu_2 \neq 0$. $\beta \equiv 0$ is then impossible as will subsequently be proved. If we use (4.7) to introduce s as a new t into (1.3) we get

$$(1 - \tau_2\varepsilon^2 + \ldots)\dot{\underline{v}} + \beta\underline{v} = \underline{L}_{t,\varepsilon}(\underline{v}).$$

Also, we have (with the new t)

$$\underline{v} = \underline{v}_0(t) + \varepsilon\underline{v}_1(t) + \varepsilon^2\underline{v}_2(t) + \ldots$$

where all the $\underline{v}_i$ have the same period T_0. If we introduce the power series for μ, β, $\underline{v}$, $\underline{y}$, it follows (dropping the subscript zero on the operators as before) that

$$\dot{\underline{v}}_0 = \underline{L}(\underline{v}_0), \tag{5.5}$$

$$\dot{\underline{v}}_1 = \underline{L}(\underline{v}_1) + 2\underline{Q}(\underline{y}_0,\underline{v}_0), \tag{5.6}$$

$$\begin{aligned} \beta_2\underline{v}_0 - \tau_2\dot{\underline{v}}_0 + \dot{\underline{v}}_2 &= \underline{L}(\underline{v}_2) + \mu_2\underline{L}'(\underline{v}_0) \\ &\quad + 2\underline{Q}(\underline{y}_1,\underline{v}_0) + 2\underline{Q}(\underline{y}_0,\underline{v}_1) + 3\underline{K}(\underline{y}_0,\underline{y}_0,\underline{v}_0). \end{aligned} \tag{5.7}$$

These equations have the trivial solution

$$\beta_i = 0, \quad v_i = \dot{y}_i \quad (i = 0,1,\ldots). \tag{5.8}$$

Thus, one has

$$\ddot{\underline{y}}_1 = \underline{L}(\dot{\underline{y}}_1) + 2\underline{Q}(\underline{y}_0,\dot{\underline{y}}_0). \tag{5.9}$$

$$\begin{aligned} -\tau_2\ddot{y}_0 + \ddot{\underline{y}}_2 &= \underline{L}(\dot{\underline{y}}_2) + \underline{\mu}_2 L'(\dot{\underline{y}}_0) \\ &\quad + 2\underline{Q}(\underline{y}_1,\dot{\underline{y}}_0) + 2\underline{Q}(\underline{y}_0,\dot{\underline{y}}_1) + 3\underline{K}(\underline{y}_0,\underline{y}_0,\dot{\underline{y}}_0). \end{aligned} \tag{5.10}$$

Since we may assume that $\underline{v}_0 \neq 0$

$$\underline{v}_0 = \rho \underline{y}_0 + \sigma \dot{\underline{y}}_0,^{*} \tag{5.11}$$

where at least one of the coefficients is not 0. If we set

$$\underline{v}_1 - 2\rho \underline{y}_1 - \sigma \underline{y}_1 = \underline{w}, \tag{5.12}$$

it follows from (4.10), (5.6) and (5.9) that $\dot{\underline{w}} = \underline{L}(\underline{w})$, thus

$$\underline{w} = \rho' \underline{y}_0 + \sigma' \dot{\underline{y}}_0. \tag{5.13}$$

If one forms the combination

$$(5.7) - \rho(4.11) - \sigma(5.10),$$

in which $\underline{L}'$ cancels out, and sets

$$\underline{v}_2 - \rho \underline{y}_2 - \sigma \dot{\underline{y}}_2 = \underline{u},$$

then, using (5.11) and (5.12), one obtains:

$$\beta_2 \underline{v}_0 + \dot{\underline{u}} = \underline{L}(\underline{u}) + 2\rho(2\underline{Q}(\underline{y}_0, \underline{y}_1) + \underline{K}(\underline{y}_0, \underline{y}_0, \underline{y}_0)) + \underline{R} \tag{5.14}$$

with

$$\underline{R} = 2\underline{Q}(\underline{y}_0, \underline{w}).$$

If we now apply the bracket criterion of the previous section to (4.10) and (5.9), it follows from (5.13) that

$$[\underline{R}]_1 = [\underline{R}]_2 = 0.$$

If we apply it to (5.14), in which $\underline{u}$ has the period T_0, it follows from (4.6) (with $\underline{z} = \underline{y}_0$) that

$$\rho \beta_2 = 2\rho[2\underline{Q}(\underline{y}_0, \underline{y}_1) + \underline{K}(\underline{y}_0, \underline{y}_0, \underline{y}_0)]_1.$$

---

* The ρ and σ in (5.11) are unrelated to the symbols ρ and σ as used in Section 2.

Hence, by (4.16),

$$\rho\beta_2 = -2\rho\mu_2 \underline{R}(\alpha').$$

Likewise, it follows that

$$\sigma\beta_2 = -2\rho(\tau_2 + \mu_2 \frac{\mathrm{Im}(\alpha')}{\alpha}).$$

From this, either β_2 is given by (1.4) (and then β_2 is not zero since $\mu_2 \neq 0$) or else $\beta_2 = 0$. In either case $\rho:\sigma$ is completely determined (in the second case $\rho = 0$).

To check that the first case really occurs we must undertake a somewhat longer consideration. One may think of the process as schematized in the following manner. The equation for β and $\underline{v}$ (namely the equation which follows equation (5.4)) should be divided by the factor in parenthesis. It is then once again of the form

$$\dot{\underline{v}} + \beta\underline{v} = \underline{L}_{t,\varepsilon}(\underline{v})$$

with

$$\underline{L}_{t,\varepsilon} = \underline{L}_0 + \varepsilon\underline{L}_1 + \varepsilon^2\underline{L}_2 + \cdots,$$

where $\underline{L}_0$ is a constant operator, while L_i, $i > 0$, depend on t with the period T_0. The coefficients of $1, \varepsilon$ are not altered by the division. Introduction of the power series leads to

$$\begin{aligned} \dot{\underline{v}}_0 &= \underline{L}_0(\underline{v}_0), \\ \dot{\underline{v}}_1 &= \underline{L}_0(\underline{v}_1) + \underline{L}_1(\underline{v}_0),^* \end{aligned} \tag{5.15}$$

$$\beta_2\underline{v}_0 + \dot{\underline{v}}_2 = \underline{L}_0(\underline{v}_2) + \underline{L}_1(\underline{v}_1) + \underline{L}_2(\underline{v}_0),$$

---

* One does not really have to assume $\beta_1 = 0$. From the bracket criterion this is a consequence of (5.17).

$$\beta_3\underline{v}_0 + \beta_2\underline{v}_1 + \dot{\underline{v}}_3 = \underline{L}_0(\underline{v}_3) + \underline{L}_1(\underline{v}_2) + \underline{L}_2(\underline{v}_1) + \underline{L}_3(\underline{v}_0)$$

and so forth. The situation is the following. For $\varepsilon = 0$ there are two solutions $\underline{z}$, $\dot{\underline{z}}$ with period T_0. Furthermore

$$\underline{v}_0 = \rho\underline{z} + \sigma\dot{\underline{z}} \tag{5.16}$$

and

$$[\underline{L}_1(\underline{z})] = [\underline{L}_1(\dot{\underline{z}})] = 0 \tag{5.17}$$

for both bracket subscripts. It follows that

$$\underline{v}_1 = \rho\underline{g} + \sigma h + \rho'\underline{z} + \sigma'\underline{z} \tag{5.18}$$

with fixed periodic $\underline{g}$ and $\underline{h}$. For the third equation of (5.15), the bracket criterion gives

$$\begin{aligned} \beta_2\rho &= A_1\rho + B_1\sigma \\ \beta_2\sigma &= A_2\rho + B_2\sigma \end{aligned} \tag{5.19}$$

with

$$\begin{aligned} A_i &= [\underline{L}_1(\underline{g}) + \underline{L}_2(\underline{z})]_i, \\ B_i &= [\underline{L}_1(\underline{h}) + \underline{L}_2(\dot{\underline{z}})]_i, \end{aligned} \tag{5.20}$$

while (5.17) implies that ρ', σ' drop out. The situation now is that the equations (5.19) with the unknowns β_2, ρ, σ have two distinct real solutions β_2.* To them belong two linearly independent pairs (ρ,σ). Each of the two solution systems leads now to a unique determination of the β_i and $\underline{v}_i$ through the recursion formulas, if one suitably normalizes

---

*In the general case, that is if the special condition (5.17) is not fulfilled, the splitting into two cases occurs already at the second equation (5.15). The solution of the problem in this case is found in F. R. Houlton, Periodic Orbits. Compare Chapter 1, particularly pages 34 and 40.

$\underline{v}$. To this end choose a constant vector $\underline{a} \neq 0$ in such a way that $\underline{v}_0 \cdot \underline{a} = 1$ $(t = 0)$ for both pairs (ρ,σ) in (5.16). One concludes then that

$$\underline{v} \cdot \underline{a} = 1, \quad t = 0,$$

that is, $\underline{v}_i \cdot \underline{a} = 0$ at $(t = 0)$ for $i > 0$. Let

$$\underline{z} \cdot \underline{a} = C, \quad \underline{\dot{z}} \cdot \underline{a} = D \quad (t = 0).$$

Then, for either of the two values of β_2, the system of equations

$$\begin{aligned} (A_1 - \beta_2)\rho + B_1\sigma &= 0 \\ A_2\rho + (B_2 - \beta_2)\sigma &= 0 \\ C\rho + D\sigma &= 1 \end{aligned} \tag{5.21}$$

uniquely determines the unknowns ρ and σ. Up to now β_2, ρ, σ, $\underline{v}_0$ are determined. Using the definition of $\underline{g}$, $\underline{h}$ and (5.18), one obtains from the third equation of (5.15)

$$\underline{v}_2 = \rho'\underline{g} + \sigma'\underline{h} + \rho''\underline{z} + \sigma''\underline{\dot{z}} + \cdots, \tag{5.22}$$

where the terms omitted are already known. From the fourth equation of (5.15) one obtains the equations

$$\begin{aligned} \rho\beta_3 - (A_1 - \beta_2)\rho' - B_1\sigma' &= \cdots, \\ \sigma\beta_3 - A_2\rho' - (B_2 - \beta_2)\sigma' &= \cdots \end{aligned}$$

by using (5.18), (5.20), (5.22) and the bracket criterion. Since $\underline{v}_1 \cdot a = 0$ $(t = 0)$, we add to these equations the equation

$$C\rho' + D\sigma' = \cdots$$

Through the three equations, the three quantities β_2, ρ', σ' are now uniquely determined. With the help of (5.21), the determinant is found to be

$$A_1 + B_2 - 2\beta_2.$$

It is not equal to zero, since by hypothesis, (5.19) has two distinct solutions β_2. From this β_3, ρ', σ' and $\underline{v}_1$ are determined.

It is now easy to see that at the next step β_4, ρ'', σ'' are determined by equations with exactly the same left hand side, and that by the further analogous steps everything is determined.

We return now to the special problem which interests us, and assume that by suitable normalization two different formal power series pairs $(\beta,\underline{v})$ exist which solve the equation

$$(1 - \tau_2\varepsilon^2 + \ldots)\dot{\underline{v}} + \beta\underline{v} = \underline{L}_{t,\varepsilon}(\underline{v}).$$

On the other hand it was previously demonstrated that under the assumption $\beta \not\equiv 0$, two actual solutions exist, of which one is known, namely (5.8). Under this assumption the second (normalized) solution can thus be represented by the power series and the formula (1.4) for β_2 does in fact hold. To dispose of this completely we must still show that $\beta \equiv 0$ cannot occur if $\mu_2 \neq 0$. We show this also in terms of the schematic problem. Since (5.19) has the solution $\beta_2 = \rho = 0$ and the second $\beta_2 \neq 0$,

$$B_1 = B_2 = 0, \quad A_1 \neq 0. \tag{5.23}$$

If β were $\equiv 0$, then (5.4) would have a solution with the properties given there.

Setting in the power series for $\underline{v}$, $\underline{w}$ gives

$$\begin{aligned} \underline{v}_0 + \dot{\underline{w}}_0 &= \underline{L}_0(\underline{w}_0) \\ \underline{v}_1 + \dot{\underline{w}}_1 &= \underline{L}_0(\underline{w}_1) + \underline{L}_1(\underline{w}_0) \\ \underline{v}_2 + \dot{\underline{w}}_2 &= \underline{L}_0(\underline{w}_2) + \underline{L}_1(\underline{w}_1) + \underline{L}_2(\underline{w}_0). \end{aligned} \tag{5.24}$$

We have

$$\underline{w}_0 = \rho\underline{z} + \sigma\dot{\underline{z}}. \tag{5.25}$$

Since v_0 is also of this form, according to the bracket criterion $\underline{v}_0$ must be equal to zero. By (5.17), it follows analogously that $\underline{v}_1 = 0$. Similarly, as in (5.18), we find

$$\underline{w}_1 = \rho\underline{g} + \sigma\underline{h} + \rho'\underline{z} + \sigma'\dot{\underline{z}}.$$

It has been demonstrated above that $\dot{\underline{v}} = \underline{L}_{t,\varepsilon}(\underline{v})$ has a solution $(\underline{y})$ of period T_0, unique up to a factor. Thus we certainly have

$$\underline{v}_2 = \lambda\dot{\underline{z}}.$$

As above, using (5.20), application of the bracket rule to (5.24) gives the equations

$$0 = A_1\rho + B_1\sigma,$$

$$\lambda = A_2\rho + B_2\sigma.$$

(in which ρ', σ' once again fall out). According to (5.23) it thus follows that $\rho = \lambda = 0$, and from this $\underline{v}_2 = 0$. According to (5.25) $\underline{w}_0 = \sigma\dot{z}$. If one subtracts from the second equation of (5.4) the solution $\sigma\dot{\underline{y}}$ of $\dot{\underline{w}} = \underline{L}(\underline{w})$ and divides by ε, then the whole process can be repeated, and we find successively that the $\underline{v}_i = 0$, and thus $\underline{v} = 0$. With this it is demonstrated that β cannot be equal to zero.

The verification of the formula (1.4) is thus complete under the assumption $\mu_2 \neq 0$. This assumption could be replaced by $\mu \not\equiv 0$. The considerations would be changed only in that in the calculation of the coefficients the case of splitting will occur later.

The difficulties of these considerations could be avoided in the following manner. One first calculates purely formally as above the coefficient of the power series for β and $\underline{v}$ and then shows the convergence directly by a suitable application of the method of majorants. This would correspond to our intention of facilitating the application to partial differential systems. But one can also carry out the discussion of the case of splitting and the proof of (1.4) exclusively with determinants.[††††]

[††††] See editorial comments in Section 5A.

SECTION 5A

EDITORIAL COMMENTS

BY L. N. HOWARD AND N. KOPELL

(†) 1. Hopf's argument can be considerably simplified. After "blowing up" the equation (1.1) to (2.3), one wishes to show that for each sufficiently small ε there is a $\mu(\varepsilon)$, a period $T(\varepsilon)$ and initial conditions $\underline{y}^0(\varepsilon)$ (suitably normalized), so that (2.16) holds; the family of solutions to (1.1) asserted in the theorem is then $\underline{x}(t,\varepsilon) = \varepsilon\underline{y}(t,\mu(\varepsilon),\varepsilon,\underline{y}^0)$. Now (2.16) is satisfied if $\mu = \varepsilon = 0$, $\underline{y}^0 = \underline{z}^0$. Hence, the existence of the functions $\mu(\varepsilon)$, $T(\varepsilon)$, $\underline{y}^0(\varepsilon)$ follows from an implicit function theorem argument, provided that the $n \times n$ matrix

$$\left(\frac{\partial \underline{y}}{\partial t}, \frac{\partial \underline{y}}{\partial \mu}, \frac{\partial \underline{y}}{\partial \underline{y}^0}\right)\Bigg|_{t = T_0,\ \mu = 0,\ \varepsilon = 0,\ \underline{y}^0 = \underline{z}^0}$$

has maximal rank. (Here $\frac{\partial \underline{y}}{\partial \underline{y}^0}$ is an $n \times (n-2)$ matrix representing the derivative of $\underline{y}$ with respect to $(n-2)$ initial directions; there are two restrictions on the initial conditions from the normalization.) We show below how the rank of this matrix may be computed more easily.

Let $\underline{r}$ and $\underline{\ell}$ be the right and left eigenvectors corresponding to a pure imaginary eigenvalue of L_0; by rescaling time, we may assume that this eigenvalue is i. ($\overline{\underline{r}}$ and $\overline{\underline{\ell}}$ are eigenvectors for $-i$.) We may also assume that $\underline{\ell} \cdot \underline{r} = 1$. Let $L' = \frac{d}{d\mu} L_\mu \big|_{\mu = 0}$.

We note that hypothesis (1.2) may be rephrased: $\mathrm{Re}(\underline{\ell} \cdot L'\underline{r}) \neq 0$. (To see this, let $\underline{e}(\mu)$ be the eigenvector of L_μ which corresponds to the eigenvalue $\alpha(\mu)$ near a pure imaginary eigenvalue, normalized by $\underline{\ell} \cdot \underline{e} = 1$, so $\underline{e}(0) = \underline{r}$. Differentiating $L_\mu \underline{e} = \alpha(\mu)\underline{e}$ with respect to μ at $\mu = 0$, we get

$$L_0 \frac{d\underline{e}}{d\mu} + L'\underline{r} = \alpha'(0)\underline{r} + \alpha(0)\frac{d\underline{e}}{d\mu} \tag{5A.1}$$

Now $\underline{\ell} \cdot \frac{d\underline{e}}{d\mu} = 0$ and $\underline{\ell} L_0 = \alpha(0)\underline{\ell}$. Hence, if (5A.1) is multiplied by $\underline{\ell}$ on the left, we get $\underline{\ell} \cdot L'\underline{r} = \alpha'(0)$.)

Let $\underline{y}$ be defined by $\underline{y} = \underline{x}/\varepsilon$. t is replaced by $s = t/(1+\tau)$, where τ is to be adjusted (for each ε) so that the period in s is 2π. Then (1.1) becomes

$$\frac{d\underline{y}}{ds} = (1+\tau)[L_\mu \underline{y} + \varepsilon S(\underline{y},\varepsilon,\mu)].$$

For each ε, τ and μ we construct the solution with initial condition $\underline{y}(0)$, normalized by requiring $\underline{y}(0) = \frac{1}{2}(\underline{r}+\overline{\underline{r}}) + \underline{z}$, where $\underline{\ell} \cdot \underline{z} = \overline{\underline{\ell}} \cdot \underline{z} = 0$. (Hence, the initial conditions are parameterized by points in the $n-2$ dimensional space $W = (\underline{\ell} \oplus \overline{\underline{\ell}})^\perp$. Note that by the simplicity of the imaginary eigenvalues, W is transverse to $\underline{r} \oplus \overline{\underline{r}}$.) This solution we denote by $\underline{y}(s,\tau,\mu,\underline{z},\varepsilon)$.

Let $\underline{V}(\tau,\mu,\underline{z},\varepsilon) = \underline{y}(2\pi,\tau,\mu,\underline{z},\varepsilon) - \underline{y}(0,\tau,\mu,\underline{z},\varepsilon)$. At $\mu = \tau = \varepsilon = 0$, $\underline{z} = \underline{0}$, we have $\underline{y}(s) \equiv \underline{y}_0(s) = \text{Re}(\underline{r}e^{is})$ and $v = \underline{0}$. To show that there is a family of 2π-periodic functions with $\tau = \tau(\varepsilon)$, $\mu = \mu(\varepsilon)$, $\underline{z} = \underline{z}(\varepsilon)$, it suffices to show that $\partial\underline{V}/\partial(\tau,\mu,\underline{z})$ has rank n at $\mu = \tau = \varepsilon = 0$, $\underline{z} = \underline{0}$.

Let $\underline{y}_\tau(s) = \frac{\partial\underline{y}}{\partial\tau}(s,0,0,\underline{0},0)$. Then $\underline{y}_\tau$ satisfies the variational equation

$$\frac{d}{ds}\underline{y}_\tau = L_0\underline{y}_\tau + L_0\underline{y}_0$$

with initial condition $\underline{y}_\tau(0) = \underline{0}$. The solution to this equation is $\underline{y}_\tau = s\frac{d\underline{y}_0}{ds}$ which implies that $\frac{\partial\underline{V}}{\partial\tau} = 2\pi\frac{i}{2}(\underline{r}-\underline{\bar{r}}) = -2\pi\text{Im}\,\underline{r}$.

We next calculate $\underline{y}_\mu(s) = \frac{\partial\underline{y}}{\partial\mu}(s,0,0,\underline{0},0)$, which satisfies the variational equation

$$\frac{d}{ds}\underline{y}_\mu = L_0\underline{y}_\mu + L'\underline{y}_0$$

with initial condition $\underline{y}_\mu(0) = \underline{0}$. Since $L'\underline{y}_0 = \text{Re}\,L'\underline{r}e^{is}$, $\underline{y}_\mu$ is the real part of $\underline{\eta}$, where $\underline{\eta}$ satisfies

$$\frac{d}{ds}\underline{\eta} - L_0\underline{\eta} = L'\underline{r}e^{is} \tag{5A.2}$$

with initial condition $\underline{\eta}(0) = \underline{0}$.

A particular solution to (5A.2) is $\underline{y} = s(\underline{\ell}\cdot L'\underline{r})\underline{r}e^{is} + \underline{b}e^{is}$, where $\underline{b}$ is any complex vector which satisfies:

$$(i - L_0)\underline{b} + (\underline{\ell}\cdot L'\underline{r})\underline{r} = L'\underline{r}. \tag{5A.3}$$

Now $(i-L_0)$ is singular, but (5A.3) may be solved for $\underline{b}$ since $\underline{\ell}\cdot(L'\underline{r} - (\underline{\ell}\cdot L'\underline{r})\underline{r}) = 0$. The solutions $\underline{b}$ all have the form $\underline{b}_0 + k\underline{r}$ for any k. There is a unique such value of $\underline{b}$ for which $\underline{\ell}\cdot\text{Re }\underline{b} = \overline{\underline{\ell}}\cdot\text{Re }\underline{b} = 0$. (Take $k = -\underline{\ell}\cdot(\underline{b}_0 + \overline{\underline{b}}_0)$.) We use this value of $\underline{b}$. The solution to (5A.2) satisfying $\underline{\eta}(0) = 0$ then has real part

$$\underline{\eta}_\mu = \text{Re}\{(s(\underline{\ell}\cdot L'\underline{r})\underline{r} + \underline{b})e^{is}\} + \underline{\gamma}$$

where $\frac{d\underline{\gamma}}{ds} = L_0\underline{\gamma}$ and $\underline{\gamma}(0) = -\text{Re }\underline{b}$. Hence $\frac{\partial \underline{V}}{\partial \mu} = 2\pi\,\text{Re}(\underline{\ell}\cdot L'\underline{r})\underline{r} + \underline{\gamma}(2\pi) - \underline{\gamma}(0)$. We note that $\underline{\gamma}(2\pi) - \underline{\gamma}(0) = \underline{\gamma}_1$ satisfies $\underline{\ell}\cdot\gamma_1 = \overline{\underline{\ell}}\cdot\gamma_1 = 0$. (This follows since $\frac{d}{ds}(\underline{\ell}\cdot\underline{\gamma}) = \underline{\ell}\cdot L_0\underline{\gamma} = i\underline{\ell}\cdot\underline{\gamma}$. Since $\underline{\ell}\cdot\underline{\gamma}(0) = -\underline{\ell}\cdot\text{Re }\underline{b} = 0$, $\underline{\ell}\cdot\underline{\gamma}(s) \equiv 0$. Similarly, $\overline{\underline{\ell}}\cdot\underline{\gamma}(s) \equiv 0$.)

Finally, we compute $\frac{\partial \underline{V}}{\partial \underline{z}}$. Let $\delta\underline{y}$ be the variation in $\underline{y}$ due to the variation $\delta\underline{z}$ in initial conditions. Then $\delta\underline{y}(s)$ satisfies $\frac{d}{ds}(\delta\underline{y}) = L_0(\delta\underline{y})$, $\delta\underline{y}(0) = \delta\underline{z}$, and $\underline{\ell}\cdot\delta\underline{z} = \overline{\underline{\ell}}\cdot\delta\underline{z} = 0$. This implies that $\frac{\partial \underline{V}}{\partial \underline{z}}(\delta\underline{z}) = (e^{2\pi L_0} - I)\delta\underline{z}$. Now $\delta\underline{z}$ is in the subspace W orthogonal to $\underline{\ell}$ and $\overline{\underline{\ell}}$. Since there are no other pure imaginary eigenvalues for L_0 (in particular, no integer multiples of $\pm i$), the matrix $(e^{2\pi L_0} - I)$ is invertible on W. Hence $\frac{\partial \underline{V}}{\partial \underline{z}}$ has rank $n-2$.

Now R^n is the direct sum of W and the span of $\text{Re }\underline{r}$ and $\text{Im }\underline{r}$. (This follows from the simplicity of the pure imaginary eigenvalues.) The range of $(e^{2\pi L_0} - I)$ is W, so $\partial\underline{V}/\partial(\tau,\mu,\underline{z})$ has rank n if and only if $\text{Im }\underline{r}$ and $\text{Re}(\underline{\ell}\cdot L'\underline{r})\underline{r}$ are independent. This is true if $\text{Re}(\underline{\ell}\cdot L'\underline{r}) \neq 0$.

2. The argument in this section does not require analyticity; it merely sets up the hypotheses of an implicit function theorem. Hence this argument provides a proof for

a C^r version of this theorem. More specifically, suppose that $\underline{F}(\underline{x},\mu)$ is r times differentiable in $\underline{x}$ and μ. Then the right hand side of (2.3) is r times differentiable in $\underline{x}$ and μ, but only C^{r-1} in ε. The function $\underline{V}(\tau,\mu,\underline{z},\varepsilon)$ defined above is C^{r-1} in ε and at least C^r in the other variables. Hence, the implicit function theorem says that the functions $\tau(\varepsilon)$, $\mu(\varepsilon)$, $\underline{z}(\varepsilon)$ and $\underline{y}(s,\varepsilon) \equiv \underline{y}(s,\mu(\varepsilon),\varepsilon,\underline{z}(\varepsilon))$ are all C^{r-1}. The periodic solutions to (1.1), namely $x(t,\varepsilon) = y(\frac{t}{1+\tau(\varepsilon)},\varepsilon)$, are C^r.

(††) The uniqueness proved in this argument is weaker than that of Theorem 3.15 of these notes. That is, it is not proved in Hopf's paper that the periodic solutions which are found are the only ones in some neighborhood of the critical point. For example, Hopf's argument does not rule out a sequence of periodic functions $x_k(t)$ such that $\max|x_k(t)| \to 0$, the associated $\mu_k \to 0$, and the periods $T_k \to \infty$. Such behavior *is* ruled out by the center manifold theorem, which says that any point not on the center manifold must eventually leave a sufficiently small neighborhood (at least for a while) or tend to the center manifold as $t \to \infty$. Thus the center manifold contains all sufficiently small closed orbits; since the center manifold is three dimensional (including the parameter dimension), the uniqueness of the periodic solutions is a consequence of the uniqueness for the two-dimensional theorem.

(†††) Formulas equivalent to Hopf's but somewhat easier to apply can be obtained in a simpler manner. The main point is to use the "e^{is}" form of the solutions more explicitly and thereby avoid introducing the bracket criterion.

We again assume that time has been scaled so that the pure imaginary eigenvalues of L_0 are $\pm i$, and we use the notation introduced in (†). Following Hopf we further rescale time by $t = (1+\tau(\varepsilon))s$, $\tau(0) = 0$, and let $\underline{y} = \varepsilon\underline{x}$. Then (1.1) becomes

$$\dot{\underline{y}} = [1+\tau(\varepsilon)][L_0\underline{y} + \mu L'\underline{y} + \varepsilon Q(\underline{y},\underline{y}) + \varepsilon^2 K(\underline{y},\underline{y},\underline{y}) + \ldots] \quad (5A.4)$$

where Q and K are respectively the quadratic and cubic terms when $\mu = 0$.

Let the 2π-periodic solution of (5A.4) be $\underline{y}(s,\varepsilon) = \underline{y}_0 + \varepsilon\underline{y}_1 + \varepsilon^2\underline{y}_2 + \ldots$, where, as before, $\underline{y}_0 = \mathrm{Re}(e^{is}\underline{r})$, and the $\underline{y}_i$ are 2π-periodic with $\underline{\ell}\cdot\underline{y}_i(0) = \overline{\underline{\ell}}\cdot\underline{y}_i(0) = 0$ for $i \geq 1$. (Since the $\underline{y}_i$ are real, we may simply require $\underline{\ell}\cdot\underline{y}_i(0) = 0$.)

To get recursive equations for the $\underline{y}_i$, the series for $\underline{y}(s,\varepsilon)$ is inserted in (5A.4) and like powers of ε are collected. We use the fact that $\tau_1 = \mu_1 = 0$, so $\tau = \varepsilon^2\tau_2 + \ldots$ and $\mu = \varepsilon^2\mu_2 + \ldots$. We find that $\underline{y}_1$ should satisfy $\dot{\underline{y}}_1 = A\underline{y}_1 + \underline{Q}(\underline{y}_0,\underline{y}_0)$, and $\underline{Q}(\underline{y}_0,\underline{y}_0) = \frac{1}{2}\underline{Q}(\underline{r},\overline{\underline{r}}) + \frac{1}{2}\mathrm{Re}[e^{2is}\underline{Q}(\underline{r},\underline{r})]$. A periodic solution to this equation is $\underline{a} + \mathrm{Re}(\underline{c}e^{2is})$ where $\underline{a}$ and $\underline{c}$ are constant vectors satisfying

$$-L_0\underline{a} = \frac{1}{2}\underline{Q}(\underline{r},\overline{\underline{r}}) \qquad (5A.5)$$

$$(2i - L_0)\underline{c} = \frac{1}{2}\underline{Q}(\underline{r},\underline{r}).$$

(Since $-L_0$ and $2i - L_0$ are non-singular, these formulas do determine $\underline{a}$ and $\underline{c}$.) Thus $\underline{y}_1 = \underline{a} + \mathrm{Re}(\underline{c}e^{2is}) + \mathrm{Re}(C_1\underline{r}e^{is})$, where the complex number C_1 is to be chosen so that $\underline{\ell}\cdot\underline{y}_1(0) = 0$; using equations (5A.5), one readily finds that

$$C_1 = \frac{1}{2}\, \underline{\ell}\cdot[\underline{Q}(\underline{r},\underline{\bar{r}}) - \frac{1}{2}\underline{Q}(\underline{r},\underline{r}) + \frac{1}{6}\underline{Q}(\underline{\bar{r}},\underline{\bar{r}})].$$

Now $\underline{y}_2$ is a periodic solution of

$$\dot{\underline{y}}_2 = L_0\underline{y}_2 + \mu_2 L'\underline{y}_0 + 2\underline{Q}(\underline{y}_0,\underline{y}_1) + \underline{K}(\underline{y}_0,\underline{y}_0,\underline{y}_0) + \tau_2 L_0\underline{y}_0$$

Hence

$$\begin{aligned}\dot{\underline{y}}_2 - L_0\underline{y}_2 &= \mathrm{Re}\,\mu_2 L'\underline{r}e^{is} + \mathrm{Re}\underline{Q}(\underline{r}e^{is}+\underline{\bar{r}}e^{-is},\ \underline{a}+\underline{c}e^{2is}+C_1\underline{r}e^{is}) \\ &\quad + \frac{1}{4}\mathrm{Re}[\underline{K}(\underline{r},\underline{r},\underline{r})e^{3is} + 3\underline{K}(\underline{r},\underline{r},\underline{\bar{r}})e^{is}] + \tau_2\mathrm{Re}[i\underline{r}e^{is}] \\ &= \mathrm{Re}\{C_1\underline{Q}(\underline{\bar{r}},\underline{r}) + e^{is}[(\mu_2 L'+i\tau_2)\underline{r} + 2\underline{Q}(\underline{r},\underline{a}) + \underline{Q}(\underline{r},\underline{c}) \\ &\quad + \frac{3}{4}\underline{K}(\underline{r},\underline{r},\underline{\bar{r}})] + e^{2is}C_1\underline{Q}(\underline{r},\underline{r}) + e^{3is}[\underline{Q}(\underline{r},\underline{c})+\frac{1}{4}\underline{K}(\underline{r},\underline{r},\underline{r})]\}.\end{aligned}$$

These equations have a periodic solution if and only if there is no resonance, which requires that the coefficient of e^{is} in this last formula should be orthogonal to $\underline{\ell}$ (the bracket criteria in disguise). Thus

$$\mu_2(\underline{\ell}\cdot L'\underline{r}) + i\tau_2 = -2\underline{\ell}\cdot\underline{Q}(\underline{r},\underline{a}) - \underline{\ell}\cdot\underline{Q}(\underline{\bar{r}},\underline{c}) - \frac{3}{4}\,\underline{\ell}\cdot\underline{K}(\underline{r},\underline{r},\underline{\bar{r}}) \equiv B.$$

Hence we get the formulas for μ_2 and τ_2:

$$\begin{aligned}\mu_2\ \mathrm{Re}(\underline{\ell}\cdot L'\underline{r}) &= -\mathrm{Re}\ B \\ \tau_2 &= -\mu_2\mathrm{Im}(\underline{\ell}\cdot L'\underline{r}) - \mathrm{Im}\ B\end{aligned} \qquad (5A.6)$$

where $\underline{a}$ and $\underline{c}$ are the solutions of (5A.5). [These formulas are unchanged if the eigenvalue is $i\omega$ instead of i, except that, instead of the second equation (5A.5), $\underline{c}$ is the solution of $(2i\omega - L_0)\underline{c} = \frac{1}{2}\underline{Q}(\underline{r},\underline{r})$. Also the value of C_1 given above should be divided by ω.]

The formulas (5A.6) are equivalent to Hopf's (4.16). The determination of the left eigenvector $\underline{\ell}$ and the solution of the linear equations (5A.5) for $\underline{a}$ and $\underline{c}$ takes the place of finding the adjoint eigenfunctions and evaluating the integrals implied by the bracket symbols.

(††††) 1. The translators must admit that they have found this section somewhat less transparent than the rest of the paper. In their article, Joseph and Sattinger [1] point out an apparent circularity in a part of Hopf's argument; they also show there that it can be rectified rather easily.

2. The relationship of β, the Floquet exponent near zero (of the periodic solution), to the coefficient μ_2 can be found with relatively little calculation, as follows. The argumented system

$$\begin{aligned} \dot{\underline{x}} &= F_\mu(\underline{x}) \\ \dot{\mu} &= 0 \end{aligned} \tag{5A.7}$$

has the origin as a critical point. There are three eigenvalues of this critical point with zero real part; a zero eigenvalue with the μ-axis as eigenvector, and the conjugate pair of imaginary eigenvalues $\pm i$ (after suitably rescaling the time variable) with eigenvectors $\underline{r}$ and $\overline{\underline{r}}$. All other eigenvalues are off the imaginary axis, so this critical point has a 3 dimensional center manifold. This center manifold must contain the μ-axis, the periodic solutions given by Hopf's Theorem, and any trajectories of (5A.7) which for all time remain close to the origin; it is tangent to the linear space generated by the μ-axis and the real and imaginary parts of $\underline{r}$. Let us set $\underline{x} = \varepsilon(\zeta\underline{r}+\overline{\zeta\underline{r}}) + \underline{x}_2$

where $\underline{\ell}\cdot x_2 = \underline{\overline{\ell}}\cdot x_2 = 0$. $\varepsilon \geq 0$ is regarded as a replacement for μ, given by the function $\mu(\varepsilon)$ of Hopf's Theorem: $\mu = \mu(\varepsilon) = \mu_2\varepsilon^2 + \ldots$, where we are now assuming that $\mu_2 \neq 0$. Thus we may think of the real and imaginary parts of ζ, and ε, as parameters on the center manifold. For any $(\underline{x},\mu)$ on this manifold $\underline{x}_2 = 0(\varepsilon^2)$ since it is at least quadratic in $\underline{\ell}\cdot\underline{x} = \varepsilon\zeta$ and $\underline{\overline{\ell}}\cdot x = \varepsilon\overline{\zeta}$. Thus we may write the equations of the center manifold as $\mu = \mu(\varepsilon)$, $\underline{x}_2 = \varepsilon^2\underline{g}(\zeta,\overline{\zeta},\varepsilon)$, where $\underline{\ell}\cdot\underline{g} = \underline{\overline{\ell}}\cdot\underline{g} = 0$ and $\underline{g}$ is at least quadratic in ζ and $\overline{\zeta}$. For any trajectory on the center manifold we then have, with the notations of (†) and (†††),

$$\begin{aligned}\underline{r}\dot{\zeta} + \underline{\overline{r}}\,\dot{\overline{\zeta}} + \varepsilon(\underline{g}_\zeta\dot{\zeta} + \underline{g}_{\overline{\zeta}}\dot{\overline{\zeta}}) &= i\zeta\underline{r} - i\overline{\zeta}\,\underline{\overline{r}} \\ &+ \varepsilon L_0\underline{g} + \mu_2\varepsilon^2 L'(\zeta\underline{r} + \overline{\zeta}\,\underline{\overline{r}}) + \varepsilon Q(\zeta\underline{r} + \overline{\zeta}\,\underline{\overline{r}}) \\ &+ 2\varepsilon^2 Q(\zeta\underline{r} + \overline{\zeta}\,\underline{\overline{r}},\underline{g}) + \varepsilon^2 C(\zeta\underline{r} + \overline{\zeta}\,\underline{\overline{r}}) + 0(\varepsilon^3).\end{aligned} \tag{5A.8}$$

By multiplying on the left with $\underline{\ell}$, we obtain

$$\begin{aligned}\dot{\zeta} &= i\zeta + \mu_2\varepsilon^2\underline{\ell}\cdot L'(\zeta\underline{r} + \overline{\zeta}\,\overline{r}) \\ &+ \varepsilon\underline{\ell}\cdot[\zeta^2 Q(\underline{r},\underline{r}) + 2\zeta\overline{\zeta}Q(\underline{r},\underline{\overline{r}}) + \overline{\zeta}^2 Q(\underline{\overline{r}},\underline{\overline{r}})] \\ &+ \varepsilon^2\gamma(\zeta,\overline{\zeta}) + 0(\varepsilon^3)\end{aligned} \tag{5A.9}$$

where γ is cubic in ζ and $\overline{\zeta}$.

We now introduce the function

$$I(\zeta,\overline{\zeta}) = \frac{1}{2}\zeta\overline{\zeta} + \varepsilon\mathrm{Re}[\frac{i\zeta^3}{3}\underline{\overline{\ell}}\cdot Q(\underline{r},\underline{r}) + i\zeta^2\overline{\zeta}(2\underline{\overline{\ell}}\cdot Q(\underline{r},\underline{\overline{r}}) + \underline{\ell}\cdot Q(\underline{r},\underline{r})].$$

As we will see below, $I(\zeta,\overline{\zeta})$ is approximately invariant along trajectories lying on the center manifold. For any trajectory on the center manifold we have, using (5A.9) and its complex conjugate,

$$\frac{dI}{dt} = \text{Re}\{\bar{\zeta}\dot{\zeta} + \varepsilon[i\zeta^2\dot{\zeta}\underline{\bar{\ell}}\cdot Q(\underline{r},\underline{r}) + i(\zeta^2\dot{\bar{\zeta}} + 2\zeta\bar{\zeta}\,\dot{\zeta})\cdot$$

$$(2\underline{\bar{\ell}}\cdot Q(\underline{r},\underline{\bar{r}}) + \underline{\ell}\cdot Q(\underline{r},\underline{r}))]\}$$

$$= \text{Re}\{i\zeta\bar{\zeta} + \varepsilon\underline{\ell}\cdot[\bar{\zeta}\zeta^2 Q(\underline{r},\underline{r}) + 2\zeta\bar{\zeta}^2 Q(\underline{r},\underline{\bar{r}}) + \bar{\zeta}^3 Q(\underline{\bar{r}},\underline{\bar{r}})]$$

$$+ \varepsilon[-\zeta^3\underline{\bar{\ell}}\cdot Q(\underline{r},\underline{r}) + (\zeta^2\bar{\zeta} - 2\zeta^2\bar{\zeta})(2\underline{\bar{\ell}}\cdot Q(\underline{r},\underline{\bar{r}}) + \underline{\ell}\cdot Q(\underline{r},\underline{r}))]$$

$$+ \mu_2\varepsilon^2\bar{\zeta}\underline{\ell}\cdot L'(\zeta\underline{r} + \bar{\zeta}\,\underline{\bar{r}}) + \varepsilon^2\bar{\zeta}\gamma(\zeta,\bar{\zeta}) + \varepsilon^2\delta(\zeta,\bar{\zeta})\} + O(\varepsilon^3)$$

where δ is quartic in ζ and $\bar{\zeta}$. The terms of order ε in the above are $\varepsilon\text{Re}\{\bar{\zeta}^3\underline{\ell}\cdot Q(\underline{\bar{r}},\underline{\bar{r}}) - \zeta^3\underline{\bar{\ell}}\cdot Q(\underline{r},\underline{r}) + \bar{\zeta}\zeta^2\underline{\ell}\cdot Q(\underline{r},\underline{r}) - \bar{\zeta}\zeta^2\underline{\ell}\cdot Q(\underline{r},\underline{r}) + 2\zeta\bar{\zeta}^2\underline{\ell}\cdot Q(\underline{r},\underline{\bar{r}}) - 2\bar{\zeta}\zeta^2\underline{\bar{\ell}}\cdot Q(\underline{r},\underline{\bar{r}})\} \equiv 0$. Thus

$$\frac{dI}{dt} = \mu_2\varepsilon^2\text{Re}[\bar{\zeta}\underline{\ell}\cdot L'(\zeta r + \bar{\zeta}\,\underline{\bar{r}})] + \varepsilon^2\delta_1(\zeta,\bar{\zeta}) + O(\varepsilon^3) \quad (5A.10)$$

where δ_1 is also quartic in ζ and $\bar{\zeta}$. Thus $\frac{dI}{dt}$ is of order ε^2.

($I_0 = \frac{1}{2}|\zeta|^2$ is also an approximate invariant, but $\frac{dI_0}{dt}$ is $O(\varepsilon)$ whereas $\frac{dI}{dt}$ is only $O(\varepsilon^2)$. If we consider a trajectory on the center manifold starting at $t = 0$ with $\zeta = \bar{\zeta} = c$, we see from (5A.9) that it is given to $O(\varepsilon)$ by $\zeta = ce^{it}$; thus, after a time of approximately 2π, it must once again return to $\text{Im}\,\zeta = 0$. This arc is a circle to $O(\varepsilon)$, but is more accurately described (to $O(\varepsilon^2)$) as a curve of constant I.)

We see from (5A.10) that the change in I in going once around this way is given, to $O(\varepsilon^2)$, by

$$\Delta I = 2\pi[\mu_2\varepsilon^2 c^2\text{Re}(\underline{\ell}\cdot L'\underline{r}) + \varepsilon^2 c^4\delta_2] \quad (5A.11)$$

where $\delta_2 = \frac{1}{2\pi}\int_0^{2\pi} \delta_1(e^{it}, e^{-it})dt$. However, we know that if $c = 1$ we get the periodic solution, for which $\Delta I = 0$; consequently $\delta_2 = -\mu_2 \mathrm{Re}(\underline{\ell}\cdot L'\underline{r}) = -\mu_2 \mathrm{Re}(\alpha'(0))$ as noted in (†). Thus, in general,

$$\Delta I = 2\pi\varepsilon^2 \mu_2 \mathrm{Re}(\alpha'(0)(c^2 - c^4) + 0(\varepsilon^3). \tag{5A.12}$$

Since $c = 1$ gives the periodic solution, the $0(\varepsilon^3)$ part is also divisible by $(c-1)$. Thus, for c near 1, (5A.12) may be written

$$\Delta I = 2\pi\varepsilon^2 (c-1)[-2\mu_2 \mathrm{Re}\ \alpha'(0) + 0(c-1)]. \tag{5A.13}$$

For small ε, any trajectory on the center manifold with $\zeta = 0(1)$ must keep going around an approximate circle. However, it cannot be periodic unless it passes through $\zeta = 1$. Hence, it is apparent from (5A.12) that, when $\mu_2 \mathrm{Re}\ \alpha'(0) > 0$, all trajectories on the center manifold (at a given ε, i.e., μ) which are inside the periodic solution must spiral out towards it as $t \to +\infty$ (or as $t \to -\infty$ if $\mu_2 \mathrm{Re}\ \alpha'(0) < 0$). Since I is approximately $\frac{1}{2}|\zeta|^2$, $\Delta I \cong (\zeta(2\pi) - c)c$. Thus (5A.13) implies that these trajectories asymptotically approach the periodic solution with exponential rate $\beta = -2\varepsilon^2 \mu_2 \mathrm{Re}\ \alpha'(0) + 0(\varepsilon^3)$, and this must thus be the numerically smallest non-zero Floquet exponent.

3. Equation (5A.12) actually tells us more; it implies that we may approximately describe the trajectories on the center manifold as slowly expanding (or contracting if $\mu_2 \mathrm{Re}\ \alpha'(0) < 0$) circles whose radius c varies according to the formula

$$c^2 = \frac{1}{2}(1 + \tanh(\varepsilon^2\mu_2 \mathrm{Re}\ \alpha'(0)(t-t_1)))$$

where t_1 is the time at which $c^2 = 1/2$.

4. The function I is also of some use in relating the above to the "vague attractor" hypothesis. If we set $\mu = 0$, the n-dimensional system $\dot{\underline{x}} = F_0(x)$ has a two-dimensional center manifold for its critical point at 0, tangent to the linear space spanned by the real and imaginary parts of $\underline{r}$. As in a previous paragraph, we set $\underline{x} = \varepsilon(\zeta\underline{r} + \overline{\zeta}\,\overline{\underline{r}}) + \underline{x}_2$, where $\underline{\ell}\cdot\underline{x}_2 = \overline{\underline{\ell}}\cdot\underline{x}_2 = 0$. On this center manifold $\underline{x}_2 = 0(\varepsilon^2)$ and is at least quadratic in ζ and $\overline{\zeta}$; ε is now an arbitrary scaling parameter. For any trajectory on this center manifold one obtains the same formula (5A.9) except for the omission of the $\mu_2 L'$ term - the other terms written down all come from the μ-dependent pairs of F_μ. If we then consider the function I for a trajectory on this center manifold, we obtain (5A.10) again, with the μ_2 term omitted, but the same δ_1; integrating this around, we get (5A.11) without the μ_2 term but the same δ_2. Since $\delta_2 = -\mu_2 \mathrm{Re}\ \alpha'(0)$ we have for trajectories in the center manifold at $\mu = 0$, to order ε^2,

$$\Delta(\frac{1}{2}\ c^2) = -2\pi\ \varepsilon^2\mu_2 \mathrm{Re}\ \alpha'(0)c^4,$$

or

$$\Delta(\varepsilon c) = -2\pi\mu_2 \mathrm{Re}\ \alpha'(0)(\varepsilon c)^3.$$

Since $\Delta(\varepsilon c)$ is $V(x_1)$ (the Poincaré map minus identity), where $\varepsilon c = x_1$ is the coordinate $\mathrm{Re}(\underline{\ell}\cdot\underline{x})$, we see that $V'''(0) = -2\pi\mu_2 \mathrm{Re}(\alpha'(0))\cdot 6 = -2\pi \mathrm{Re}(\alpha'(0))\cdot 3\mu''(0)$. This relates the calculations here to the stability calculations done in §4.

SECTION 6

THE HOPF BIFURCATION THEOREM FOR DIFFEOMORPHISMS

Let X be a vector field and let γ be a closed orbit of the flow ϕ_t of X. Let P be a Poincaré map associated with γ. (See §2B). Suppose there is a circle σ that is invariant under P. Then it is clear that $\bigcup_t \phi_t(\sigma)$ is an invariant torus for the flow of X (see Figure 6.1).

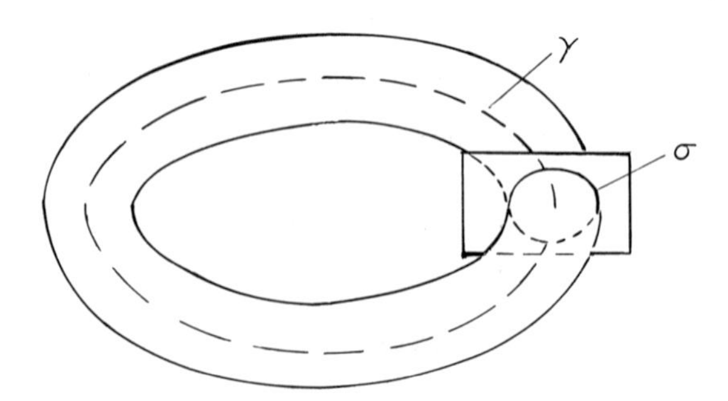

Figure 6.1

If we have a one parameter family of vector fields and closed orbits X_μ and γ_μ, it is quite conceivable that for small μ, γ_μ might be stable, but for large μ it might become unstable and a stable invariant torus take its place. Recall that γ_μ is stable (unstable) if the eigenvalues of the derivative of the Poincaré map P_μ have absolute value < 1 (> 1). (See §2B). The Hopf Bifurcation Theorem for diffeomorphisms gives conditions under which we may expect bifurcation to stable invariant tori after loss of stability of γ_μ. The theorem we present is due to Ruelle-Takens [1]; we follow the exposition of Lanford [1] for the proof.

In order to apply these ideas, one needs to know how to compute the spectrum of the Poincaré map P. Fortunately this can be done because, as we have remarked earlier, the spectrum of the time τ map of the flow is that of $P \cup \{1\}$. (See §2B).

Reduction to Two Dimensions

We thus turn our attention to bifurcations for diffeomorphisms. The first thing to do is to reduce to the two dimensional case.* This is done by means of the center manifold theorem exactly as we did in the previous case; i.e., assume we have a one parameter family of diffeomorphisms $\Phi_\mu: Z \to Z$, $\Phi_\mu(0) = 0$ and assume a single complex conjugate pair of simple non-real eigenvalues crosses the unit circle as μ increases past zero. Then the center manifold theorem applied to $\Psi: (x,\mu) \to (\Phi_\mu(x),\mu)$ yields a locally invariant three manifold M; the μ slices M_μ then give a family of

---

*As remarked before, for partial differential equations, P may become a diffeomorphism only after the reduction, and be only a smooth map before.

manifolds which we can identify by some fixed coordinate chart. Then on M_μ we have induced a family of diffeomorphisms containing all the recurrence.

Thus we are reduced to the following case: (modulo "global" stability problems as in the last section).

We have a one parameter family $\Phi_\mu: \mathbb{R}^2 \to \mathbb{R}^2$ of diffeomorphisms satisfying:

a) $\Phi_\mu(0,0) = (0,0)$

b) $d\Phi_\mu(0,0)$ has two non-real eigenvalues $\lambda(\mu)$ and $\overline{\lambda(\mu)}$ such that for $\mu < 0$ $|\lambda(\mu)| < 1$ and for $\mu > 0$ $|\lambda(\mu)| > 1$

c) $\left.\dfrac{d|\lambda(\mu)|}{d\mu}\right|_{\mu=0} > 0.$

We can reparametrize so that the eigenvalues of $d\Phi_\mu(0,0)$ are $(1+\mu)e^{\pm i\theta(\mu)}$. By making a smooth μ-dependent change of coordinates, we can arrange that:

$$d\Phi_\mu(0,0) = (1+\mu)\begin{pmatrix} \cos\theta(\mu) & -\sin\theta(\mu) \\ \sin\theta(\mu) & \cos\theta(\mu) \end{pmatrix}.$$

<u>The Canonical Form</u>

The next step is to make a further change of coordinates to bring Φ_μ approximately into appropriate canonical form. To be able to do this, we need a technical assumption:*

$$e^{im\theta(0)} \neq 1 \quad m = 1,2,3,4,5. \tag{6.1}$$

(6.1) <u>Lemma</u>. <u>Subject to Assumption (6.1), we can make a smooth μ-dependent change of coordinates bringing</u> Φ_μ

---

*As D. Ruelle has pointed out, only m = 1,2,3,4 is needed for the bifurcation theorems as can be seen from the proof in §6A.

into the form:

$$\Phi_\mu(x) = N\Phi_\mu(x) + 0(|x|^5)$$

where, in polar coordinates,

$$N\Phi_\mu: (r,\phi) \mapsto ((1+\mu)r - f_1(\mu)r^3, \quad \phi + \theta(\mu) + f_3(\mu)r^2).$$

The proof of this proposition uses standard techniques and may be obtained, for example, from §23 of Siegel and Moser [1].* We give a straightforward and completely elementary proof in Section 6A. As indicated above, we think of $N\Phi_\mu$ as an approximate canonical form for Φ_μ. Note two special features of $N\Phi_\mu$:

i) The new r depends only on the old r, not on ϕ.

ii) The new ϕ is obtained from the old ϕ by an r-dependent rotation. We now add a final assumption:

$$f_1(0) > 0. \qquad (6.2)$$

This assumption implies that for small positive μ, $N\Phi_\mu$ has an invariant circle of radius r_0, where r_0 is obtained by solving

$$(1+\mu)r_0 - f_1(\mu)r_0^3 = r_0,$$

i.e., $r_0^2 = \dfrac{\mu}{f_1(\mu)}$.

---

*The canonical form is important in celestial mechanics for proving the existence and stability of closed orbits near a given one; i.e. for finding fixed points or invariant circles for the Poincaré map. In the Hamiltonian case this map is symplectic (see Abraham-Marsden [1]) so Birkhoff's theorem applies, as are the results of Kolmogorov, Arnold and Moser if it is a "twist mapping".

We shall verify shortly that this circle is attracting for $N\Phi_\mu$. Since Φ_μ differs only a little from $N\Phi_\mu$, it is not surprising that Φ_μ has a nearby invariant circle.

The Main Result

(6.2) Theorem. (Ruelle-Takens, Sacker, Naimark). *Assume (6.1) and (6.2)*. Then for all sufficiently small positive μ, Φ_μ has an attracting invariant circle.*

Before giving the proof, let us look at what happens if (6.2) is replaced by the assumption $f_1(0) < 0$. Then, for a small positive μ, $N\Phi_\mu$ has no invariant sets except $\{0\}$ and $\mathbb{R}^2$. For $\mu < 0$, $N\Phi_\mu$ does have an invariant circle, but it is repelling rather than attracting. By applying the result of Ruelle and Takens to Φ_μ^{-1}, we prove that, in this case, Φ_μ has a nearby invariant circle. Thus, we again find the usual situation that supercritical branches are stable and subcritical branches unstable. If

$$\Phi_\mu\colon (y,\phi) \to ((1-2\mu)y - \mu(3y^2+y^3) + \mu^2 0(1),$$

$$\phi + \theta(\mu) + \mu\frac{f_3(\mu)}{f_1(\mu)}(1+y)^2 + \mu^2 0(1)).$$

Finally, we scale y again by putting

$$y = \sqrt{\mu}\, z;$$

then

---

*It would be interesting to explicitly compute $f_1(0)$ in terms of Φ_μ directly as we did in §4 for the bifurcation to invariant circles. However the labor involved in the earlier calculations, and the promise of a harder, if not impossible computation has left the present authors sufficiently exhausted to leave this one to the ambitious reader. The calculation of $f_1(0)$ in terms of Φ_μ rather than X_μ is not so hard and has been done by Wan [preprint] and Iooss [6].

$$\Phi_\mu: (z,\phi) \to ((1-2\mu)z - \mu^{3/2}(3z^2 + \mu^{1/2}z^3) + \mu^{3/2}0(1),$$
$$\phi + \theta(\mu) + \mu \frac{f_3(\mu)}{f_1(\mu)}(1 + \mu^{1/2}z)^2 + \mu^2 0(1);$$

we rewrite this last formula as

$$(z,\phi) \to ((1-2\mu)z + \mu^{3/2}H_\mu(z,\phi), \phi + \theta_1(\mu) + \mu^{3/2}K_\mu(z,\phi)).$$

The functions $H_\mu(z,\phi)$, $K_\mu(z,\phi)$ are smooth in z, ϕ, μ on

$$-1 \le z \le 1, \quad 0 \le \phi \le 2\pi, \quad 0 \le \mu \le \mu_0$$

for some sufficiently small μ_0; the region $-1 \le z \le 1$, $0 \le \phi \le 2\pi$ corresponds to an annulus of width $0(\mu)$ about the invariant circle for $N\Phi_\mu$ (which has radius $0(\mu)$). We are going to produce an invariant circle inside this annulus.

The qualitative behavior of Φ_μ is now easy to read off: Φ_μ can be written as

$$(z,\phi) \mapsto ((1+2\mu)z, \phi + \theta_1(\mu))$$

plus a small perturbation. The approximate Φ is simply a rotation in the ϕ direction and a contraction in the z direction. Note, however, that the strength of the contraction goes to zero with μ. If this were not the case, we could simply invoke known results about the persistence of attracting invariant circles under small perturbations. As it is, we need to make a slightly more detailed argument, exploiting the fact that the size of the perturbation goes to zero faster than the strength of the contraction.

We are going to look for an invariant manifold of the form

$$\{z = u(\phi)\},$$

where

i) $u(\phi)$ is periodic in ϕ with period 2π

ii) $|u(\phi)| \leq 1$ for all ϕ

iii) $u(\phi)$ is Lipschitz continuous with Lipschitz constant 1 (i.e., $|u(\phi_1) - u(\phi_2)| \leq |\phi_1 - \phi_2|$).

The space of all functions u satisfying i), ii) and iii) will be denoted by U.

We shall give a proof based on the contraction mapping principle. In outline, the argument goes as follows: We start with a manifold

$$M = \{z = u(\phi)\},$$

with $u \in U$, and consider the new manifold $\Phi_\mu M$ obtained by acting on M with Φ_μ. We show that, for μ sufficiently small, $\Phi_\mu M$ again has the form $\{z = \hat{u}(\phi)\}$ for some $\hat{u} \in U$. Thus, we construct a non-linear mapping $\mathcal{F}$ of U into itself by

$$\mathcal{F}u = \hat{u}.$$

We then prove, again for small positive μ, that $\mathcal{F}$ is a contraction on U (with respect to the supremum norm) and hence as a unique fixed point u^*. The manifold $\{z = u^*(\phi)\}$ is the desired invariant circle. As a by-product of the proof of contractivity, we prove this manifold is attracting in the following sense: Pick a starting point (z,ϕ) with $|z| \leq 1$, and let (z_n,ϕ_n) denote $\Phi_\mu^n(z,\phi)$. Then

$$\lim_{n\to\infty} z_n - u^*(\phi_n) = 0.$$

It is not hard to see that the domain of attraction is much

larger than the annulus $|z| \leq 1$. In particular, it contains everything inside the annulus except the fixed point at the center, but we shall not pursue this point.

To carry out the argument outlined above, we must first construct the non-linear mapping $\mathscr{F}$. To find $\mathscr{F}u(\phi)$, we should proceed as follows:

A) Show that there is a unique $\tilde{\phi}$ such that the ϕ-component of $\Phi_\mu(u(\tilde{\phi}),\tilde{\phi})$ is ϕ, i.e., such that

$$\phi \equiv \tilde{\phi} + \theta_1(\mu) + \mu^{3/2}K_\mu(u(\tilde{\phi}),\tilde{\phi})(2\pi).^* \tag{6.3}$$

and

B) Put $\mathscr{F}u(\phi)$ equal to the z-component of $\Phi_\mu(u(\tilde{\phi}),\tilde{\phi})$, i.e.,

$$\mathscr{F}u(\phi) = (1-2\mu)u(\tilde{\phi}) + \mu^{3/2}H_\mu(u(\tilde{\phi}),\tilde{\phi}). \tag{6.4}$$

In the estimates we are going to make, it will be convenient to introduce

$$\lambda = \sup_{\substack{0\leq\phi\leq 2\pi \\ -1 \\ -1\leq z\leq 1}} \{|H_\mu| \vee |K_\mu| \vee \frac{|\partial H_\mu|}{\partial z} \vee \frac{|\partial K_\mu|}{\partial z} \vee \frac{|\partial H_\mu|}{\partial \phi} \vee \frac{|\partial K_\mu|}{\partial \phi}\};$$

so defined, λ depends on μ but remains bounded as $\mu \to 0$.

We now prove that (6.3) has a unique solution. To do this, it is convenient to denote the right-hand side of (6.3) temporarily by $x(\tilde{\phi})$:

$$x(\tilde{\phi}) = \tilde{\phi} + \theta_1(\mu) + \mu^{3/2}K_\mu(u(\tilde{\phi}),\tilde{\phi}).$$

We want to show that, as $\tilde{\phi}$ runs from 0 to 2π, $x(\tilde{\phi})$ runs

---

* i.e., ϕ differs from $\tilde{\phi} + \theta_1(\mu) + \mu^{3/2}K_\mu(u(\tilde{\phi}),\tilde{\phi})$ by an integral multiple of 2π.

exactly once over an interval of length 2π. From the periodicity of $u(\tilde{\phi})$, $K_\mu(z,\tilde{\phi})$ in $\tilde{\phi}$, it follows that

$$x(2\pi) = x(0) + 2\pi.$$

We therefore only have to show that x is strictly increasing. Let $\tilde{\phi}_1 < \tilde{\phi}_2$. Then

$$x(\tilde{\phi}_2)-x(\tilde{\phi}_1) = \tilde{\phi}_2 - \tilde{\phi}_1 + \mu^{3/2}[K_\mu(u(\tilde{\phi}_2),\tilde{\phi}_2)-K_\mu(u(\tilde{\phi}_2),\tilde{\phi}_2)].$$

Now

$$|K_\mu(u(\tilde{\phi}),\tilde{\phi})-K_\mu(\tilde{\phi}_1),\tilde{\phi}_1)| \leq \lambda[|u(\tilde{\phi}_2)-u(\tilde{\phi}_1)| + |\tilde{\phi}_2-\tilde{\phi}_1|]$$

$$\leq 2\lambda|\tilde{\phi}-\tilde{\phi}_1| = 2\lambda(\tilde{\phi}_2-\tilde{\phi}_1).$$

(The second inequality follows from the Lipschitz continuity of u.) Thus

$$x(\tilde{\phi}_2) - x(\tilde{\phi}_1) \geq (1 - 2\lambda\mu^{3/2})(\tilde{\phi}_2-\tilde{\phi}_1), \quad \text{so, provided}$$

$$1 - 2\lambda\mu^{3/2} > 0, \tag{6.5}$$

x is strictly increasing and (6.3) has a unique solution. We thus get $\tilde{\phi}$ as a function of ϕ, and it follows from our above estimates that $\tilde{\phi}$ is Lipschitz continuous:

$$|\tilde{\phi}(\phi_1) - \tilde{\phi}(\phi_2)| \leq (1 - 2\lambda\mu^{3/2})^{-1}|\phi_1-\phi_2|. \tag{6.6}$$

The definition (6.4) of $\mathscr{F}u$ therefore makes sense, and we next have to check that $\mathscr{F}u \in U$. Condition i), corresponds to 6.7 is immediate. For (ii), note that

$$|\mathscr{F}u(\phi)| \leq (1-2\mu)|u(\tilde{\phi})| + \mu^{3/2}|H_\mu(u(\tilde{\phi}),\tilde{\phi})|$$

$$\leq 1 - 2\mu + \mu^{3/2} .$$

Thus, $|\mathscr{F}u(\phi)| \leq 1$ for all ϕ provided

$$2\mu - \mu^{3/2}\lambda \geq 0. \tag{6.7}$$

Finally,

$$\begin{aligned} |\mathscr{F}u(\phi_1) - \mathscr{F}u(\phi_2)| &\leq (1-2\mu)|u(\tilde{\phi}_1) - u(\tilde{\phi}_2)| \\ &\quad + \mu^{3/2}\lambda[|u(\tilde{\phi}_1) - u(\tilde{\phi}_2)| + |\tilde{\phi}_1 - \tilde{\phi}_2|] \\ &\leq (1-2\mu + 2\mu^{3/2}\lambda)|\tilde{\phi}_1 - \tilde{\phi}_2| \end{aligned}$$

by the Lipschitz continuity of u. Inserting estimate (6.6) for $|\tilde{\phi}_1 - \tilde{\phi}_2|$, we get

$$|\mathscr{F}u(\phi_1) - \mathscr{F}u(\phi_2)| \leq (1-2\mu+2\mu^{3/2}\lambda)(1-2\mu^{3/2}\lambda)^{-1}|\phi_1 - \phi_2|,$$

so $\mathscr{F}u$ is Lipschitz continuous with Lipschitz constant 1 provided

$$(1-2\mu+2\mu^{3/2}\lambda)(1-2\mu^{3/2}\lambda)^{-1} \leq 1. \tag{6.8}$$

Evidently (6.8) holds for all sufficiently small positive μ, so (iii) holds.

The next step is to prove that $\mathscr{F}$ is a contraction. Thus, let $u_1, u_2 \in U$, choose ϕ, and let $\tilde{\phi}_1$, $\tilde{\phi}_2$ denote the solutions of

$$\phi = \tilde{\phi}_1 + \theta_1(\mu) + \mu^{3/2}K_\mu(u_1(\tilde{\phi}_1), \tilde{\phi}_1)$$

$$\phi = \tilde{\phi}_2 + \theta_1(\mu) + \mu^{3/2}K_\mu(u_2(\tilde{\phi}_2), \tilde{\phi}_2),$$

respectively. Subtracting these equations, transposing, and taking absolute values yields

$$|\tilde{\phi}_1-\tilde{\phi}_2| \le \mu^{3/2}|K_\mu(u_1(\tilde{\phi}_1),\tilde{\phi}_1) - K_\mu|u_2(\tilde{\phi}_2),\tilde{\phi}_2)| \qquad (6.9)$$
$$\le \mu^{3/2}\lambda[|u_1(\tilde{\phi}_1)-u_2(\tilde{\phi}_2)| + |\tilde{\phi}_1-\tilde{\phi}_2|].$$

Now

$$|u_1(\tilde{\phi}_1)-u_2(\tilde{\phi}_2)| \le |u_1(\tilde{\phi}_1)-u_2(\tilde{\phi}_1)| + |u_2(\tilde{\phi}_1)-u_2(\tilde{\phi}_2)|$$
$$\le ||u_1-u_2|| + |\tilde{\phi}_1-\tilde{\phi}_2|.$$

Inserting this inequality into (6.9), collecting all the terms involving $|\tilde{\phi}_1-\tilde{\phi}_2|$ on the left, and dividing yields

$$|\tilde{\phi}_1-\tilde{\phi}_2| \le (1-2\mu^{3/2}\lambda)^{-1}\mu^{3/2}\lambda\cdot||u_1-u_2||. \qquad (6.10)$$

Now we use the definition (6.4) of $\mathscr{F}u$:

$$|\mathscr{F}u_1(\phi)-\mathscr{F}u_2(\phi)| \le (1-2\mu)|\tilde{u}_1(\tilde{\phi}_1)-\tilde{u}_2(\tilde{\phi}_2)|$$
$$+ \mu^{3/2}|H_\mu(u_1(\tilde{\phi}_1),\tilde{\phi}_1) - H_\mu(u_2(\tilde{\phi}_2),\tilde{\phi}_2)|$$
$$\le (1-2\mu)[||u_1-u_2|| + |\tilde{\phi}_1-\tilde{\phi}_2|]$$
$$+ \mu^{3/2}\lambda[||u_1-u_2|| + 2|\tilde{\phi}_1-\tilde{\phi}_2|]$$
$$\le ||u_1-u_2||\{(1-2\mu)(1+\mu^{3/2}\lambda(1-2\mu^{3/2}\lambda)^{-1}]$$
$$+ \mu^{3/2}\lambda[1 + 2\mu^{3/2}\lambda(1-2\mu^{3/2}\lambda)^{-1}]\}.$$

Let α denote the expression in braces. Then

$$\alpha = 1 - 2\mu + 0(\mu^{3/2}),$$

so we can make $\alpha < 1$ by making μ small enough. If this is done, we have

$$||\mathscr{F}u_1- \mathscr{F}u_2|| \le \alpha\cdot||u_1-u_2|| \quad \text{with} \quad \alpha < 1, \qquad (6.11)$$

i.e., $\mathscr{F}$ is a contraction on U and hence has a unique fixed point u^*.

To prove that the invariant manifold $\{z = u^*(\phi)\}$ is attracting, we pick a point (z,ϕ) in the annulus $|z| \leq 1$, and we let (z_1,ϕ_1) denote $\phi_\mu(z,\phi)$. Note that

$$|z_1| \leq (1-2\mu)|z| + \mu^{3/2}\lambda \leq 1 - 2\mu + \mu^{3/2}\lambda \leq 1$$

(by (6.7)), so (z_1,ϕ_1) is again in the annulus. Now let $\tilde{\phi}_1$ denote the solution of

$$\phi_1 = \tilde{\phi}_1 + \theta_1(\mu) + \mu^{3/2}K_\mu(u^*(\tilde{\phi}_1),\tilde{\phi}_1).$$

The definition of ϕ_1, on the other hand, needs

$$\phi_1 = \phi + \theta_1(\mu) + \mu^{3/2}K_\mu(z,\phi).$$

Subtracting these equations and then estimating and re-arranging as in the proof of (6.8), we get

$$|\tilde{\phi}_1-\phi| \leq \mu^{3/2}\lambda(1-2\mu^{3/2}\lambda)^{-1}|z-u^*(\phi)|.$$

Now subtract the equations

$$u^*(\phi_1) = \mathscr{F}u^*(\phi_1) = (1-2\mu)u^*(\tilde{\phi}_1) + \mu^{3/2}H_\mu(u^*(\tilde{\phi}_1),\tilde{\phi}_1)$$

$$z_1 = (1-2\mu)z + \mu^{3/2}H_\mu(z,\phi)$$

and again imitate the proof that $\mathscr{F}$ is a contraction to get

$$|z_1-u^*(\phi_1)| \leq \alpha\cdot|z-u(\phi)|,$$

with the same α as in (6.11). By induction,

$$|z_n-\phi^*(\phi_n)| \leq \alpha^n|z-u(\phi)| \to 0 \quad \text{as} \quad n \to \infty.$$

In our proof, we used only the continuity of H_μ, K_μ

and their first derivatives, and we obtained a Lipschitz continuous u^*. Closer examination of the argument shows that we needed only Lipschitz continuity of H_μ, K_μ. If we have more differentiability of H_μ, K_μ, we would expect to obtain more differentiability for u^*. This is indeed the case. Specifically, let U_k denote the set of periodic functions $u(\phi)$ of class C^k satisfying

i) $|u^{(j)}(\phi)| \leq 1$, $j = 0,1,\ldots,k$; all ϕ.

ii) $|u^{(k)}(\phi)|$ is Lipschitz continuous with Lipschitz constant one.

If H_μ, K_μ have Lipschitz continuous k^{th} derivatives, a straightforward generalization of the estimates we have given shows that for μ sufficiently small, $\mathcal{F}$ maps U_k into itself. It may be shown that U_k is complete in the supremum norm (as in the proof of the center manifold theorem), so the fixed point of $\mathcal{F}$ must be in U_k, i.e., u^* has Lipschitz continuous k^{th} derivative. If we make the weaker assumption that H_μ, H_μ have continuous k^{th} derivatives, slightly more complicated arguments show that u^* also has a continuous k^{th} derivative; we proceed with the proof by showing that the set of u's, whose k^{th} derivatives have an appropriately chosen modulus of continuity, is mapped into itself by $\mathcal{F}$.

SECTION 6A

THE CANONICAL FORM

We shall give here, following Lanford [1], an elementary and straightforward derivation of the canonical form for the mapping $\Phi_\mu: \mathbb{R}^2 \to \mathbb{R}^2$. Recall that we have already arranged things so that

$$\Phi_\mu\begin{pmatrix} x \\ y \end{pmatrix} = (1+\mu)\begin{pmatrix} \cos\theta(\mu) & -\sin\theta(\mu) \\ \sin\theta(\mu) & \cos\theta(\mu) \end{pmatrix}\begin{pmatrix} x \\ y \end{pmatrix} + 0(r^2).$$

We want to organize the second, third, and fourth degree terms by making further coordinate changes. It will be convenient to identify R^2 with the complex plane by writing

$$z = x + iy.$$

Then

$$\Phi_\mu(z) = \lambda(\mu)z + 0(|z|^2), \quad \lambda(\mu) = (1+\mu)e^{i\theta(\mu)}.$$

From now on we shall leave μ out of our notation as much as possible.

The higher-order terms in the Taylor series for Φ may

be written as polynomials in z and $\overline{z}$, i.e.,

$$\Phi(z) = \lambda z + A_2(z) + A_3(z) + \dots,$$

where, for example,

$$A_2(z) = \sum_{j=0}^{2} a_j^2 z^j \overline{z}^{2-j}.$$

Let us begin in a pedestrian way with A_2. We choose a new coordinate $z' = z + \gamma(z)$, where γ is homogeneous of degree 2, i.e., has the same form as A_2. We can invert the relation between z and z' as

$$z = z' - \gamma(z') + \text{higher order terms}.$$

Since for the moment we are only concerned with terms of degree 2 or lower, we calculate module terms of degree 3 and higher and replace equality signs by congruance signs ($\equiv$). Thus we have

$$z \equiv z' - \gamma(z') = (I-\gamma)(z').$$

In terms of the new coordinate we have

$$\begin{aligned}
\Phi'(z') &\equiv (I+\gamma)\Phi(z'-\gamma(z')) \\
&\equiv (I+\gamma)[\lambda z' - \lambda\gamma(z') + A_2(z'-\gamma(z'))] \\
&\equiv (I+\gamma)[\lambda z' - \lambda\gamma(z') + A_2(z')] \\
&\equiv \lambda z' - \lambda\gamma(z') + A_2(z') + \gamma(\lambda z'-\lambda\gamma(z')+A_2(z')) \\
&\equiv \lambda z' + A_2(z') + \gamma(\lambda z') - \lambda\gamma(z').
\end{aligned}$$

Now

$$\gamma(z') = \gamma_2 z'^2 + \gamma_1 z'\overline{z'} + \gamma_0 \overline{z'}^2$$

$$\gamma(\lambda z') - \lambda\gamma(z') = \gamma_2(\lambda^2-\lambda)z'^2 + \gamma_1(|\lambda|^2-\lambda)z'\overline{z'}+\gamma_0(\overline{\lambda}^2-\lambda)\overline{z'}^2.$$

On the other hand,

$$A_2(z') = a_2^2 z'^2 + a_1^2 z'\overline{z'} + a_0^2 \overline{z'}^2,$$

so, if we put

$$\gamma_2 = \frac{-a_2^2}{\lambda^2 - \lambda}, \quad \gamma_1 = \frac{-a_1^2}{|\lambda|^2 - \lambda}, \quad \gamma_0 = \frac{-a_0^2}{\overline{\lambda}^2 - \lambda},$$

we get

$$\Phi'(z') = \lambda z' + O(|z'|^3).$$

We must, of course, make sure that the denominators in our expressions for the γ_i do not vanish. Since $|\lambda| = 1+\mu$, there is no problem for $\mu \neq 0$, but we want our μ-dependent coordinate change to be well-behaved as $\mu \to 0$. This will be the case provided

$$e^{2i\theta(0)} \neq e^{i\theta(0)}, \quad 1 \neq e^{i\theta(0)}, \quad e^{-2i\theta(0)} \neq e^{i\theta(0)},$$

i.e., provided

$$e^{i\theta(0)} \neq 1, \quad e^{3i\theta(0)} \neq 1.$$

Thus, if these conditions hold, we can make a smooth μ-dependent coordinate change, bringing A_2 to zero. We assume that we have made this change and drop the primes:

$$\Phi(z) = \lambda z + A_3(z) + \cdots .$$

(A_3 is <u>not</u> the original A_3) and see what we can do about A_3.

This time, we take a new coordinate $z' = z + \gamma(z)$, γ homogeneous of degree 3, and we calculate modulo terms of degree 4 and higher. Just as before, we have

$$\begin{aligned}\Phi'(z') &\equiv (I + \gamma)\Phi(z' - \gamma(z')) \\ &\equiv \lambda z' + A_3(z') + \gamma(\lambda z') - \lambda\gamma(z').\end{aligned}$$

Again, we write out

$$\gamma(z') = \gamma_3 a'^3 + \gamma_2 z^2 \bar{z}' + \gamma_1 z\bar{z}'^2 + \gamma_0 \bar{z}'^2$$

$$\gamma(\lambda z') - \lambda\gamma(z') = \gamma_3(\lambda^3 - \lambda)z'^3 + \gamma_2(|\lambda|^{-} - 1)\lambda z'^2\bar{z}'$$

$$+ \gamma_1(|\lambda|^2\bar{\lambda} - \lambda)z'\bar{z}'^2 + \gamma_0(\bar{\lambda}^3 - \lambda)\bar{z}'^3$$

$$A_3(z') = a_3^3 z'^3 + a_2^3 z'^2\bar{z}' + a_1^3 z'\bar{z}'^2 + a_0^3 \bar{z}'^3.$$

By an appropriate choice of γ_3, γ_1, γ_0, we can cancel the a_3^3, a_1^3, a_0^3 terms provided

$$e^{2i\theta(0)} \neq 1, \quad e^{4i\theta(0)} \neq 1.$$

The a_2^3 term presents a new problem. For $\mu \neq 0$, we can, of course, cancel it by putting

$$\gamma_3 = \frac{-a_2^3}{\lambda(|\lambda|^2 - 1)}.$$

This expression, however, diverges as $\mu \to 0$, independent of the value of $\theta(0)$. For this reason, we shall not try to adjust this term and simply put $\gamma_3 = 0$. Then, in the new coordinates (dropping the primes)

$$\Phi(z) = (\lambda + a_2^3|z|^2) + 0(|z|^4).$$

We next set out to cancel the 4th degree terms by a coordinate change $z' = z + \gamma(z)$, γ homogeneous of degree 4. A straightforward calculation of a by now familiar sort shows that such a coordinate change does not affect the terms of degree ≤ 3 and that all the terms of degree 4 can be cancelled provided

$$e^{5i\theta(0)} \neq 1.$$

Thus we get

$$\Psi(z) = (\lambda + a_2^3|z|^2)z + 0(|z|^3).$$

This is still not quite the desired form. To complete the argument, we write

$$\lambda + a_2^3|z|^2 = (1+\mu)e^{i\theta(\mu)}[1 - \frac{f_1(\mu)}{1+\mu}|z|^2 + if_3(\mu)|z|^2]$$

(where f_1, f_3 are real)

$$= (1+\mu-f_1(\mu)|z|^2)e^{i[\theta(\mu)+f_3(\mu)|z|^2]} + 0(|z|^4).$$

Thus

$$\Phi(z) = (1+\mu-f_1(\mu)|z|^2)e^{i[\theta(\mu)+f_3(\mu)|z|^2]}z + 0(|z|^5);$$

when we translate back into polar coordinates, we get exactly the desired canonical form.

SECTION 7

BIFURCATIONS WITH SYMMETRY

by

Steve Schecter

In this section we investigate what happens in the bifurcation theorems if a symmetry group is present. This is a non-generic condition, so special situations and degeneracies are encountered. (See Exercises 1.16 and 4.3).

At the end of the section we shall briefly discuss how the ideas presented here apply to Couette flow (we thank D. Ruelle for a communication on this subject).

Parts 0-4 of this section are based on Ruelle [3]. A related reference is Kopell-Howard [3]. See also Sattinger [6].

In the first parts of the section we confine ourselves to diffeomorphisms. There are entirely analogous results for vector fields.

0. Introduction.

Let E be a real Banach space with C^{ℓ} norm, $\ell \geq 3$.

(This means that the map $x \mapsto ||x||$ is C^ℓ on $E - \{0\}$.) Let G be a Lie group and Λ_G a (smooth) representation of G as a group of linear isometries of E. We denote the elements of Λ_G by Λ_g, where $g \in G$. We wish to consider diffeomorphisms $f: E \to E$ that commute with Λ_G, i.e., that satisfy $f \circ \Lambda_g = \Lambda_g \circ f$ for all $g \in G$.

If we know $f(x)$ and f commutes with Λ_G, then $f(y)$ is determined for all y in the orbit of x under the action of Λ_G. For example, if $E = R^2$ with the Euclidean norm and Λ_G is the group of all rotations of the plane (a representation of $G = SO(2)$), then if f commutes with Λ_G it follows that f takes circles to circles.

Notice:

(0.1) If f and h are diffeomorphisms of E that commute with Λ_G then for all $a,b \in R$, $af + bh$ also commutes with Λ_G.

(0.2) If f commutes with Λ_G and $f(0) = 0$, then $Df(0)$ commutes with Λ_G.

(0.3) If $\phi: E \to R$ satisfies $\phi\Lambda_g(x) = \phi(x)$ and all $g \in G$, for all $x \in E$, and f commutes with Λ_G, then ϕf commutes with Λ_G.

Because E has a C^ℓ norm, we can construct C^ℓ bump functions $\phi: E \to R$ satisfying $\phi = 1$ on a neighborhood of 0, $\phi = 0$ outside a larger neighborhood of 0, and ϕ is constant on each sphere centered at 0. These ϕ satisfy the conditions of (0.3). Given a C^ℓ diffeomorphism $f: E \to E$ and a linear isomorphism $A: E \to E$ that commutes with Λ_G and is sufficiently close to $Df(0)$, these bump functions together with properties (0.1) - (0.3) allow us to perturb f to a new diffeomorphism h C^{ℓ}-close to f such that h commutes with Λ_G and $Dh(0) = A$.

Given $1 \leq k \leq \ell$, let $\mathcal{F}$ denote the space of level-preserving maps $f: E \times (-1,1) \to E \times (-1,1)$ satisfying:

(a) Each f_{μ} is a C^{ℓ} diffeomorphism of E.

(b) f is C^k in the second variable, μ.

(c) $f_{\mu}(0) = 0$ for all μ.

(d) f_{μ} commutes with Λ_G for all μ.

(e) $Df_0(0)$ has a finite number of isolated eigenvalues on $|z| = 1$, each of finite multiplicity. (Since these eigenvalues are isolated, the rest of the spectrum is bounded away from $|z| = 1$.)

We wish to study how the qualitative picture of f_{μ} near the origin changes as μ passes 0, for generic $f \in \mathcal{F}$. In fact, we will study an open dense subset of $\mathcal{F}$, to be defined in Section 2.*

1. Reduction to Finite Dimensions.

Let E^0 be the finite dimensional subspace of E corresponding to the eigenvalues of $Df_0(0)$ on $|z| = 1$.

---

* $\mathcal{F}$ is given the appropriate Whitney topology. A basic open set $B(f,\phi)$ containing $f \in \mathcal{F}$ is given by specifying a strictly positive C^0 function $\phi: E \times (-1,1) \to \mathbb{R}$ and setting $B(f,\phi) = \{h \in \mathcal{F} \mid ||h(x,\mu)-f(x,\mu)|| < \phi(x,\mu)$ for all (x,μ); $||\frac{\partial^i}{\partial x^i} h(x,\mu) - \frac{\partial^i}{\partial x^i} f(x,\mu)|| < \phi(x,\mu)$ for all (x,μ) and $1 \leq i \leq \ell$; and $||\frac{\partial^i}{\partial \mu^i} h(x,\mu) - \frac{\partial^i}{\partial \mu^i} f(x,\mu)|| < \phi(x,\mu)$ for all (x,μ) and $1 \leq i \leq k\}$.

(7.1) Proposition. (1) $\Lambda_g E^0 = E^0$ for all $g \in G$.

(2) E^0 can be given a Hilbert space structure so that $\Lambda_G | E^0 = \Lambda_G^0$ remains a group of isometries.

Proof. (1) Let C be a simple closed curve in $\mathbb{C}$ such that $C = \bar{C}$ (where $\bar{C}$ denotes the complex conjugate of C), $\text{Spec } Df_0(0) \cap C = \emptyset$ and $\text{Spec } Df_0(0) \cap \text{Int } C = \text{Spec } Df_0(0) \cap \{z: |z| = 1\}$. In $E \otimes \mathbb{C}$, the complexification of E, let $P = \frac{1}{2\pi i} \int_C (zI - Df_0(0) \otimes I)^{-1} dz$. P commutes with $\Lambda_G \otimes I$ because it is the limit of operators that do. By the Real Spectral Splitting Theorem, P is the complexification of a real operator $Q: E \to E$. Clearly Q commutes with Λ_G, so $\text{Im } Q$ is invariant under Λ_G. But $\text{Im } Q = E^0$.

(2) Let Γ be the closure of Λ_G^0, a compact group. The desired inner product on E^0 is $(x,y) = \int_\Gamma d\gamma \langle x, \gamma y \rangle$, where $d\gamma$ is Haar measure on Γ and $\langle\ ,\ \rangle$ is any inner product on E^0. $\square$

(7.1) Theorem. $f \in \mathscr{F}$ has a C^k local center manifold V near $(0,0) \in E \times (-1,1)$, tangent to $E^0 \times (-1,1)$, satisfying:

(1) Each V_μ $(V_\mu = V \cap (E \times \{\mu\}))$ is C^ℓ and Λ_G-invariant.

(2) There is a level-preserving chart $\phi: V \to E^0 \times (-1,1)$ satisfying $\phi\Lambda_g = \Lambda_g\phi$ for all $g \in G$.

All the local recurrence of f_μ near 0 takes place in V_μ.

Proof. (1) The point here is that the construction of a center manifold in the center manifold theorem can be done in a Λ_G-invariant manner. In order that V be Λ_G-invariant, it suffices that the first "trial center manifold" used in its construction be Λ_G-invariant. (Consult Section 2 and the center manifold theorem for flows.) Take the first trial center manifold to be the subspace of E corresponding to $\operatorname{Spec} Df_0(0) \cap \{z: |z| \geq 1\}$, which is Λ_G-invariant by the argument of Proposition 1 (1).

(2) $\phi = (Q|E^0) \times I$, where Q is as defined in the proof of Proposition 7.1 (1). □

2. Generic Behavior.

From now on, because of Theorem 7.1, we will think of $f|V$ as acting on a neighborhood of $(0,0)$ in $E^0 \times (-1,1)$, say $U \times J$, and we will let A_μ stand for $Df_\mu(0)\, E^0$. Also, Λ_G^0 is now a group of isometries of the Hilbert space E^0. We will now exhibit a tractable generic subset of $\mathscr{F}$.

The characteristic polynomial of A_0 is a product of factors of the form $(x-1), (x+1)$, and $(x^2-2\operatorname{Re}\lambda x + \lambda\bar{\lambda})$, with $|\lambda| = 1$. Let λ be a complex eigenvalue of A (a similar argument applies if $\lambda = \pm 1$) and let F be the null space of $(A_0^2 - 2\operatorname{Re}\lambda A_0 + \lambda\bar{\lambda}I)^k$, where k is the exponent of $(x^2 - 2\operatorname{Re}\lambda x + \lambda\bar{\lambda})$ in the characteristic polynomial of A_0. Then F is Λ_G-invariant because it is the null space of an operator that commutes with Λ_G, and $F^\perp$ is Λ_G-invariant because F is and the Λ_g preserve the inner product.

Let $B: E^0 \to E^0$ be orthogonal projection onto $F^\perp$. Then $A_0 + \varepsilon A_0 B$ commutes with Λ_G and leaves F and $F^\perp$ invariant. $A_0 + \varepsilon A_0 B|F$ has only the eigenvalues λ and $\bar{\lambda}$; and, for all $\varepsilon \neq 0$, $A_0 + \varepsilon A_0 B|F^\perp$ has no eigenvalues on

$|z| = 1$. Using (0.1) and (0.3) it is now easy to perturb $f|U \times J$ so that the new $D(f|U \times J)_0(0) = A_0 + \varepsilon A_0 B$ has only the one pair of eigenvalues λ, $\bar{\lambda}$ on $|z| = 1$. Since E^0 has a Λ_G-invariant complement in E $(E = E^0 \oplus \ker Q)$, it is easy to extend this perturbation of $f|U \times J$ to a perturbation of f such that the new $Df_0(0)$ has only the eigenvalues λ and $\bar{\lambda}$ on $|z| = 1$, each with finite multiplicity. This perturbation reduces the dimension of E^0. If it is not the case that now Spec A_μ consists of exactly one pair of complex eigenvalues λ_μ, $\bar{\lambda}_\mu$ for each μ near 0, then we may make another perturbation to further reduce the dimension of E^0. Thus we see that by arbitrarily small perturbations of f we can eventually ensure that Spec A_μ consists of either one real eigenvalue λ_μ for all small μ, or one pair of complex eigenvalues λ_μ, $\bar{\lambda}_\mu$ for all small μ, with C^{k-1} dependence on μ.

Case 1. Spec A_μ consists of one real eigenvalue for all small μ, with $\lambda_0 = \pm 1$. Write $A_0 = S + T$, S symmetric, T antisymmetric. Using the fact that the Λ_g are isometries of Hilbert space, one checks that S and T commute with Λ_G. We choose an orthonormal basis for E^0 with respect to which S is diagonal; then with respect to that basis, $T_{ij} = -T_{ji}$.

Lemma 1. *If* $T \neq 0$, *then for arbitrarily small* $\varepsilon > 0$, Spec$(A_0+\varepsilon T)$ *has more than one point*.

Proof. Suppose Spec$(A_0+\varepsilon T)$ consists of exactly one point for all small $\varepsilon > 0$. Then for all small $\varepsilon > 0$,

$$\det(zI-(A_0+\varepsilon T)) = (z-\lambda(\varepsilon))^n \tag{2.1}$$

where $\text{Spec}(A_0+\varepsilon T) = \{\lambda(\varepsilon)\}$ and $n = \dim E^0$. By considering the coefficient of z^{n-1} on each side of (2.1) (the diagonal entries of εT are 0!) one sees that $\text{Tr } A_0 = n\lambda(\varepsilon)$. But $\text{Tr } A_0 = n\lambda(0)$. Therefore $\lambda(\varepsilon) = (0)$ for all small $\varepsilon > 0$. This implies that $\det(zI-(A_0+\varepsilon T))$ is constant for small $\varepsilon > 0$. But the coefficient of z^{n-2} in this polynomial is $\sum_{i<j} S_{ii}S_{jj} + \sum_{i<j} (T_{ij}+\varepsilon T_{ij})^2$, which is not a constant function of ε unless $T = 0$. $\square$

Since T commutes with Λ_G, if $T \neq 0$ Proposition 2 allows us to make small perturbations h of $f|U \times J$ for which $\text{Spec } Dh_0(0)$ contains more than one point. This allows us to make another perturbation further reducing the dimension of E^0. Therefore for some perturbation of f we must have $T = 0$. In this situation $A_0 = \pm I$. In fact, for some perturbation of f we must have $A_\mu = \lambda_\mu I$ for all small μ, $\lambda_\mu \in \mathbb{R}$, $\lambda_0 = \pm 1$.

Furthermore, for some such perturbation of f, Λ_G^0 is "irreducible of real type," i.e., the only elements of $\text{Hom } E^0$ that commute with Λ_G^0 are the multiples of I. To see this, suppose $R \in \text{Hom } E^0$, R commutes with Λ_0^G, and $R \neq \lambda I$. Then $A_0 + \varepsilon R$ commutes with Λ_0^G for all ε and is not a multiple of I. Hence one can construct a perturbation h of $f|U \times J$ such that $Dh_0(0) = A_0 + \varepsilon R$ and then another perturbation reducing $\dim E^0$.

Therefore for some arbitrarily small perturbation of f we have

(1) Λ_G^0 is irreducible of real type and

(2) $A_\mu = \lambda_\mu I$ for all small μ, with $\lambda_0 = \pm 1$.

But the set $f \in \mathscr{F}$ satisfying (1) and (2) is open. Therefore (1) and (2) hold generically.

Case 2. Spec A_μ consists of one pair of complex eigenvalues for all small μ. For each μ we have the following commutative diagram:

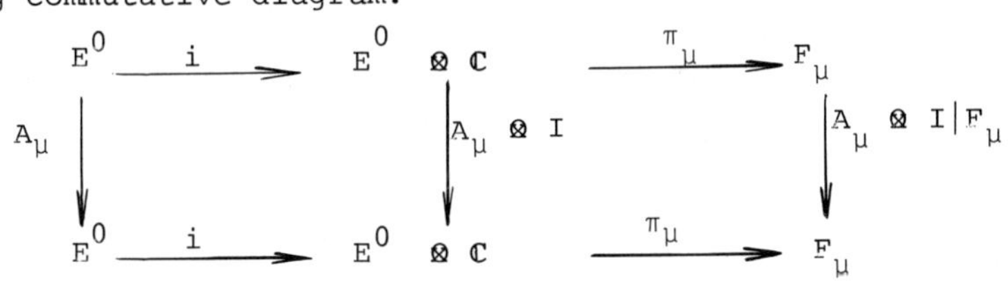

Here F_μ is the null space of $[(A_\mu \otimes I) - \lambda_\mu I]^{n/2}$ in $E^0 \otimes \mathbb{C}$, a complex subspace of complex dimension $\frac{n}{2}$, $i(x) = x \otimes 1$, and π_μ is orthogonal projection. The inner product on E^0 induces a complex inner product on $E^0 \otimes \mathbb{C}$ with respect to which $\Lambda_G^0 \otimes I$ is a group of unitary operators. F_μ is invariant under $\Lambda_G^0 \otimes I$ because $A_\mu \otimes I$ commutes with $\Lambda_G^0 \otimes I$. $A_\mu \otimes I|F_\mu$ is conjugate to A_μ and commutes with $\Lambda_G^0 \otimes I|F_\mu$. We now work with the $A_\mu \otimes I|F_\mu$ in analogy to Case 1.

We see that generically

(1) $\Lambda_G^0 \otimes I|F_\mu$ is irreducible and

(2) $A_\mu = \lambda_\mu I$ for all small μ, where I denotes the identity operator in $F_\mu = \mathbb{C}^{n/2}$ and λ_μ is complex with $|\lambda_0| = 1$. In other words, generically E^0 may be regarded as a complex inner product space and Λ_G^0 as an irreducible

group of unitary operators on E^0, while $A_\mu = \lambda_\mu I$ for all small μ, with λ_μ complex and $|\lambda_0| = 1$.

3. Results in the Case $Df_0(0)$ has One Pair of Complex Eigenvalues on $|z| = 1$.

Ruelle's main result, which we state without proof, is Theorem 7.2 below, which helps one find invariant manifolds in Case 2 of Section 2. We now think of $f|U \times J$ as defined on a neighborhood of $(0,0)$ in $F \times J$, where F is a finite-dimensional complex inner product space and J is an interval in R containing 0.

(7.2) *Lemma. Suppose* $f|U \times J$ *is* C^ℓ *for fixed* μ, C^k *in* μ, $1 \leq k \leq \ell$, $k \geq 3$. *Suppose also that* $Df_0(0) = \lambda_\mu I$ *with* $|\lambda_0| = 1$ *but* $\lambda_0^3 \neq 1$ *and* $\lambda_0^4 \neq 1$ *(a generic assumption). By a level-preserving change of coordinates that is* C^{k-3} *in* μ *and* C^∞ *for fixed* μ *and commutes with* Λ_G^0, *we can put* f *in the form*

$$f'_\mu(z) = \lambda_\mu z + P_\mu(z) + Q_\mu(z)$$

where P_μ *is a homogeneous polynomial of degree* 2 *in* z *and* 1 *in* $\bar{z}$, *and* Q_μ *is* $o(|z|^3)$. *In fact*, $|Q_\mu(z)| \leq c(|z|)|z|^3$ *and* $|DQ_\mu(z)| \leq c(|z|)|z|^2$ *where* $c(\cdot)$ *is independent of* μ *and* $\lim_{u \to 0} c(u) = 0$.

If $z \in \mathbb{C}^2$, $z = (z_1, z_2)$, each $z_i \in \mathbb{C}$, then a "homogeneous polynomial of degree 2 in z and 1 in $\bar{z}$" on $\mathbb{C}^2$ is one of the form

$$P(z) = (Az_1^2\bar{z}_1 + Bz_1z_2\bar{z}_1 + Cz_2^2\bar{z}_1 + Dz_1^2\bar{z}_2 + Ez_1z_2\bar{z}_2 + Fz_2^2\bar{z}_2,$$

$Qz_1^2\bar{z}_1 + Rz_1z_2\bar{z}_1 + Sz_2^2\bar{z}_1 + Tz_1^2\underline{z}_2 + Uz_1z_2\bar{z}_2 + Vz_2^2\bar{z}_2$.)

By way of motivation one might note that for such $P(z)$ and for $\lambda \in \mathbb{C}$ with $|\lambda| = 1$, one has $P(\lambda z) = \lambda P(z)$.

(7.2) Theorem. *Let $\Phi_\mu: \mathbb{C}^n \to \mathbb{C}^n$ be a one-parameter family of C^ℓ diffeomorphisms, $1 \le \ell < \infty$, depending on a real parameter μ varying in an interval about 0. Suppose $\Phi_\mu(z) = \lambda_\mu z + P_\mu(z) + Q_\mu(z)$, where $\mu \mapsto \lambda_\mu$ is a continuous complex function; P_μ is a homogeneous polynomial of degree 2 in z and 1 in $\bar{z}$ with coefficients continuous in μ; $|Q_\mu(z)| \le c(|z|)|z|^3$ and $|DQ_\mu(z)| \le c(|z|)|z|^2$ where $c(\cdot)$ is independent of μ and $\lim_{u\to 0} c(u) = 0$. We also assume $|\lambda_0| = 1$ and $|\lambda_\mu| > 1$ for $\mu > 0$. Suppose the vector field $z \mapsto z + \lambda_0^{-1}P_0(z)$ leaves the compact manifold S invariant and is normally hyperbolic*to it. Suppose S is also invariant under the transformations $z \mapsto ze^{i\sigma}$ (all real σ). Then for small $\mu > 0$ there exist maps $\theta_\mu \in C^\ell(S,\mathbb{C}^n)$ and manifolds $S_\mu \subset \mathbb{C}^n$ such that*

(1) θ_μ is a diffeomorphism of S onto S_μ.

(2) S_μ is invariant under Φ_μ and Φ_μ is normally hyperbolic to S_μ.

(3) $S_\mu \to 0$ as $\mu \to 0$.

(4) If Λ is a group of unitary transformations of $\mathbb{C}^n$ such that Φ_μ commutes with Λ for all μ and $\Lambda S = S$,

---

*Normally hyperbolic is defined as follows. Let S be a differentiable submanifold of a normed vector space E invariant under a diffeomorphism $f: E \to E$. Let $B \to S$ be a subbundle of $TE|S$ that is invariant under Df. Define $\rho(Df|B) = \limsup_{n\to\infty} \sup_{x\in S} ||Df^n(x)|B||^{1/n}$. f is said to be *normally hyper-*

<u>then</u> $\Lambda S_\mu = S_\mu$. <u>In fact, each</u> θ_μ <u>commutes with</u> Λ.

(5) <u>If</u> $\mu \to \Phi_\mu$ <u>is continuous from</u> R <u>to</u> C^k, $k \leq \ell$, <u>then</u> $\mu \to \theta_\mu$ <u>is continuous from</u> $\{\mu: 0 < \mu < \mu_0\}$ <u>to</u> $C^k(S,\mathbb{C}^n)$.

Theorem 7.2 gives information on invariant manifolds for $\mu > 0$. If we assume $|\lambda_\mu| < 1$ for $\mu < 0$, then we can obtain similar information for $\mu < 0$ by applying Theorem 2 to $\Phi^{-1}_{-\mu}$. It is easy to check that

$$\Phi^{-1}_{-\mu}(z) = \lambda^{-1}_{-\mu}z - \lambda^{-3}_{-\mu}\bar{\lambda}^{-1}_{-\mu}P_{-\mu}(z) + Q'_{-\mu}(z).$$

Therefore one should look for invariant manifolds of the vector field $z \mapsto z - \lambda_0^{-1}P_0(z)$.

4. <u>Examples</u>.

1. Let Λ^0_G be the full orthogonal group of $E^0 = R^n$ and let $\lambda_0 = 1$ be the only eigenvalue of $Df_0(0)$ on $|z| = 1$. Let $h = f|V$, V the center manifold for f of Proposition 7.1. According to Theorem 7.1 we may think of h as defined on a neighborhood of $(0,0)$ in $R^n \times R$, where:

(1) Each h_μ commutes with Λ^0_G.

(2) $h_\mu(0) = 0$ for all μ.

(3) $Dh_\mu(0) = \lambda_\mu I$, $\lambda_0 = 1$, $\lambda_\mu \in R$ for all μ.

<u>bolic to</u> <u>S</u> if there is a splitting of $TE|S$, $TE|S = N_+ \oplus TS \oplus N_-$, such that $\rho(Df|N_-) < \min\{1,\rho(Df|TS)\}$ and $\rho(Df^{-1}|N_+) < \min\{1,\rho(Df^{-1}|TS)\}$. This means that iteration Df contracts every vector of N_- more than Df contracts any vector of TS, and under iteration Df expands every vector of N_+ more than it expands any vector of TS.

Because Λ_G^0 contains all the reflections, it is easy to see that $h_\mu(x)$ is a multiple of x. Also, h_μ takes spheres about 0 to spheres about 0. Therefore $h_\mu(x) = \lambda_\mu x + p_\mu(|x|)x$, where for each μ, $u \mapsto p_\mu(u)$ is a real-valued $C^{\ell-1}$ function of u with $p_\mu(0) = 0$. Write $p_\mu(|x|) = a_\mu|x|^2 + o(|x|^2)$. The non-wandering set of h_μ near 0 consists of exactly those spheres that are taken into themselves; each such sphere consists entirely of fixed points. If f is at least C^3 in μ and $a_0 < 0$ (i.e., $(0,0)$ is a vague attractor), we can apply exactly the analysis of to see that for small $\mu > 0$ there is a one-parameter family of such spheres, one for each small $\mu > 0$, with the spheres converging to 0 as $\mu \to 0$.

2.* We think of $0(2)$ as generated by the complex numbers α with $|\alpha| = 1$ and a reflection r. Suppose we are in the situation of Case 2, Section 2, with $F = \mathbb{C}^2$ and Λ_G^0 the irreducible representation of $0(2)$ on $\mathbb{C}^2$ given by

$$\Lambda_\alpha^0(z_1,z_2) = (\alpha z_1,\alpha^{-1}z_2)$$

$$\Lambda_r^0(z_1,z_2) = (z_2,z_1).$$

We think of $f|V$ as acting on a neighborhood of $(0,0)$ in $\mathbb{C}^2 \times R$ and denote this map by h. According to Lemma 7.2, after a change of coordinates we have a new map h' with

$$h'_\mu(z) = \lambda_\mu z + P_\mu(z) + Q_\mu(z)$$

where $P_\mu(z)$ is homogeneous of degree 2 in z and 1 in

---

*David Fried explained this example to us. It is a reworking of Section 4.9 of Ruelle's paper.

$\bar{z}$, and $Q_\mu(z)$ is $o(|z|^3)$ uniformly in μ. $|\lambda_0| = 1$, and we assume $|\lambda_\mu| > 1$ for $\mu > 1$, $|\lambda_\mu| < 1$ for $\mu > 1$, and $\lambda_0^3 \neq 1$, $\lambda_0^4 \neq 1$.

It is easy to see that each $P_\mu(z)$ is Λ_G^0-invariant. Let $P(z)$ be a Λ_G^0-invariant homogeneous polynomial of degree 2 in z and 1 in $\bar{z}$. Write $P(z) = (P_1(z), P_2(z))$, where

$$P_1(z) = Az_1^2\bar{z}_1 + Bz_1z_2\bar{z}_1 + Cz_2^2\bar{z}_1 + Dz_1^2\bar{z}_2 + Ez_1z_2\bar{z}_2 + Fz_2^2\bar{z}_2$$

and

$$P_2(z) = Qz_1^2\bar{z}_1 + Rz_1z_2\bar{z}_1 + Sz_2^2\bar{z}_1 + Tz_1^2\bar{z}_2 + Uz_1z_2\bar{z}_2 + Vz_2^2\bar{z}_2.$$

Since $P(\alpha z_1, \alpha^{-1}z_2) = (\alpha P_1(z), \alpha^{-1}P_2(z))$ for all α with $|\alpha| = 1$, we see that $B = C = D = F = Q = S = T = U = 0$. Since $P(z_2, z_1) = (P_2(z), P_1(z))$, it follows that $A = V$ and $E = R$. Thus

$$P(z) = (Az_1^2\bar{z}_1 + Ez_1z_2\bar{z}_2,\ Az_2^2\bar{z}_2 + Ez_1z_2\bar{z}_1).$$

One more calculation shows we can write $P(z)$ in the form

$$P(z) = a(|z_1|^2+|z_2|^2)(z_1,z_2) + b(|z_1|^2-|z_2|^2)(z_1,-z_2); \quad a,b \in \mathbb{C}.$$

Therefore

$$h'_\mu(z) = \lambda_\mu z + a_\mu(|z_1|^2+|z_2|^2)z + b_\mu(|z_1|^2-|z_2|^2)(z_1,-z_2)+Q_\mu(z).$$

According to Theorem 2 we should now study the differential equation

$$\frac{dz}{dt} = z \pm \lambda_0^{-1}\{a_0(|z_1|^2+|z_2|^2)(z_1,z_2) + b_0(|z_1|^2-|z_2|^2)(z_1,-z_2)\}. \tag{4.1}$$

We will look for invariant manifolds S of this differential equation which are also invariant under the action of Λ^0_G and under the maps $z \mapsto \alpha z$, $|\alpha| = 1$. Now suppose $q\colon \mathbb{C}^2 \to \mathbb{C}$ is a polynomial in z and $\bar{z}$ invariant under Λ^0_G and the maps $z \mapsto \alpha z$, $|\alpha| = 1$ (i.e., $q(z) = q(\Lambda^0_g z)$ for all $g \in G$ and $q(z) = q(\alpha z)$ for all α with $|\alpha| = 1$). Suppose further that S is exactly the union of the orbits of (any) one of its points under the actions of Λ^0_G and the group of maps $z \mapsto \alpha z$, $|\alpha| = 1$. Then it follows that $\frac{d}{dt} q(z) = 0$ for $z \in S$. With this motivation, we will search for the manifolds S by studying these polynomials $q(z)$.

For $|\alpha| = 1$, $q(z_1,z_2) = q(\alpha z_1, \alpha^{-1} z_2) = q(\alpha^2 z_1, z_2)$. Hence for fixed z_2, $q(z_1,z_2)$ depends only on $|z_1|^2$. Similarly, for fixed z_1, $q(z_1,z_2)$ depends only on $|z_2|^2$. Since $q(z_1,z_2) = q(z_2,z_1)$, q is symmetric in z_1 and z_2. Now let $s(z_1,z_2) = |z_1|^2 + |z_2|^2$ and let $d(z_1,z_2) = 2|z_1 z_2|$. Then s and d^2 are a basis over $\mathbb{C}$ for the polynomials in which we are interested.

Let $\alpha = \mathrm{Re}\, \lambda_0^{-1} a_0$ and let $\beta = \mathrm{Re}\, \lambda_0^{-1} b_0$. Let $z^1, z^2 \in \mathbb{C}^2$ with $z^1 = (z_1^1, z_2^1)$, $z^2 = (z_1^2, z_2^2)$, where the $z_j^i \in \mathbb{C}$. Let $[z^1, z^2] = z_1^1 z_1^2 + z_2^1 z_2^2$. Then $s(z) = [z, \bar{z}]$ and

$$\begin{aligned}
\frac{ds}{dt} &= [z, \frac{d\bar{z}}{dt}] + [\frac{dz}{dt}, \bar{z}] \\
&= [z, \bar{z} \pm \lambda_0 \{\bar{a}_0(|z_1|^2+|z_2|^2)(\bar{z}_1,\bar{z}_2) \\
&\qquad\qquad + \bar{b}_0(|z_1|^2-|z_2|^2)(\bar{z}_1,-\bar{z}_2)\}] \\
&\quad + [z \pm \lambda_0^{-1}\{a_0(|z_1|^2+|z_2|^2)(z_1,z_2) \\
&\qquad\qquad + b_0(|z_1|^2-|z_2|^2)(z_1,-z_2)\}, \bar{z}] \\
&= 2s \pm (2\alpha s^2 + 2\beta(s^2-d^2)).
\end{aligned}$$

Also, since $d^2 = 4|z_1|^2|z_2|^2 = 4z_1\bar{z}_1z_2\bar{z}_2$, one calculates that $\frac{d}{dt} d = 2(d \pm sd)$. Therefore:

$$\begin{aligned} \frac{1}{2}\frac{ds}{dt} &= s \pm (\alpha s^2 + \beta(s^2-d^2)) \\ \frac{1}{2}\frac{d}{dt} d &= d \pm \alpha sd. \end{aligned} \tag{4.2}$$

We make the generic assumption that α, β, and $\alpha + \beta$ are all nonzero. Recalling that d and s are non-negative functions, we see that we have found three possibilities for invariant manifolds:

(1) $d = 0,\ s = 0$

(2) $d = 0,\ s = \mp(\alpha+\beta)^{-1}$

(3) $d = s = \mp \alpha^{-1}$.

Now we have the following commutative diagram

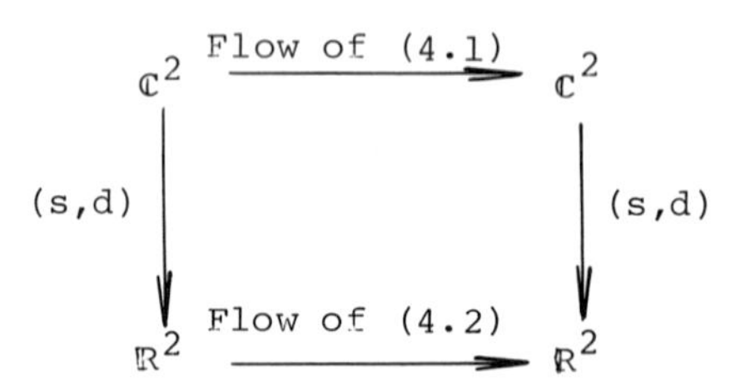

Because the vector field (4.2) is Lipschitz, its zeros represent fixed points of its flow. Hence their inverse images in $\mathbb{C}^2$ are invariant under the flow of (4.1). Therefore we have found the following invariant manifolds of (4.1):

(1) $S^{(1)} = \{0\}$

(2) $S^{(2)} = \{z \in \mathbb{C}^2 : d = 0,\ s = \mp(\alpha+\beta)^{-1}\}$

$\quad = \{z \in \mathbb{C}^2 : z_1 = 0,\ |z_2|^2 = \mp(\alpha+\beta)^{-1}\} \cup$

$\{z \in \mathbb{C}^2: z_2 = 0, |z_1|^2 = \mp(\alpha+\beta)^{-1}\}$

= two circles or $\emptyset$.

(3) $S^{(3)} = \{z \in \mathbb{C}^2: s = d = \mp \frac{1}{\alpha}\} = \{z: |z_1|^2 = |z_2|^2 = \mp \frac{1}{2\alpha}\}$ = torus or $\emptyset$.

The vector field (4.1) is normally hyperbolic to each of these manifolds:

(1) The derivative of (4.1) at 0 is I.

(2) The derivative of (4.2) on $S^{(2)}$, with respect to the variables s, d is $\begin{pmatrix} -2 & 0 \\ 0 & 2\beta/\alpha+\beta \end{pmatrix}$.

(3) The derivative of (.) on $S^{(3)}$, with respect to the variable s, d is $\begin{pmatrix} -2(\alpha+2\beta/\alpha) & 4\beta/\alpha \\ -2 & 0 \end{pmatrix}$, which has eigenvalues -2 and $\frac{-4\beta}{\alpha}$.

In this case one can see that the vector field (4.1) contain no other compact invariant manifolds by examining the flow of (4.2) in the sd-plane. Because of the definition of s and d, one needs only consider the region $0 \leq d \leq s$.

Now for definiteness suppose $f_\mu: E \to E$ and Spec $Df_\mu(0)$ is contained in $|z| < 1$ except for the eigenvalues $\lambda_\mu, \overline{\lambda_\mu}$. Assume $|\lambda_\mu| < 1$ for $\mu < 0$ and $|\lambda_\mu| > 1$ for $\mu > 0$. Also suppose $\alpha < 0$ and $\alpha + \beta < 0$ (a "weak attractor" condition). Then for each $\mu > 0$ we have the following invariant manifolds:

$S_\mu^{(1)} = \{0\}$, non-attracting for $\mu > 0$.

$S_\mu^{(2)}$ = two circles, invariant under f and the connected component of the identity in Λ_G^0, interchanged by reflections. They are attracting if $\beta < 0$, non-attracting if

$\beta > 0$.

$S_\mu^{(3)}$, a torus, attracting if $\beta > 0$, non-attracting if $\beta < 0$. $S_\mu^{(3)}$ can be analyzed further. The subspace $\Pi_\alpha = \{(z_1,z_2) \in \mathbb{C}^2: z_2 = \alpha z_1\}$ is pointwise fixed by $\Lambda_r^0 \circ \Lambda^0 \in \Lambda_G^0$ (this is the operator $z_1 \mapsto \alpha z_2$, $z_2 \mapsto \alpha^{-1} z_1$). Since the map θ_μ of Theorem 7.2 commutes with Λ_G^0, we have that $\theta_\mu(S^{(3)} \cap \Pi_\alpha) \subset \Pi_\alpha$. Since $h = f|V$ also commutes with Λ_G^0, it follows that $h(S_\mu^{(3)} \cap \Pi_\alpha) = S_\mu^{(3)} \cap \Pi_\alpha$. Thus $S_\mu^{(3)}$ is a disjoint union of circles of the form $S_\mu^{(3)} \cap \Pi_\alpha$, each invariant under h, interchanged by the elements of Λ_G^0.

It seems that in this example "two Hopf bifurcations take place at once," resulting in invariant sets, for each $\mu > 0$, of the form (point $\cup$ circle) $\times$ (point $\cup$ circle).

5. Flow Between Two Cylinders. Let us recall the set-up for Couette flow: Suppose we have a viscous incompressible fluid between two concentric cylinders. Let R_1 be the radius of the inner cylinder and R_2 the radius of the outer cylinder. Suppose we forcibly rotate the two cylinders. Let Ω_1 be the angular velocity of the inner cylinder and Ω_2 the angular velocity of the outer cylinder. Assume $\Omega_1 > 0$ and $\Omega_2 > 0$ (i.e., we rotate both cylinders counterclockwise). For small values of Ω_1 and Ω_2 one observes a steady horizontal laminar flow, called Couette flow. In fact, for arbitrary Ω_1, Ω_2 Couette flow is an explicit solution of the Navier Stokes equations, which, in cylindrical coordinates (r,ϕ,z), is given as follows: (see §1)

$$v_\theta = \frac{\Omega_2 R_2^2 - \Omega_1 R_1^2}{R_2^2 - R_1^2} r + \frac{(\Omega_1 - \Omega_2) R_1^2 R_2^2}{R_2^2 - R_1^2} \frac{1}{r} .$$

To get this solution one must ignore special phenomena that occur at the ends of the cylinders. We will do this by identifying the ends of the cylinders, so that the space A in which our fluid sits is an annulus crossed with the circle.

Let E be the space of C^r vector fields Y on A satisfying $\operatorname{div} Y = 0$ and $Y|\partial A = 0$, with the C^r topology. We will think of Ω_1 as small fixed and Ω_2 as varying: Ω_2 will be our bifurcation parameter μ. For each μ let Z_μ be the corresponding Couette vector field on A. Then for each μ the vector fields Y on A satisfying

$$\operatorname{div} Y = 0$$
$$Y \text{ does not slip at } \partial A$$

form a space E_μ which we may think of as $Z_\mu + E$. We will identify each E_μ with E in the obvious way. We will think of the Navier-Stokes equations as defining, for each μ, a vector field X_μ on E. We make the false assumption that the vector field $X\colon E \times \mathbb{R} \to E$ given by $X(Y,\mu) = X_\mu(Y)$ satisfies the conditions of Ruelle's paper.* Note that the point $(0,\mu)$ of $E \times \mathbb{R}$ corresponds to the Couette vector field Z_μ. Because Couette flow is a steady solution of the Navier-Stokes equations, we have $X_\mu(0) = 0$ for all μ.

Let $G = SO(2) \times O(2)$, the rotations of the annulus crossed with the full orthogonal group of the circle. This is the natural symmetry group of the situation. Notice that the Couette vector fields Z_μ are G-invariant. For each $g \in G$

---

*To make this really work, one can use the methods of sections 8,9.

define $\Lambda_g: E \to E$ by $(\Lambda_g Y)(x) = DgY(g^{-1}x)$, where $x \in A$. Because each g is an isometry of A and $D^k g = D^k g^{-1} = 0$ for $k > 1$, one can check that each Λ_g is a linear isometry of E. Thus $\Lambda_G = \{\Lambda_g: g \in G\}$ is a group of linear isometries of E. The physical symmetries of the situation imply that $X_\mu \Lambda_g = \Lambda_g X_\mu$ for all $g \in G$.

We assume that for small μ, 0 is an attracting fixed point for X_μ. (Actually this can be proved; cf. Serrin [1,2] and Sections 2A and 9.) This is consistent with the experimental observation that Couette flow is stable for small Ω_1, Ω_2. Assume that at the first bifurcation point $\mu_0 > 0$ only a finite number of eigenvalues of $DX_{\mu_0}(0)$ reach the imaginary axis, each of finite multiplicity. Then generically the stable manifold of the bifurcation $V \subset E \times \mathbb{R}$ must be tangent at $(0,\mu_0)$ to $E^0 \times \mathbb{R}$, where E^0 is a subspace of E on which Λ_G acts irreducibly.

Let us assume:

(1) $E^0 \cong \mathbb{R}^2$

(2) If $g \in SO(2) \times 1$, then Λ_g fixes E^0 pointwise.

(3) If $g \in 1 \times O(2)$, then Λ_g acts on E^0 exactly like the corresponding element of the full orthogonal group of $\mathbb{R}^2$.

Thus Λ_G^0 is essentially the full orthogonal group of the plane. If $(0,\mu_0)$ is a "vague attractor" for X_{μ_0}, then according to Section 4, Example 1 (modified for vector fields), for each small $\mu > \mu_0$ we expect the following invariant sets near the origin of $E \times \{\mu\}$ (See Figure 7.1):

(1) The origin, a zero of X_μ, non-attracting.

(2) A circle S_μ of zeroes of X_μ in V_μ. Each S_μ is invariant under Λ_G, $S_\mu \to 0$ as $\mu \to 0$, and each S_μ as a set is attracting in $E \times \{\mu\}$.

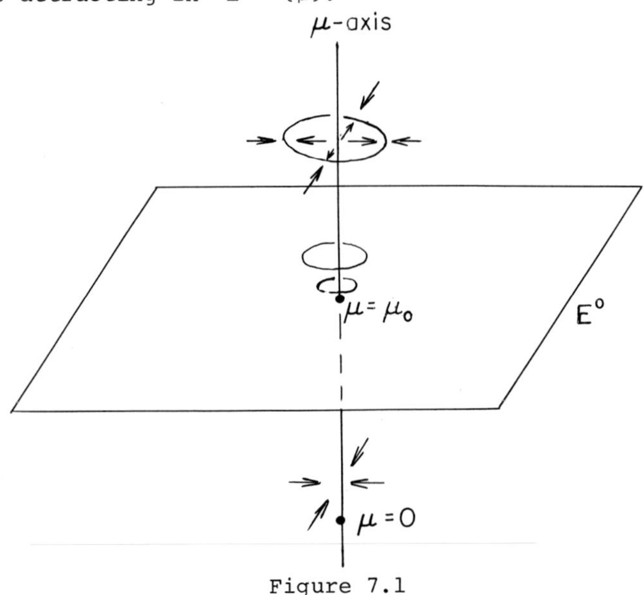

Figure 7.1

Suppose $Y_1, Y_2 \in S_\mu$ and $Y_1 = \Lambda_g Y_2$, $g \in 1 \times 0(2)$. Then $Y_1(x) = DgY_2(g^{-1}x)$. This means that Y_1 and Y_2 differ only by a vertical rotation and/or flip of A. Also, since $\Lambda_{SO(2) \times 1}$ is the identity on E^0, we see that each $Y \in S_\mu$ is fixed by the elements of $\Lambda_{SO(2) \times 1}$, i.e., each $Y \in S_\mu$ exhibits horizontal rotational symmetry.

This description of the S_μ is consistent with the experimentally observed phenomenon of Taylor cells. When μ reaches some critical value one commonly observes that Couette flow breaks up into cells within which the fluid moves radically from inner cylinder to outer cylinder and back while it continues its circular motion about the vertical axis (see Figure 7.2).

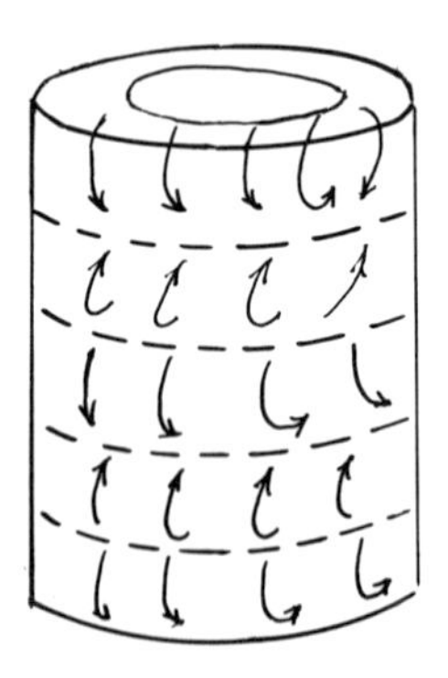

Figure 7.2

Taylor cells are a stable solution of the Navier-Stokes equations with horizontal rotational symmetry. Since any Taylor cell picture is no more likely to occur than its vertical translation by any distance (assuming the ends of the cylinder identified), we see why a whole circle of fixed points, interchanged by $\Lambda_{1 \times 0(2)}$, blossoms out from Couette flow. As μ increases, S_μ recedes from $(0,\mu)$ in our model; this corresponds to the Taylor cells becoming "stronger," i.e., stronger radial movement. Note that for $\mu > \mu_0$ Couette flow is still a zero of X_μ, but an unstable one.

(The above does not account for the fact that a Taylor cell vector field is invariant under a finite number of vertical translations, or, in our model, under a nontrivial subgroup of $\Lambda_{1 \times S0(2)}$. To get this result one should assume that for $g \in 1 \times 0(2)$, Λ_g acts on $E^0 \cong \mathbb{R}^2$ as follows: View $0(2)$ as generated by numbers θ, $0 \leq \theta < 2\pi$, where θ represents rotation through θ degrees, and by a flip r.

We then represent $1 \times O(2)$ in $O(2)$, the isometries of $\mathbb{R}^2$, by the homomorphism $(1,\theta) \mapsto n\theta$, $(1,r) \mapsto r$. Because this representation is onto, we get the S_μ as above. Now, however, each $Y_\mu \in S_\mu$ is invariant under Λ_H, where $H = SO(2) \times \{0, \frac{2\pi}{n}, \frac{4\pi}{n}, \ldots, \frac{2(n-1)\pi}{n}\}$. This additional invariance under the new Λ_G^0 would be preserved in the following.)

We now study the second bifurcation of X_μ. The situation is complicated by the circumstance that for $\mu > \mu_0$ the zeroes of X_μ in which we are interested occur in one-dimensional sets. In what follows we assume we have made a change of coordinates so that $S_\mu \subset E^0 \times \{\mu\}$.

Let $Y_\mu \in S_\mu$. For small $\mu > \mu_0$, $DX_\mu(Y_\mu)$ has an eigenvalue 0 of multiplicity 1 and the rest of its spectrum lies in $\text{Re } z < 0$. Notice that if $Y'_\mu \in S_\mu$ also, then $Y'_\mu = \Lambda_g Y_\mu$ for some $g \in 1 \times O(2)$, so $X(Y'_\mu) = X\Lambda_g(Y_\mu) = \Lambda_g X(Y_\mu)$. Therefore $DX(Y'_\mu)\cdot\Lambda_g = \Lambda_g \cdot DX(Y_\mu)$, so $DX(Y'_\mu)$ and $DX(Y_\mu)$ are conjugate. Hence $\text{Spec } DX(Y_\mu) = \text{Spec } DX(Y'_\mu)$.

Assuming $\Lambda_{1 \times O(2)}$ acts on E^0 like the full orthogonal group of the plane, Y_μ is fixed by just two elements of $\Lambda_{1 \times O(2)}$, namely $\Lambda_{(1,1)}$ and $\Lambda_{(1,r)}$, where $\Lambda^0_{(1,r)}$ is just the flip about the line determined by Y_μ. Since E^0 is pointwise fixed by $\Lambda_{SO(2) \times 1}$, the subgroup of Λ_G that fixes Y_μ is $\Lambda_{SO(2)\times\{1,r\}}$. Therefore $DX(Y_\mu)\cdot\Lambda_g = \Lambda_g \cdot DX(Y_\mu)$ for all $g \in SO(2) \times \{1,r\}$.

Suppose for $\mu_0 < \mu < \mu_1$ and $Y_\mu \in S_\mu$, $\text{Spec } DX_\mu(Y_\mu)$ is contained in $\text{Re } z < 0$ except for one eigenvalue 0 of multiplicity 1. Suppose also that for $Y_{\mu_1} \in S_{\mu_1}$,

Spec $DX_{\mu_1}(Y_{\mu_1}) \cap \text{Re } z = 0$ consists of a finite number of isolated eigenvalues (including 0), each of finite multiplicity. Fix $Y_{\mu_1} \in S_{\mu_1}$ and let r now denote the unique element of $O(2)$ such that Y_{μ_1} is fixed by $\Lambda_{SO(1)\times\{1,r\}}$. X has a center manifold W at Y_{μ_1}, tangent to $T_{Y_{\mu_1}} S_{\mu_1} \oplus E^1 \oplus \mu$-axis, where E^1 is a finite-dimensional space and $T_{Y_{\mu_1}} \oplus E^1$ is invariant under $\Lambda_{SO(2)\times\{1,r\}}$. Since W contains all the local recurrence of X near Y_{μ_1}, W contains all the points of $\cup_\mu S_\mu$ in a neighborhood of Y_{μ_1}.

Let $\Pi: E \times \mathbb{R} \to E$ be projection onto the first factor. Then for each μ near μ_1 there is a unique $Y_\mu \in S_\mu$ such that $\Pi Y_\mu / ||Y_\mu|| = \Pi Y_{\mu_1} / ||Y_{\mu_1}||$. Take this now as the definition of Y_μ. Then the subgroup of Λ_G that fixes Y_μ is exactly $\Lambda_{SO(2)\times\{1,r\}}$.

As in Section 1 we may identify a neighborhood of Y_{μ_1} in W with a neighborhood U of $(Y_{\mu_1},0,0)$ in $T_{Y_{\mu_1}} S_{\mu_1} \oplus E^1 \oplus \mu$-axis and $X|V$ with a vector field on U, still called X, in such a way that:

(1) Y_μ corresponds to $(Y_{\mu_1},0,\mu)$.

(2) $\Lambda_{SO(2)\times\{1,r\}}$ acts as a group of Hilbert isometries of $T_{Y_{\mu_1}} S_{\mu_1} \oplus E^1$.

(3) X on U commutes with $\Lambda_{SO(2)\times\{1,r\}}$. In particular, $DX_\mu(Y_{\mu_1},0,\mu)$ commutes with $\Lambda_{SO(2)\times\{1,r\}}$.

Generically we have one of two situations:

(1) Spec $DX_\mu(Y_{\mu_1},0,\mu)$ consists of the eigenvalue 0

with eigenspace $T_{Y_{\mu_1}} S_{\mu_1}$ and a single real eigenvalue λ_μ with $\lambda_{\mu_1} = 0$.

(2) Spec $DX_\mu(Y_{\mu_1},0,\mu)$ consists of the eigenvalue 0 with eigenspace $T_{Y_{\mu_1}} S_{\mu_1}$ and a single pair of complex conjugate eigenvalues $\lambda_\mu, \overline{\lambda_\mu}$ with $\text{Re}\,\lambda_\mu = 0$.

Let us assume (2) holds. Then for each μ near μ_1 there is a unique subspace E^2_μ of $T_{(Y_{\mu_1},0,\mu)}U$ such that $DX(Y_{\mu_1},0,\mu)$ leaves E^2_μ invariant and $DX(Y_{\mu_1},0,\mu)|E^2_\mu$ has only the eigenvalues $\lambda_\mu, \overline{\lambda_\mu}$. If we now regard E^2_μ as a subset of $T_{Y_{\mu_1}} S_{\mu_1} \oplus E^1 \oplus \mu$-axis, then E^2_μ is invariant under the action of $\Lambda_{SO(2)\times\{1,r\}}$ because $DX(Y_{\mu_1},0,\mu)$ is.

Generically $\Lambda_{SO(2)\times\{1,r\}}$ acts irreducibly on $E^2_{\mu_1}$. Assume $E^2_{\mu_1} \cong R^2$, so $E^1 \cong R^2$ and each $E^2_\mu \cong R^2$. Assume $\Lambda_{SO(2)\times\{1,r\}}$ acts on $E^2_{\mu_1}$ like the rotations of the plane. In other words, we assume $\Lambda_{(1,r)}$ fixes $E^2_{\mu_1}$ pointwise and $\Lambda_{SO(2)\times 1}$ acts on $E^2_{\mu_1}$ like the rotations of the plane. Since $\Lambda_{(1,r)}$ fixes $E^2_{\mu_1}$ pointwise, it follows that $E^2_{\mu_1}$ is invariant under X_{μ_1}. Because E^2_μ is almost parallel to $E^2_{\mu_1}$ and is $\Lambda_{SO(2)\times\{1,r\}}$-invariant, we can conclude that E^2_μ is in fact parallel to $E^2_{\mu_1}$, hence E^2_μ is X_μ-invariant.

Because of the last paragraph we see that we should consider the bifurcation of a vector field X' defined on $E^2_{\mu_1} \times \mu$-axis, where $X'_\mu = X|E^2_\mu$ (here we identify E^2_μ with $E^2_{\mu_1} \times \{\mu\}$). We have that X'_μ commutes with the rotations of $E^2_{\mu_1} \cong R^2$ and Spec $DX'_\mu(0,\mu) = \{\lambda,\overline{\lambda}\}$ where $\text{Re}\,\lambda < 0$ for $\mu < \mu_1$, $\text{Re}\,\lambda = 0$ for $\mu = \mu_1$, and we assume $\text{Re}\,\lambda > 0$ for

$\mu > \mu_1$. If we assume $(0,\mu_1)$ is a vague attractor for X'_{μ_1}, we get a "Hopf bifurcation with symmetry": the closed orbits of X'_μ that appear for $\mu > \mu_1$ are geometric circles (they are invariant under rotations) and the motion on these circles is with constant velocity.

Globally, the circle of fixed points S_{μ_1} has bifurcated into an attracting torus T_μ, for $\mu > \mu_1$, where T_μ is composed of closed orbits,* one bifurcating off each fixed point of S_{μ_1}. The closed orbits are invariant under $\Lambda_{SO(2)\times\{1,r\}}$ and are interchanged by the elements of $\Lambda_{1\times O(2)}$. This bifurcation corresponds to the following experimental observation: As μ increases, Taylor cells commonly become "doubly periodic," i.e., a horizontal wave pattern is introduced, and this wave pattern rotates horizontally with constant velocity (see Figure 7.3).

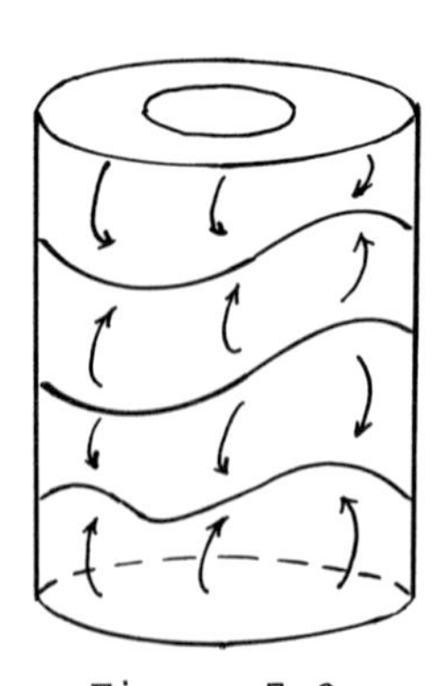

Figure 7.3

This argument does not account for the horizontal periodicity commonly observed in the wave pattern. This

---

*A similar invariant set occurs for flow behind a cyclinder.

periodicity can be explained by assuming the representation of $SO(2) \times 1$ in $O(2)$, the isometry group of $E^2_{\mu_1} \cong \mathbb{R}^2$, is of the form $(\theta,1) \mapsto n\theta$. A more serious problem is that our argument predicts that a vertical flip symmetry should still be present after the second bifurcation-- the elements of T_μ are invariant under $\Lambda_{(1,r)}$. This symmetry is not observed experimentally. The whole situation deserves further study. The methods discussed here seem very fruitful towards this end.

SECTION 8

BIFURCATION THEOREMS FOR PARTIAL DIFFERENTIAL EQUATIONS

As we have seen in earlier sections, there are two methods generally available for proving bifurcation theorems. The first is the original method of Hopf, and the second is using invariant manifold techniques to reduce one to the finite (often two) dimensional case.

For partial differential equations, such as the Navier-Stokes equations (see Section 1) the theorems as formulated by Hopf (see Section 5) or by Ruelle-Takens (see Sections 3,4) do not apply as stated. The difficulty is precisely that the vector fields generating the flows are usually not smooth functions on any reasonable Banach space.

For partial differential equations, Hopf's method can be pushed through, provided the equations are of a certain "parabolic" type. This was done by Judovich [11], Iooss [3], Joseph and Sattinger [1] and others. In particular, the methods do apply to the Navier-Stokes equations. The result is that if the spectral conditions of Hopf's theorem are fulfilled, then indeed a periodic solution will develop, and moreover, the stability analysis given earlier, applies. The crucial hypothesis needed in this method is analyticity of the solution in t.

Here we wish to outline a different method for obtaining results of this type. In fact, the earlier sections were written in such a way as to make this method fairly clear: instead of utilizing smoothness of the generating vector field, or t-analyticity of the solution, we make use of smoothness of the flow F_t^μ. This seems to have technical advantages when one considers the next bifurcation to invariant tori; analyticity in t is not enough to deal with the Poincaré map of a periodic solution (see Section 2B).

It is useful to note that there are general results applicable to concrete evolutionary partial differential equations which enable the determination of the smoothness of their flows on convenient Banach spaces. These results are found in Dorroh-Marsden [1]. We have reproduced some of the relevant parts of this work along with useful background material in Section 8A for the reader's convenience.

We shall begin by formulating the results in a general manner and then in Section 9 we will describe how this procedure can be effected for the Navier-Stokes equations. In the course of doing this we shall establish basic existence, uniqueness and smoothness results for the Navier-Stokes equations by using the method of Kato-Fujita [1] and results of Dorroh-Marsden [1] (§8A).

It should be noted that bifurcation problems for partial

differential equations other than the Navier-Stokes equations are fairly common. For instance in chemical reactions (see Kopell-Howard [1,2,5]) and in population dynamics (see Section 10). Problems in other subjects are probably of a similar type, such as in electric circuit theory and elastodynamics (see Stern [1], Ziegler [1] and Knops and Wilkes [1]). It seems likely that the real power of bifurcation theorems and periodic solutions is only beginning to be realized in applications.

The General Set-Up and Assumptions.

We shall be considering a system of evolution equations of the general form

$$dx/dt = X_\mu(x), \quad x(0) \quad \text{given},$$

where X_μ is a densely defined nonlinear operator on a suitable function space E, a Banach space, and depends on a parameter μ. For example, X_μ may be the Navier-Stokes operator and μ the Reynolds number (see Section 1). This system is assumed to define unique local solutions $x(t)$ and thereby a semiflow F_t which maps $x(0)$ to $x(t)$, for μ fixed, $t \geq 0$.

The key thing we need to know about the flow F_t of our system is that, for each fixed t,μ, F_t is a C^∞ mapping on the Banach space E (F_t is only locally defined in general). We note (see Section 8A at the end of this section) the properties that one usually has for F_t and which we shall assume are valid:

(a) F_t is defined on an open subset of

$$R^+ \times E, \quad R^+ = \{t \in R | t \geq 0\};$$

(b) $F_{t+s} = F_t \circ F_s$ (where defined);

(c) $F_t(x)$ is separately (hence jointly [§ 8A]) continuous in $t, x \in R^+ \times E$.

We shall make two standing assumptions on the flow. The first of these is

(8.1) <u>Smoothness Assumption</u>. <u>Assume that for each fixed</u> t, F_t <u>is a</u> C^∞ <u>map of</u> (<u>an open set in</u>) E <u>to</u> E.

This is what we mean by a smooth semigroup. Of course we cannot have smoothness in t since, in general, the generator X_μ of F_t will only be densely defined and is not a smooth map of E to E. However, as explained in Section 8A it is not unreasonable to expect smoothness in μ, t if $t > 0$. (This is the nonlinear analogue of "analytic semigroups" and holds for "parabolic type" equations). We shall need this below.

In Section 9 we shall outline how one can check this assumption for the Navier-Stokes equations by using general criteria applicable to a wide variety of systems. (For systems such as nonlinear wave equations, this is well known through the work of Segal [1] and others.)

The second condition is

(8.2) <u>Continuation Assumption</u>. <u>Let</u> $F_t(x)$, <u>for fixed</u> x <u>lie in a bounded set in</u> E <u>for all</u> t <u>for which</u> $F_t(x)$

<u>is defined</u>. <u>Then</u> $F_t(x)$ <u>is defined for all</u> $t \geq 0$.

This merely states that our existence theorem for F_t is strong enough to guarantee that the only way an orbit can fail to be defined is if it tends to infinity in a finite time. This assumption is valid for most situations and in particular for the Navier-Stokes equations.

Suppose we have a fixed point of F_t, which we may assume to be $0 \in E$; i.e., $F_t(0) = 0$ for all $t \geq 0$. Letting DF_t denote the Fréchet derivative of F_t for fixed t, $G_t = DF_t(0)$ is clearly a linear semigroup on E. Its generator, which is formally $DX(0)$, is therefore a densely defined closed linear operator which represents the linearized equations.* Our hypotheses below will be concerned with the spectrum of the linear semigroup G_t, which, under suitable conditions (Hille-Phillips [1]) is the exponential of the spectrum of $DX(0)$. (Compare Section 2A).

The third assumption is:

(8.3) <u>Hypotheses on the Spectrum</u>. <u>Assume we have a family</u> F_t^μ <u>of smooth nonlinear semigroups defined for</u> μ <u>in an interval about</u> $0 \in \mathbb{R}$. <u>Suppose</u> $F_t^\mu(x)$ <u>is jointly smooth in</u> t, x, μ, <u>for</u> $t > 0$. <u>Assume</u>:

(a) 0 is a fixed point for F_t^μ;

(b) for $\mu < 0$, the spectrum of G_t^μ is contained in

---

*Even if a semigroup is not smooth, it may make sense to linearize the equations and the flow. For example the flow of the Euler equations is C^1 from H^s to H^{s-1}, but the derivative extends to a bounded operator on H^{s-1}; cf. Dorroh-Marsden [1].

$D = \{z \in \mathbb{C}: |z| < 1\}$, where $G_t^\mu = D_x F_t^\mu(x)|_{x=0}$;

(c) for $\mu = 0$ (resp. $\mu > 0$) the spectrum of G_1^μ at the origin has two isolated simple eigenvalues $\lambda(\mu)$ and $\overline{\lambda(\mu)}$ with $|\lambda(\mu)| = 1$ (resp. $|\lambda(\mu)| > 1$) and the rest of the spectrum is in D and remains bounded away from the unit circle.

(d) $(d/dt)|\lambda(\mu)|\big|_{\mu=0} > 0$ (the eigenvalues move steadily across the unit circle).

Under these conditions, bifurcation to periodic orbits takes place. For their stability we make:

(8.4) <u>Stability Assumption</u>. <u>The condition</u> $V'''(0) < 0$ <u>holds, where</u> $V'''(0)$ <u>is calculated according to the procedures of Section 4</u> (<u>see Section 4A</u>).

This calculation may be done directly on the vector field X, since the computations are finite dimensional; unboundedness of the generator X causes no problems.

<u>Bifurcation to Periodic Orbits</u>.

Let us recap the result:

(8.5) <u>Theorem</u>. <u>Under the above hypotheses, there is a fixed neighborhood</u> U <u>of</u> 0 <u>in</u> E <u>and an</u> $\varepsilon > 0$ <u>such that</u> $F_t^\mu(x)$ <u>is defined for all</u> $t \geq 0$ <u>for</u> $\mu \in [-\varepsilon, \varepsilon]$ <u>and</u> $x \in V$. <u>There is a one-parameter family of closed orbits for</u> F_t^μ <u>for</u> $\mu > 0$, <u>one for each</u> $\mu > 0$ <u>varying continuously with</u> μ. <u>They are locally attracting and hence stable</u>. <u>Solutions near them are defined for all</u> $t \geq 0$. <u>There is a neighborhood</u> U <u>of the origin such that any closed orbit in</u> U <u>is</u>

one of the above orbits.

Note especially that near the periodic orbit, solutions are defined for all $t \geq 0$. This is an important criterion for global existence of solutions (see also Sattinger [1,2]).

Of course one can consider generalizations: for instance, when the system depends on many parameters with multiple eigenvalues crossing or to a system with symmetry as was previously described. Also, the bifurcation of periodic orbits to invariant tori can be proved in the same way.

Proof of Theorem (Outline). From our work in Section 2 we know that the center manifold theorem applies to flows. Thus, for the smooth flow $F_t(x,\mu) = (F_t^{\mu}(x),\mu)$ we can deduce the existence of a locally invariant center manifold C; a three-manifold tangent to the μ axis and the two eigendirections of $G_t^0(0)$. (The invariant manifold is attracting and contains all the local recurrence, but F_t still is only a local flow on this center).

Now there is a remarkable property of smooth semiflows (going back to Bochner-Montgomery [1]; cf. Chernoff-Marsden [2]) which is proved in Section 8A: this is that the semiflow F_t is generated by a C^{∞} vector field on the finite dimensional manifold C; i.e., the original X restricts to a C^{∞} vector field (defined at all points) on C.

This trick then immediately reduces us to the Hopf theorem in two dimensions and the proof can then be referred back to Section 3. □

Bifurcation to Invariant Tori.

This can be carried out exactly as in Section 6. However, as explained in Section 2B, we need to know that $F_t^{\mu}(x)$ is smooth in t, μ, x for $t > 0$. Then the Poincaré map for the closed orbit will be well defined and smooth and after we reduce to finite dimensions via the center manifold theorem as in Section 6, it will be a diffeomorphism by the corollary on p. 265. Therefore we can indeed use exactly the same bifurcation theorems as in Section 6 for bifurcation to tori. To check the hypothesis of smoothness, one uses results of Section 8A and Section 9.

SECTION 8A

NOTES ON NONLINEAR SEMIGROUPS

In this section we shall assemble some tools which are useful in the proofs of bifurcation theorems. We begin with some general properties of flows and semiflows ($\equiv$ nonlinear groups and nonlinear semigroups) following Chernoff-Marsden [1,2]. These include various important continuity and smoothness properties. Next, we give a basic criterion for when a semiflow consists of smooth mappings following Dorroh-Marsden [1].

Flows and Semiflows.

(8A.1) Definitions. Let D be a set. A flow on D is a collection of maps $F_t: D \to D$ defined for all $t \in \mathbb{R}$ such that:

1) $F_0 = \text{Identity}$

and

2) $F_{t+s} = F_t \circ F_s$ for all $t, s \in \mathbb{R}$.

Note that for fixed t, F_t is one to one and onto, since $F_{-t} \circ F_t = \mathrm{Id}$, and $F_t \circ F_{-t} = \mathrm{Id}$; i.e., $F_t^{-1} = F_{-t}$.

A semiflow on D is a collection of maps $F_t: D \to D$ defined for $t \geq 0$, also satisfying 1) and 2) for $t,s \geq 0$.

Warning: a semiflow need not consist of bijections.

(8A.2) Definition. Let N be a topological space and $D \subset N$. A local flow on D is a map $F: \mathcal{D} \subset \mathbb{R} \times D \to D$, where $\mathcal{D}$ is open in the N topology induced on $\mathbb{R} \times D$, such that for all $x \in D$, $(0,x) \in \mathcal{D}$ and if $\mathcal{D}_t = \{x \in D \mid (t,x) \in \mathcal{D}\}$ so we can define $F_t: \mathcal{D}_t \to D$, then F_t satisfies (1) and (2) where defined. The flow is maximal if $(t,x) \in \mathcal{D}$, $(s,F_t(x)) \in \mathcal{D} \Rightarrow (s+t,x) \in \mathcal{D}$. Similarly one defines a local semiflow and a maximal semiflow.

Now let N be a Banach manifold. A vector field with domain D is a map $X: D \to T(N)$ such that $X(x) \in T_x(N)$ for all $x \in D$. (T_xN is the tangent space to N at $x \in D \subset N$.) An integral curve for X is a curve c: $(a,b) \subset \mathbb{R} \to D$ such that c is differentiable as a map from (a,b) to N and $c'(t) = X(c(t))$. A flow (resp. semiflow, local flow) for X is a flow (resp. semiflow, local flow) on D such that for all $x \in D$, the map $t \mapsto F_t(x)$ is an integral curve of X.

If F_t is a flow on N such that $F: \mathbb{R} \times N \to N$ via $(t,x) \mapsto F_t(x)$ is a C^0 map, we say F is a C^0 flow on N.

If F_t is a flow for X and F_t extends to a continuous map $F_t: N \to N$ and the extension is a C^0 flow on M, then F_t is a C^0 flow for X.

If F_t is a C^0 flow on N and for all t fixed

$F_t\colon N \to N$ is of class C^k (resp. T^k), then F_t is a <u>flow of class C^k</u>. (resp. T^k) Here $F_t\colon N \to N$ is T^k means that the jth tangent map $T^j F_t\colon T^j(N) \to T^j(N)$ exists and is continuous for $j \le k$; $F_t\colon N \to N$ is C^k means that in each chart, the map $x \to d_x^j F$ $j \le k$ is continuous in the norm topology. ($d_x^j F$ is the jth total derivative of F at x. It is a j-multilinear map on the model space for N.) The T^k case differs from the C^k case in that norm continuity is replaced by strong continuity.

<u>Warning</u>. A C^k flow is not assumed to be C^k in the t-variable, only in the x-variable. A flow will be C^k in the t-variable only if it is generated by a smooth everywhere defined vector field; see however the Bochner-Montgomery theorem stated below.

<u>Separate and Joint Continuity</u>.

(8A.3) <u>Theorem</u>. (<u>Chernoff-Marsden</u> [2]). <u>Let</u> N <u>be a Banach manifold</u>. <u>Let</u> F_t <u>be a flow (or local flow) on</u> N, <u>and let</u> F <u>be separately continuous in</u> x <u>and</u> t (<u>i.e.</u>, $t \mapsto F_t(x)$ <u>is continuous for fixed</u> x <u>and</u> $x \mapsto F_t(x)$ <u>is continuous for fixed</u> t), <u>then</u> F_t <u>is a</u> C^0 <u>flow</u>; <u>i.e.</u> F_t <u>is jointly continuous</u>.

For the proof, we shall use the following.

(8A.4) <u>Lemma</u>. (Bourbaki [1] Chapter 9, page 18; Choquet, [1] Vol. 1, page 127). <u>Let</u> E <u>be a Baire space</u>. <u>Let</u> F, G <u>be metric spaces</u>. <u>Let</u> $\phi\colon E \times F \to G$ <u>be separately continuous</u>, <u>then for all</u> $f \in F$, <u>there is a dense set</u>

$S_f \subseteq E$ whose complement is first category such that if $e \in S_f$, then ϕ is continuous at (e,f).

Proof of the Theorem (8A.3). Since this is a local theorem, we may work in a chart. Therefore, we may assume that N is a Banach space and F_t is a local flow on N. We let $E = \mathbb{R}$, $F = G = N$. Let $x \in U \subset N$, $t \in (-\varepsilon,\varepsilon)$. There is a dense set of $t_x \in (-\varepsilon,\varepsilon)$ such that F is continuous at (t_x,x). Since the domain of definition of F is assumed open in $\mathbb{R} \times N$ we can choose t_x close to t so that the various compositions are defined. Let $t_n \to t$ and $x_n \to x$, and write $F_{t_n}(x_n) = F_{t-t_x} \circ F_{t_x+t_n-t}(x_n)$. Since $t_x + t_n - t \to t_x$ and F is continuous at (t_x,x), $F_{t_x+t_n-t}(x_n) = y_n \to F_{t_x}(x)$. Since for fixed t, $x \mapsto F_t(x)$ is continuous, we have that $F_{t_n}(x_n) = F_{t-t_x}(y_n) \to F_{t-t_x}(F_{t_x}(x)) = F_t(x)$. $\square$

(8A.5) Remarks.

1) Let G be a topological group which is also a Baire space. Let $\Phi: G \times N \to N$ be a separately continuous group action of G on a metric space N, then the above argument also shows that Φ is jointly continuous.

2) Suppose that $D \subseteq N$ is dense and that F_t is a flow on D which extends by continuity to a flow on M such that $t \mapsto F_t(x)$ is continuous for each $x \in D$. Then the same is true for each $x \in N$ and the extended flow is C^0. Indeed, let $x_n \to x$, where $x_n \in D$ and $x \in N$. Then for fixed t, $F_t(x_n) \to F_t(x)$, so that $t \mapsto F_t(x)$ is the pointwise limit of continuous functions. Therefore, for each $x \in N$, there is a second category set $S_x \subseteq \mathbb{R}$ such that if

$t \in S_x$, then $t \mapsto F_t(x)$ is continuous. The argument used in the proof of Theorem 8A.3 shows that $S_x = \mathbb{R}$ for all $x \in N$.

3) Many of these results can be generalized to the case in which N is not locally metrizable; e.g. a manifold modelled on a topological vector space (e.g.: a manifold modelled on a Banach space with the weak topology; - a "weak manifold"). c.f. Ball [1].

4) The same argument also works for semiflows, at least for $t > 0$. If N is locally compact, joint continuity is also true at $t = 0$ (Dorroh [1], but one can give a more direct argument). In general, however, joint continuity may fail at $t = 0$ so it has to be postulated.

Using these methods we can obtain an interesting result on the t-continuity of the derivatives of a differentiable flow.

(8A.6) Theorem. *Let N be a Banach manifold. Let F_t be a C^0 flow (or local flow, or semiflow) on N. Let F_t be of class T^k for $k \geq 1$. Then for each $j \leq k$, $T^jF_t: T^j(N) \to T^j(N)$ is jointly continuous in $t \in \mathbb{R}$ and $x \in T^j(N)$. (Only $t > 0$ for semiflows.)*

Proof. By induction and Theorem 8A.3 we are reduced immediately to the case $k = 1$. We may also assume that we are working in a chart. Therefore, $TF_t(x,v) = (F_t(x), D_xF_t(x)\cdot v)$. By assumption, this is continuous in the space variable x, so we need to show it is continuous in t. But clearly $D_xF_t(x)\cdot v = \lim_{n\to\infty} n(F_t(x + \frac{v}{n}) - F_t(x))$. Thus $t \mapsto D_xF_t(x)\cdot v$ is the pointwise limit of continuous functions so has a dense set of points of t-continuity. The rest of the proof is as in Remark 2 of (8A.5). □

The Generalized Bochner-Montgomery Theorem.

For simplicity we will give the next result for the case of flat manifolds. But it holds for general manifolds M, as one sees by working in local charts.

(8A.7) Theorem. (Chernoff-Marsden). *Let F_t be a jointly continuous flow on a Banach space $\mathbb{E}$. Suppose that, for each t, F_t is a C^k mapping, $k \geq 1$. Assume also that, for each $x \in \mathbb{E}$, $\|DF_t(x)-I\| \to 0$ as $t \to 0$, where $\|\cdot\|$ is the operator norm. Then $F_t(x)$ is jointly of class C^k in t and x. Moreover the generator X of the flow is an everywhere-defined vector field of class C^{k-1} on $\mathbb{E}$.*

Proof. Under the stated hypotheses, we can show that $DF_t(x)$ is jointly continuous as a mapping from $R \times \mathbb{E}$ into $\mathscr{L}(\mathbb{E}, \mathbb{E})$, the latter being all bounded linear maps of $\mathbb{E}$ to $\mathbb{E}$ equipped with the norm topology. In fact, if we write $\phi(t,x)$ for $DF_t(x)$. The chain rule implies the relation

$$\phi(s+t,x) = \phi(s,F_t(x))\cdot\phi(t,x). \tag{8A.1}$$

We have separate continuity of ϕ by assumption, and then we can apply Baire's argument as in Theorem (8A.3), together with the identity (8A.1) to deduce joint continuity.

Now let $\phi(t)$ be a C^∞ function on $\mathbb{R}$ with compact support. Define $J_\phi\colon \mathbb{E} \to \mathbb{E}$ by

$$J_\phi(x) = \int_{-\infty}^{\infty}\phi(t)F_t(x)dt. \tag{8A.2}$$

By joint continuity, we can differentiate under the integral sign in (8A.2), thus obtaining

$$DJ_{\phi}(x) = \int_{-\infty}^{\infty} \phi(t) DF_t(x) dt. \tag{8A.3}$$

Now if ϕ approximates the δ-function then $||DJ_{\phi}(x)-I||$ is small; in particular $DJ_{\phi}(x)$ is invertible. By the inverse function theorem it follows that J_{ϕ} is a local C^k diffeomorphism.

Moreover,

$$\begin{aligned} J_{\phi}(F_t(x)) &= \int_{-\infty}^{\infty} \phi(s) F_{s+t}(x) ds \\ &= \int_{-\infty}^{\infty} \phi(s-t) F_s(x) ds. \end{aligned}$$

The latter is differentiable in t and x. Since J_{ϕ} is a local C^k diffeomorphism, $F_t x$ is jointly C^k for t near 0. But then the flow identity shows that the same is true for all t. □

(8A.8) Remarks.

1) The above result is a non-linear generalization of the fact well known in linear theory that a norm-continuous linear semigroup has a bounded generator (and hence is defined for all $t \in \mathbb{R}$, not merely $t \geq 0$).

Furthermore, the same argument as above applies to semiflows and to local flows. This has the amusing consequence that a semiflow which is C^k and the derivative is norm continuous in t at $t = 0$ has integral curves which are locally uniformly extendable backwards in time (since the generator is C^{k-1}). This is most significant when combined with the next remark.

2) If $\mathbb{E}$ is finite dimensional, the norm convergence of $DF_t(x)$ to I follows automatically from the smoothness hypothesis. Indeed, Theorem (8A.6) implies that $DF_t(x) \to I$ in the strong operator topology, i.e., $DF_t(x)v \to v$ for each v; but for a finite-dimensional space this is the same as norm convergence.

Accordingly if M is a finite-dimensional manifold, a flow on M which is jointly continuous and C^k in the space variable is jointly C^k. The latter is a classical result of Montgomery. There is a generalization, due to Bochner and Montgomery [1] for actions of finite dimensional Lie groups. This generalization can also be obtained by the methods used to prove Theorem (8A.7) (cf. Chernoff-Marsden [2]).

Let us summarize a consequence of remarks (8A.8) that is useful.

(8A.9) _Corollary_. _Let_ F_t _be a local_ C^k _semiflow on a Banach manifold_ N. _Suppose that_ F_t _leaves invariant a finite dimensional submanifold_ $M \subset N$. _Then on_ M, F_t _is locally reversible, is jointly_ C^k _in_ t _and_ x _and is generated by a_ C^{k-1} _vector field on_ M.

Another fact worth pointing out is a result of Dorroh [1]. Namely, under the conditions of Theorem (8A.7), F_t is actually locally conjugate to a flow with a C^k generator (rather than C^{k-1}).

Lipschitz Flows.

(8A.10) Definitions. Let F_t be a flow (or a semi-flow) on a metric space M, e.g. a Banach manifold. We say that F_t is Lipschitz provided that for each t there is a constant M_t such that

$$d(F_t x, F_t y) \leq M_t d(x,y), \quad \forall\, x,y \in M.$$

The least such constant is called the Lipschitz norm, $||F_t||_{Lip}$.

We say that F_t is locally Lipschitz provided that, for every $x_0 \in M$ and $t_0 \in \mathbb{R}$, there is a neighborhood $\mathcal{U}$ of x_0 and a number $\varepsilon > 0$, such that

$$d(F_t x, F_t y) \leq M(t_0, x_0) d(x,y)$$

for all $x,y \in \mathcal{U}$ and $t \in [t_0-\varepsilon,\ t_0+\varepsilon]$. If $\mathcal{U}$ can be taken to be any bounded set, we say that the flow F_t is semi-Lipschitz. (This term was introduced by Segal.) Note that C^1 flows are locally Lipschitz.

Let F_t be a continuous Lipschitz flow, and let $M_t = ||F_t||_{Lip}$. Then (just as in the linear case) we have an estimate of the form

$$M_t \leq Me^{\beta|t|} \qquad (8A.4)$$

where M, β are constants. Indeed, note that M_t is sub-multiplicative: $M_{s+t} \leq M_s \cdot M_t$; this is an immediate consequence of the flow identity. Moreover, we know

$$M_t = \sup_{x \neq y} d(F_t x, F_t y)/d(x,y).$$

Thus M_t is lower semicontinuous, being the supremum of a family of continuous functions. In particular, M_t is meas-

urable. But then an argument of Hille-Phillips [1, Thm. 7.6.5] shows that (8A.4) holds for some constants M, β.

Uniqueness of Integral Curves.

It is a familiar fact that integral curves of Lipschitz vector fields are uniquely determined by their initial values, but that there are continuous vector fields for which this is not the case.* On the other hand, it is known that integral curves for generators of linear semigroups are unique. The following result shows that such uniqueness is a consequence of the local Lipschitz nature of the flow (cf. van Kampen's Theorem; Hartman [1, p. 35]).

(8A.11) _Theorem_. _Let_ X _be a vector field on the Banach manifold_ M, _with domain_ D. _Assume that_ X _has a locally Lipschitz flow_ F_t. _More precisely_, _assume that_:

(a) F_t _is a group of bijections on_ D, _and_, _for each_ $x \in D$, $t \mapsto F_t x$ _is differentiable in_ M, _with_

$$\frac{d}{dt} F_t(x) = X(F_t(x))$$

(b) _For each_ $x_0 \in M$ _and_ $t_0 \in \mathbb{R}$ _there is a neighborhood_ $\mathcal{U}$ _of_ x_0 _in_ M _and an_ $\varepsilon > 0$, _such that in local charts_,

---

*The famous example is $X(x) = x^{2/3}$ on the line. In a Fréchet space E, a continuous linear vector field $S: X \to X$ may have infinitely many integral curves with given initial data; viz. $S(x_0, x_1, \ldots) = (x_1, x_2, \ldots)$ on E the space of real sequences under pointwise convergence, or may have no integral curves with given initial data; viz. $S(f) = df/dx$ on $E = C^\infty$ functions on $[0,1]$ which vanish to all orders at 0 and 1. The result (8A.11) is generalized significantly in Dorroh-Marsden [1].

$$d(F_t x, F_t y) \leq C d(x,y)$$

for $x,y \in \mathcal{U}$, _and_ $t \in [t_0-\varepsilon, t_0+\varepsilon]$. _Here the constant_ C _is supposed to be independent of_ x, y _and_ t. (_In other words, the local Lipschitz constant is supposed to be locally bounded in_ t. _This is the case for a globally Lipschitz flow, for example_.)

Conclusion: _if_ $c(t)$ _is a curve in_ D _such that_ $c'(t) = X(c(t))$, _then_ $c(t) \equiv F_t(c(0))$.

Proof. We can work in a local chart (see (8A.13)), so we assume $M = \mathbb{E}$, a Banach space. Given t_0, let $x_0 = c(t_0)$. Then choose $\varepsilon > 0$ and a neighborhood $\mathcal{U}$ of x_0 as in hypothesis (b); in addition, ε should be small enough so that $c(t) \in \mathcal{U}$ if $|t-t_0| \leq \varepsilon$.

Define $h(t) = F_{t_0-t}c(t)$. Then, for t near t_0, and τ small,

$$\begin{aligned} ||h(t+\tau) - h(t)|| &= ||F_{t_0-t-\tau}c(t+\tau) - F_{t_0-t}c(t)|| \\ &= ||F_{t_0-t-\tau}c(t+\tau) - F_{t_0-t-\tau}F_\tau c(t)|| \\ &\leq C||c(t+\tau) - F_\tau c(t)||. \end{aligned}$$

Moreover, $\frac{1}{\tau}[c(t+\tau) - F_\tau c(t)] = \frac{1}{\tau}[c(t+\tau) - c(t)] + \frac{1}{\tau}[c(t) - F_\tau c(t)] \to X(c(t)) - X(c(t)) = 0$ as $\tau \to 0$. Thus h is differentiable, and $h'(t) \equiv 0$. It follows that $h(t)$ is constant, whence $c(t) = F_{t-t_0}c(t_0)$ for t near t_0. From this the relation $c(t) = F_t c(0)$ follows easily. □

(8A.12) _Corollary_. _The conclusion of Theorem (8A.11) applies to_ C^1 _flows_ F_t.

Proof. We shall verify condition (b) of the hypothesis. In a local chart, our results (see (8A.6)) on joint continuity show that $DF_t(x) \cdot y$ is continuous jointly in t, x, and y. Hence, by the Banach-Steinhaus Theorem, for a given x_0 and t_0 there is a convex neighborhood $\mathcal{U}$ of x_0 and an $\varepsilon > 0$ so that $||DF_t(x)|| \leq C$ if $x \in \mathcal{U}$ and $|t-t_0| \leq \varepsilon$. The mean value theorem then shows that $||F_t(x) - F_t(y)|| \leq C||x-y||$ if $x, y \in \mathcal{U}$ and $|t-t_0| \leq \varepsilon$. □

The above results generalize classical theorems of Kneser and Van Kampen. They easily generalize to semi-flows.

Note. An explicit example of a continuous vector field with a jointly continuous flow F_t for which the conclusion of Theorem (8A.11) fails is the following well known example. On $\mathbb{R}$ let X be defined by

$$X(x) = \frac{3}{2} |x|^{1/3}.$$

Define $\phi(y) = |y|^{3/2} \operatorname{sgn} y$. Then ϕ is differentiable, with $\phi'(y) = \frac{3}{2} |y|^{1/2}$. It is easy to check that $F_t(x) = \phi(t+\phi^{-1}(x))$ is a flow for X. In particular $F_t(0) = |t|^{3/2} \operatorname{sgn} t$. But $c(t) \equiv 0$ is another integral curve with $c(0) = 0$. See Hartman [1] for more examples.

(8A.13) Remarks.

1) In case one wishes to work globally on M and not in charts, one should use the proper sort of metric as follows:

Definition. Let N be a Banach manifold modelled on a Banach space $\mathbb{E}$. Let d be a metric on N. We say d is compatible with the structure of N, if d gives the topology of N and if given any $x_0 \in N$, there is, about x_0, a chart $(\mathcal{U},\phi)$ and constants $\alpha(x_0)$, $\beta(x_0)$ such that for all $x,y \in \mathcal{U}$, $d(x,y) \leq \alpha||\phi(x) - \phi(y)|| \leq \beta d(x,y)$.

2) This method is not the one usually employed to prove uniqueness of integral curves, for example, in $\mathbb{R}^n$. One usually assumes that the vector field is locally Lipschitz and then uses integration to prove this result. Let us recall how this goes. Let X be a locally Lipschitz vector field on $\mathbb{R}^n$ (or any Banach space). Let $d(t)$ and $c(t)$ be two integral curves of X such that $d(0) = c(0)$. Then

$$|d(t)-c(t)| = |\int_0^t X(d(s)) - X(c(s))ds| \leq K \int_0^t |d(s)-c(s)|ds.$$

One then uses the fact (called Gronwall's inequality) that if α is such that $\alpha(t) \leq \int_0^t K\alpha(s)ds$, then $\alpha(t) \leq \alpha(0)e^{Kt}$. Therefore, we have $d(t) = c(t)$. For purposes of partial differential equations, however, it is important to have the result as stated in (8A.11) because one is often able to find Lipschitz bounds for the constructed flows, but rarely on the given generator.

3) Another method sometimes used to prove uniqueness of integral curves for a vector field X with domain $D \subseteq N$ is called the energy method. Suppose there is a smooth function $H: D \times D \to \mathbb{R}^+$ such that $H(x,y) = 0$ if and only if $x = y$ and such that for any two integral curves c and d for X, $\frac{dH(c(t),d(t))}{dt} \leq K(t,d(0),c(0))H(c(t),d(t))$ where K is locally bounded in t. Then as in Remark 2 we can conclude that X has unique integral curves. This method is directly applicable to classical solutions of the Euler and Navier-Stokes equation, for example.

Measurable flows.

Under rather general conditions, continuity of a flow in the time variable can be deduced from measurability. For example, we have the following result. See also Ball [2].

(8A.14) Theorem. *Let* M *be a separable metric space. Let* F_t *be a flow* (*or local semiflow*) *of continuous maps on* M. *Assume that, for each* $x \in M$, *the map* $t \mapsto F_t(x)$ *is Borel measurable; that is, the inverse image of any open set is a Borel subset of* $\mathbb{R}$. *Then* F_t *is jointly continuous* (*respectively, jointly continuous for* $t > 0$).

Proof. Because M is separable, the Borel function $t \mapsto F_t(x)$ is continuous when restricted to the complement of some first-category set $C \subset \mathbb{R}$ (cf. Bourbaki [1].) Given t_0 and a sequence $t_n \to t_0$, note that $\bigcup_{n=1}^{\infty} [C-(t_0-t_n)] = D$ is of the first category, hence there exists an $s \in \mathbb{R}$ with $s \notin D$; that is, $t_n - t_0 + s \notin C$ for all n. Accordingly, $F_{t_n-t_0+s}(x) \to F_s(x)$ when $n \to \infty$. Now apply the continuous map F_{s-t_0} to deduce that $F_{t_n}(x) \to F_{t_0}(x)$.

Hence $F_t(x)$ is separately continuous, and the conclusion follows from Theorem 1. $\square$

Theorems of this sort are well known for linear semigroups (see Yosida [1] for instance).

Some Results on Time Dependent Linear Evolution Equations.

In order to study smoothness criteria we shall need to make use of some results about linear evolution equations. These results are taken from Kato [1,4,5]. We begin by defining an evolution system. (This exposition is adapted from Dorroh-Marsden [1].)

(8A.15) Definition. Let X be a Banach space and $T > 0$. A subset $\{U(t,s)\ 0 \leq s \leq t < T\}$ of $B(X) = B(X,X)$ (bounded operators on X) is called an evolution system in X if

i) $U(t,t) = I$ for $0 \leq t < T$, and

ii) $U(t,s)U(s,r) = U(t,r)$ for $0 \leq r \leq s \leq t < T$.

An evolution system $\{U(t,x) | 0 \leq s \leq t < T\}$ in X is said to be strongly continuous if for each $f \in X$, the function $U(\cdot,\cdot)f$ maps $[0,T) \times [0,T)$ continuously into X. The X-infinitesimal generator of $\{U(t,s)\}$ is the collection $\{A(s) | 0 \leq s < T\}$ of operators in X defined by

$$A(s)f = \lim_{\varepsilon \downarrow 0} \varepsilon^{-1}[U(s+\varepsilon,s)f-f]$$

with $D(A(s))$ consisting of all f for which this limit exists, where the limit is taken in X.

(8A.16) Remarks. a) If $\{U(t,s)\}$ is a strongly continuous evolution system in X, then it follows from the uniform boundedness principle that $||U(t,s)||_{X,X}$ is bounded for s and t in closed and bounded intervals.

b) Let $\{U(t,s) | 0 \leq s \leq t < T\}$ be a strongly continuous evolution system in X with X-infinitesimal generator $\{A(s) | 0 \leq s < T\}$, and let $0 < a < T$. Then $\{A(s+a) | 0 \leq s < T - a\}$ is the X-infinitesimal generator of the strongly continuous evolution system $\{U(t+a, s+a) | 0 \leq s \leq t < T - a\}$.

(8A.17) Proposition. Let $\{U(t,s) | 0 \leq s \leq t < T\}$ be a strongly continuous evolution system in X with X-infinite-

simal generator $\{A(s) | 0 \leq s < T\}$. *If* $f \in D(A(s))$ *for all* $0 \leq s < T$, *and* $A(\cdot)f$ *maps* $[0,T)$ *continuously into* X, *then*

$$(\partial/\partial s)[U(t,s)f] = -U(t,s)A(s)f \tag{8A.5}$$

for $0 \leq s \leq t < T$, $t > 0$.

Proof. If $0 \leq s < t < T$, then

$$U(t,s+\varepsilon)f - U(t,s)f = U(t,s+\varepsilon)[f-U(s+\varepsilon,s)f],$$

and therefore,

$$(\partial^+/\partial s)U(t,s)f = -U(t,s)A(s)f.$$

Thus for each $t \in (0,T)$, the function $U(t,\cdot)f$ has a continuous right derivative on $[0,t)$. Thus the function $U(t,.)f$ is continuously differentiable on $[0,t)$ (see Yosida [1], p. 239), and (8A.5) holds for $0 \leq s < t < T$. Since the derivative of $U(t,\cdot)f$ has a limit from the left at t, it follows that

$$(\partial^-/\partial s)[U(t,s)f] = -U(t,s)A(s)f$$

for $0 < s = t < T$. But, because of the domain of $U(t,\cdot)f$, this is what (8A.5) means when $s = t$. □

(8A.18) *Corollary*. *Let* $\{U(t,s) | 0 \leq s \leq t < T\}$ *be a strongly continuous evolution system in* X *with* X-*infinitesimal generator* $\{A(s) | 0 \leq s < T\}$. *Let* $f(s) \in D(A(s))$ *for* $0 \leq s < T$, *suppose* f *is continuously differentiable from* $[0,T)$ *into* X, *and that* $A(\cdot)f(\cdot)$ *maps* $[0,T)$ *continuously into* X. *Then*

$$(\partial/\partial s)[U(t,s)f(s)] = U(t,s)f'(s) - U(t,s)A(s)f(s)$$

for $0 \leq s \leq t < T$, $t > 0$.

Proof. This follows from the Proposition, the strong continuity of $\{U(t,s)\}$, and the local boundedness of $||U(t,s)||_{X,X}$. □

We call (8A.5) the backward differential equation. In order that the forward differential equation

$$(\partial/\partial t)[U(t,s)f] = A(t)U(t,s)f \qquad (8A.6)$$

hold, it is necessary that $U(t,s)f \in D(A(t))$, and this is a more restrictive condition which may not be satisfied when the hypothesis of the above Proposition is satisfied.

Now suppose Y is another Banach space with Y densely and continuously embedded in X.

(8A.19) Definition. An evolution system $\{U(t,s)\}$ in X is said to be Y-regular if each transformation $U(t,s)$ maps Y continuously into Y, and $\{U(t,s)\}$ is strongly continuous in Y; i.e., if $\{U(t,s)\}$ is a strongly continuous evolution system in Y as well as in X.

Kato in [4] and [5] gives a variety of conditions on families $\{A(s)\}$ of operators in X which are sufficient for these families to be the X-infinitesimal generators of strongly continuous evolution systems in X. Some of these conditions are also sufficient for the evolution system to be Y-regular and for the forward differential equation to hold. He also gives several convergence theorems for evolution systems and upper bounds for the operator norms in terms of certain parameters of the infinitesimal generator.

Since these results bear directly on our results later, we quickly summarize the fundamental points here for reference. Kato's papers should be consulted for details and related remarks.

(8A.20) Definitions. Let $A \in G(X)$, the set of semigroup generators in X. Y is said to be *admissible with respect to* A, or simply A-*admissible*, if $\{e^{tA}\}$ leaves Y invariant and forms a semigroup of class C_0 in Y.

A subset $G(X)$ is said to be *stable* if there are constants M and β (called *constants of stability*) such that

$$\left\| \prod_{j=1}^{k} (\lambda I - A_j)^{-1} \right\| \leq M(\lambda-\beta)^{-k}$$

for $\lambda > \beta$ and $A_1,\ldots,A_k$ elements of the subset.

(8A.21) Theorem. (*Existence Theorem*). *Let* $T > 0$, *let* $A(t) \in G(X)$ *for* $0 \leq t < T$, *and assume that*

i) $\{A(t)|0 \leq t < T\}$ *is stable, say with constant* M, β;

ii) Y *is* $A(t)$-*admissible for each* t, *and if* $A^*(t) \in G(Y)$ *is the part of* $A(t)$ *within* Y, *then* $\{A^*(t)\}$ *is stable, say with constants* M^*, β^*; *and*

iii) $Y \subset D(A(t))$ *for each* t, *and* $A^-(\cdot)$ *is continuous from* $[0,T)$ *into* $B(Y,X)$, *where* $A^-(t)$ *is the restriction of* $A(t)$ *to* Y, (*called the part of* $A(t)$ *within* Y *to* X).

Then there is a unique strongly continuous evolution

<u>system</u> $\{U(t,s)\}$ <u>in</u> X <u>with</u> X-<u>infinitesimal generator ex-tending</u> $\{A^{-}(T)\}$; <u>i.e., with infinitesimal generator</u> $\{B(t)\}$ <u>such that</u> $B(t) \supset A^{-}(t)$ <u>for each</u> t. <u>Furthermore</u>,

$$||U(t,s)||_{X,X} \leq Me^{\beta(t-s)} \quad \text{for} \quad 0 \leq s \leq t < T.$$

(8A.22) <u>Remarks</u>. a) If $A(t)$ are independent of t, the stability condition for A is the condition for the Hille-Yosida theorem (Yosida [1]).

b) Actually in Theorem (8A.21) Kato shows that the X-infinitesimal generator is precisely $\{A(t)\}$.

We can add on any bounded operator to a family $\{A(t)\}$ of generators and still get generators:

(8A.23) <u>Remark</u>. Let $\{A(t) | 0 \leq t < T\}$ satisfy the hypothesis of Theorem (8A.21), let $B(t) \in B(X,X)$ for $0 \leq t < T$, and let $B(\cdot)f$ map $[0,T)$ continuously into X for each $f \in X$. Then there is a unique strongly continuous evolution system in X with X-infinitesimal generator ex-tending $\{A(t) + B(t)\}$.

In examples, it may be difficult to verify the stabil-ity condition (i). To this end we have a useful criterion given in the following proposition. First some notation: Let $G(X,M,\beta)$ denote the generators A on X with con-stants M,β: $||(\lambda-A)^{-k}|| \leq M/(\lambda-\beta)^{k}$, $\lambda > \beta$ (corresponding to the semigroup condition $||F_t|| \leq Me^{\beta t}$). In particular, if $M = 1$ we have the generator of a <u>quasi-contractive semi-group</u>, the condition being $||(\lambda-A)^{-1}|| \leq 1/(\lambda-\beta)$, $\lambda > \beta$; or on the flow $||F_t|| \leq e^{\beta t}$. Examples of this type of semigroup are common.

(8A.24) Remark. (Trotter, Feller) For a given semigroup F_t with generator $A \in G(X,M,\beta)$ the space X can be renormed so that $||F_t|| \leq e^{\beta t}$. Indeed the new norm is $|||x||| = \sup_{t \geq 0} ||e^{-\beta t} F_t(x)||$.

One should note however that it is not always possible to renorm X so that two semigroups simultaneously become quasi-contractive.

(8A.25) Theorem. For each t, let $|| \ ||_t$ be a new norm on X equivalent to the original one and vary smoothly in t; i.e.,: satisfying

$$\frac{||x||_t}{||x||_s} \leq e^{c|t-s|}, \quad x \in X, \quad 0 \leq s, \quad t \leq T.$$

For each t, let $A(t)$ be the generator of a quasi-contractive, semigroup with constant β in the norm $||\cdot||_t$. Then $\{A(t)\}$ is stable on X with $M = e^{2cT}$, $0 \leq t \leq T$, with respect to any of the norms $|| \ ||_t$.

The proof is actually a simple verification; see Kato [3, Prop. 3.4].

There is another useful criterion for the hypotheses of (8A.21) to hold as follows:

(8A.26) Theorem. Let i) and iii) of (8A.21) hold and replace ii) by

ii") There is a family $\{S(t)\}$ of isomorphisms of Y onto X such that

$$S(t)A(t)S(t)^{-1} = A(t) + B(t),$$

$B(t) \in B(X)$ *where* $B: [0,T) \to B(X)$ *is strongly continuous. Assume* $S: [0,T) \to B(Y,X)$ *is strongly* C^1.

Then the conclusions of (8A.21) hold ((ii") $\Rightarrow$ (ii)) *and moreover, the forward differential equation holds, and the evolution system is* Y-*regular*.

Two important approximation theorems follow (see Kato [5]).

(8A.27) *Theorem. Let* $\{A_n(t)\}$ *satisfy the hypotheses of (8A.21)*, $n = 0,1,2,\ldots$ *where there are uniform stability constants in* i) *and* ii). *Assume*

$$||A_0^-(t) - A_n^-(t)||_{Y,X} \to 0 \quad \text{as} \quad n \to \infty$$

uniformly in t. *Then* $U_n(t,s) \to U_0(t,s)$ *strongly in* $B(X)$, *uniformly in* $t,s \in [0,T)$, *and* $||U_n(t,s) - U_0(t,s)||_{Y,X} \to 0$ *as* $n \to \infty$.

(8A.28) *Theorem. Let* $\{A_n(t)\}$ *satisfy the hypotheses of* (8A.26), $n = 0,1,2,\ldots$ *where the primitive constants* $M, \beta, ||S||_{\infty Y,X}, ||S^{-1}||_{\infty,Y,X}, ||B||_{\infty,X,X}, ||\dot{S}||_{\infty,Y,X}$ *can be chosen independent of* n. *Assume* $||A_0(t) - A_n(t)||_{Y,X} \to 0$ *as* $n \to \infty$ *uniformly in* t, *as in (8A.26), and in addition that* $B_n(t) \to B_0(t)$ *in* $B(X)$, $S_n(t) \to S_0(t)$ *in* $B(Y,X)$, $S_n(t) \to S_0(t)$ *in* $B(Y,X)$ *uniformly in* t. *Then,*

$$U_n(t,s) \to U_0(t,s)$$

strongly in $B(Y)$ *uniformly in* $t,s \in [0,T)$.

A Criterion for Smoothness

We now give a result which tells us when a semiflow consists of smooth mappings. The result is powerful when used in conjunction with the above linear results. The present theorem is due to Dorroh-Marsden [1], to which we refer for additional results. Before proceeding, the reader should attempt Exercise 2.9 to get a feel for the situation.

We use the following notation. X and Y are Banach spaces with Y densely and continuously embedded in X. $D \subset Y$ is open and F_t is a continuous local semiflow on D. We let $G: D \to X$ be such that F_t is a semiflow for G. For $p,q \in D$, and the line segment $\{p+r(q-p) | 0 \leq r \leq 1\} \subset D$ set

$$Z(q,p) = \int_0^1 DG(p+r(q-p))dr$$

the averaged derivative of G along the segment.

Assumptions. a) $G: D \to X$ is C^1.

b) for fixed $f \in D$ and $g \in D$ sufficiently close to f, there is a strongly continuous evolution system $\{U^g(t,s) | 0 \leq s \leq t < T_g\}$ in X whose X-infinitesimal generator is an extension of $\{Z(F_s g, F_s f): 0 \leq s < T_g\}$ (here T_g denotes a time of existence for $F_s g$ and $F_s f$).

c) $||U^g(t,s) - U^f(t,s)||_{Y,X} \to 0$ as $||g-f||_Y \to 0$ (see (8A.27)).

(8A.29) Theorem. *Under these assumptions a), b), c), $F_t: Y \to X$ is Frechet differentiable at f with $DF_t(f) = U^f(t,0)$.*

(8A.30) Remarks. The proof also shows that if

$||U^g(t,0)||_{X,X}$ is uniformly bounded, as g varies F_t will be $X \to X$ Lipschitz for g near enough to f in Y.

2. A translation argument shows $DF_{t-s}(F_s(f)) = U^f(t,s)$.

Further Assumptions.

d) $U^g(t,s)$ is Y-regular

and replace c) by

c)' $U^g(t,s)$ converges strongly in Y to $U^f(t,s)$ as g converges to f along straight line intervals (see Theorem (8A.28)).

(8A.31) Theorem. *Under* a), b), c)', d), $F_t: Y \to Y$ *is Gateaux differentiable at* f *and*

$$DF_t(f) = U^f(t,0).$$

(8A.32) Remarks. 1. The proof also shows that if $||U^g(t,0)||_{Y,Y}$ is locally bounded, then $F_t: Y \to Y$ is locally Lipschitz.

2. If (8A.26) is used, we see that, in fact, $DF_t(f)$ is locally bounded in $B(Y,Y)$ for $f \in Y$. We can iterate the use of (8A.31) to get that F_t is twice Gateaux differentiable, etc. This will imply that F_t is in fact C^∞. (Gateaux differentiability and norm continuity of the derivative implies C^1 from $f(x) - f(y) = \int_0^1 Df(x+t(x-y))(x-y)dt$; also Gateau differentiability with locally bounded derivatives implies Lipschitz continuity).

3. The derivative $DF_t(f)$ in (8A.29) extends to a bounded operator $X \to X$.

Proof of (8A.29). Let $0 < T' < T_f$. For $||g-f||_Y$

sufficiently small, we define w on $[0,T']$ by $w(s) = F_s g - F_s f$. Differentiating, and using $Z(q,p)(q-p) = G(q) - G(p)$, we have

$$w'(s) = G(F_s g) - G(F_s f) = Z(F_s g, F_s f)w(s)$$

for $0 \leq s \leq T'$. If $0 \leq s \leq t \leq T'$, then by the Corollary on p. 273.

$$(\partial/\partial s)U^g(t,s)w(s) = 0,$$

so that

$$F_t g - F_t f = U^g(t,0)(g-f)$$

for $0 \leq t \leq T'$. Thus we have the estimates

$$||F_t g - F_t f - U^f(t,0)(g-f)||_X ||g-f||_Y^{-1} \leq ||U^g(t,0) - U^f(t,0)||_{Y,X},$$

and

$$||F_t g - F_t f||_X \leq ||U^g(t,0)||_{X,X} ||g-f||_X. \qquad \square .$$

<u>Proof of (8A.31)</u>. As in the proof of (8A.29), we have

$$F_t g - F_t f = U^g(t,0)(g-f),$$

so that

$$||F_t g - F_t f||_X \leq ||U^g(t,0)||_{X,X} ||g-f||_X,$$

and

$$||F_t g - F_t f||_Y \leq ||U^g(t,0)||_{Y,Y} ||g-f||_Y.$$

This establishes the claims about Lipschitz continuity. If we let $g = f + \lambda h$, then we get

$$\lambda^{-1}[F_t(f+\lambda h) - F_t f] = U^{f+\lambda h}(t,0)h,$$

from which the differentiability claim follows directly. $\square$

Using these arguments, one can also establish Y differentiability of F_t in t and differentiability in an external parameter. We consider two such results from Dorroh-Marsden [1].

(8A.33) Corollary. *Under the hypothesis of* (8A.29) *or* (8A.31) *suppose that the evolution system* $\{U^f(t,s)\}$ *with* X-*infinitesimal generator* $\{DG(F_s)\}$ *satisfies* $U^f(t,s)X \subset Y$ *for* $0 \leq s < t < T_f$. *Then for* $f \in D$, $G(F_t f) \in Y$ *for* $0 < t < T_f$. *If* $U^f(\cdot,0)g$ *is* Y-*continuous on* $(0,T_f)$ *for each* $g \in X$, *then* $F_{(\cdot)}f$ *is continuously* Y-*differentiable on* $(0,T_f)$.

Proof. Under these hypotheses one can establish a chain rule, so that by differentiating $F_{t+s}f = F_t(F_s(f))$ in s at $s = 0$, we get

$$\begin{aligned} G(F_t f) &= DF_t(f)\cdot G(f) \\ &= U^f(t,0)\cdot G(f) \in Y \end{aligned}$$

which proves the first part. The second part follows because

$$\begin{aligned} F_t f - F_s f &= \int_s^t G(F_\tau f)\,d\tau \\ &= \int_s^t U^f(\tau,0)\cdot G(f)\,d\tau. \quad \square \end{aligned}$$

The condition that $F_t f$ be Y-differentiable for $t > 0$ is a nonlinear analogue of what one has for linear analytic semigroups (see Yosida [1]).

For the dependence on a parameter, we assume $G(f,z)$, F_t^z depend on a parameter $z \in V \subset Z$ where V is open in a

Banach space. Assume at the outset that $F_t^z(f)$ is continuous in all variables, and for each z, F_t^z is as above.

To determine differentiability of $F_t^z(f)$ in (z,f) we can use a simple suspension trick. Namely, consider the semiflow H_t on $D \times V$ defined by

$$H_t(f,z) = (F_t^z(f),z).$$

The generator is $K: D \times V \to X \times Z$,

$$K(f,z) = (G(f,z),0).$$

If (8A.29) or (8A.30) applies to H_t then we can conclude differentiability* of $F_t^z(f)$ in (f,z).

One of the key ingredients in (8A.29) is hypotheses concerning the linearized equations. Here

$$DK(f,z)(g,w) = (D_1G(f,z)\cdot g + D_2G(f,z)\cdot w,0)$$

so we would be required to solve, according to (8A.29) or (8A.31), the system

$$\frac{dw}{dt} = 0 \quad \text{i.e.} \quad w = \text{constant}$$

$$\frac{dg(t)}{dt} = D_1G(F_t^z(f),z)\cdot g(t) + D_2G(F_t^z(f),z)\cdot w$$

(similarly for systems involving the averaged generators Z). This is a linear system in g with an inhomogeneous term $D_2G(F_t^z(f),z)\cdot w$. The solution can be written down in terms of the evolution system for $D_1G(F_t^z(f),z)$ via Duhamel's formula in the usual way. For systems of this type there are theorems

---

*A more direct analysis, obtaining refined results, is given by Dorroh-Marsden [1].

available to guarantee that we have an evolution system and to study its properties. Note, for example, Theorem 7.2 of Kato [4]. In this way, one can check the hypotheses of (8A.29) or (8A.31) for H_t. We would conclude, respectively, differentiability of $F_t^{(\cdot)}(\cdot)$ from $D \times V$ to $X \times V$ and (under the stronger conditions of (8A.31)), from $D \times V$ to $Y \times V$.

(For holomorphic semigroups, smooth dependence on a parameter can be analyzed directly, as in Kato [3], p. 487).

(8A.34) Remark and Application. In Aronson-Thames [1] the following system is studied:

$$\begin{aligned}
&u_{xx} - q^2 u = u_t, \quad v_{xx} - q^2 v = v_t \quad \text{in} \quad (0,1) \times R^+ \\
&\left.\begin{aligned}
u_x(1,t) &= v_x(0,t) = 0 \\
u_x(0,t) &= -pq(f \circ v)(0,t), \\
v_x(1,t) &= pq\{1-(f \circ u)(1,t)\}
\end{aligned}\right\} \quad t \geq 0
\end{aligned}$$

Here p and q are positive parameters and $f(u) = u^2/(1+u^2)$. This system is related to enzyme diffusion in biological systems. They show that the eigenvalue conditions of the Hopf theorem are met. In Dorroh-Marsden [1] it is shown, using methods described above, that the semiflow of this system is smooth. It follows at once that the Hopf theorem is valid and hence proves the existence of stable periodic solutions for this system for supercritical values of the parameters.

These equations are usually called the Glass-Kauffman equations; cf. Glass-Kauffman [1]. Recent work of the discrete analogue has also been done by Hsü.

SECTION 9

BIFURCATIONS IN FLUID DYNAMICS AND THE PROBLEM OF TURBULENCE

This section shows how the results of Sections 8 and 8A can be used to establish the bifurcation theorem for the Navier-Stokes equations. Alternatively, although conceptually harder (in our opinion) methods are described in Sections 9A, 9B and Sattinger [5] and Joseph and Sattinger [1].

The new proof of the bifurcation theorems using the center manifold theory as given in Section 8 allow one to deduce the results very simply for the Navier-Stokes equations with a minimum of technical difficulties. This includes all types of bifurcations, including the bifurcation to invariant tori as in Section 6 or directly as in Jost and Zehnder [1]. All we need to do is verify that the semiflow of the Navier-Stokes equations is smooth (in the sense of Section 8A); the rest is then automatic since the center manifold theorem immediately reduces us to the finite dimensional case (see

Section 8 for details). We note that already in Ruelle-Takens [1] there is a simple proof of the now classical results of Velte [1] on stationary bifurcations in the flow between rotating cylinders from Couette flow to Taylor cells.

The first part of this section therefore is devoted to proving that the semiflow of the Navier-Stokes equations is smooth. We use the technique of Dorroh-Marsden [1] (see Section 8A) to do this.

This guarantees then, that the same results as given in the finite dimensional case in earlier sections, including the stability calculations hold without change.

The second part of the section briefly describes the Ruelle-Takens picture of turbulence. This picture is still conjectural, but seems to be gaining increased acceptance as time goes on, at least for describing certain types of turbulence. The relationship with the global regularity (or "all time") problem in fluid mechanics is briefly discussed.

Statement of the Smoothness Theorem.

Before writing down the smoothness theorem, let us recall the equations we are dealing with. For homogeneous incompressible viscous fluids, the classical Navier-Stokes equations are, as in Section 1,

$$(\mathrm{NS})\quad \begin{cases} \dfrac{\partial v}{\partial t} + (v\cdot\nabla)v - \nu\Delta v = -\operatorname{grad} p + b_t, \quad b_t = \text{external force} \\ \operatorname{div} v = 0 \\ v = 0 \ \text{ on } \ \partial M \quad \text{(or prescribed on } M\text{, possibly depending on a parameter } \mu) \end{cases}$$

Here M is a compact Riemannian manifold with smooth boundary

∂M, usually an open set in $\mathbb{R}^3$.

The <u>Euler equations</u> are obtained by supposing $\nu = 0$ and changing the boundary condition to $v || \partial M$:

$$\text{(E)} \qquad \begin{cases} \dfrac{\partial v}{\partial t} + (v \cdot \nabla) v = -\text{grad } p + b_t \\ \text{div } v = 0 \\ v || \partial M \end{cases}$$

The pressure $p(t,x)$ is to be determined from the incompressibility condition in these equations.

The Euler equations are a <u>singular limit</u> of the Navier-Stokes equations. Taking the limit $\nu \to 0$ is very subtle and is the subject of much recent work. The sudden disappearance of the highest order term and the associated sudden change in boundary conditions is the source of the difficulties and is the reason why boundary layer theory and turbulence theory are so difficult. This point will be remarked on later.

Note that Euler equations are reversible in the sense that if we can solve them for all sets of initial data and for $t \geq 0$, then we can also solve them for $t < 0$. This is because if v_t, $t \geq 0$ is a solution, then so is $w_t = -v_{-t}$, $t < 0$.

For $s \geq 0$ an integer and $1 < p < \infty$, let $W^{s,p}$ denote the Sobolev space of functions (or vector functions) on M whose derivatives up to order s are in L_p; another way of describing $W^{s,p}$ is to complete the C^∞ functions f in the norm

$$||f||_{s,p} = \sum_{0 \leq \alpha \leq s} ||D^\alpha f||_{L_p}$$

where $D^\alpha f$ is the αth total derivative of f. Details on Sobolev spaces can be found in Friedman [1].

We point out that in the non-compact case one must deal seriously with the asymptotic conditions and many of the results we discuss are not known in that case (see Cantor [1], and McCracken [2] however).

The following result is a special case of a general result proved in Morrey [1]. For a direct proof in this case, see Bourguignon and Brezis [1].

(9.1) Lemma. Hodge Decomposition): *Let* M *be as above. Let* X *be a* $W^{s,p}$ *vector field on* M, $s \geq 0$, $p > 1$. *Then* X *has a unique decomposition as follows:*

$$X = Y + \text{grad } p$$

where $\text{div } Y = 0$ *and* $Y||_{\partial M}$. *It is also true that* $Y \in W^{s,p}$ *and* $p \in W^{s+1,p}$.

Let $\tilde{W}^{s,p} = \{$vector fields X on $M | X \in W^{s,p}$, $\text{div } X = 0$ and $X||_{\partial M}\}$. By the Hodge theorem, there is a map $P: W^{s,p} \to \tilde{W}^{s,p}$ via $X \mapsto Y$. The problem of solving the Euler equations now becomes that of finding $v_t \in \tilde{W}^{s+1,p}$ such that

$$\frac{dv_t}{dt} + P((v_t \cdot \nabla)v_t) = 0 \tag{E}$$

(plus initial data). If $s > \frac{n}{p}$, the product of two $W^{s,p}$ functions is $W^{s,p}$ so $(v \cdot \nabla)v$ is in $W^{s,p}$ if $v \in W^{s+1,p}$. This kind of equation is thus an *evolution equation* on $\tilde{W}^{s,p}$ as in Section 8A.

Let

$$\tilde{W}_0^{s,p} = \{\text{vector fields } v \text{ on } M \mid v \text{ is of class } \tilde{W}^{s,p},\ \text{div } v = 0 \text{ and } v = 0 \text{ on } \partial M\}.$$

If $s = 0$ this actually makes sense and the space is written J_p (see Ladyzhenskaya [1]).

The Navier-Stokes equations can be written: find $v_t \in \tilde{W}_0^{s,p}$ such that

$$\text{(NS)} \qquad \frac{dv_t}{dt} - \nu P \Delta v_t + P(v_t \cdot \nabla) v_t = 0$$

again an evolution equation in $\tilde{W}_0^{s,p}$. In the terminology of Section 8A, the Banach space X here is $\tilde{W}_0^{0,p} = J_p$ and $Y = \tilde{W}_0^{2,p}$. The bifurcation parameter is often $\mu = 1/\nu$, the Reynolds number.

The case $p \neq 2$ is quite difficult and won't be dealt with here, although it is very important. If $p = 2$ one generally writes

$$\tilde{H}^s = \tilde{W}^{s,2}, \quad \tilde{H}_0^s = \tilde{W}_0^{s,2} \quad \text{etc.}$$

(9.2) Theorem. The Navier-Stokes equations in dimension 2 or 3 define a smooth local semiflow on $\tilde{H}_0^2 \subset \tilde{H}^0 \equiv J$.

This semiflow satisfies conditions 8.1, 8.2, and the smoothness in 8.3 of Section 8, so the Hopf theorems apply. (The rest of the hypotheses in 8.3 and 8.4 depend on the particular problem at hand and must be verified by calculation.)

In other words, the technical difficulties related to the fact that we have partial rather than ordinary differential equations are automatically taken care of.

(9.3) Remarks. 1. If the boundaries are moving and

speed is part of the bifurcation parameter μ, the same result holds by a similar proof. This occurs in, for instance the Taylor problem.

2. The above theorem is implicit in the works of many authors. For example, D. Henry has informed us that he has obtained it in the context of Kato-Fujita. It has been proved by many authors in dimension 2 (e.g. Prodi). The first explicit demonstration we have seen is that of Iooss [3,5].

3. For the case $p \neq 2$ see McCracken [2].

4. This smoothness for the Navier-Stokes semiflow is probably false for the Euler equations on $\tilde{H}^s$ (see Kato [6], Ebin-Marsden [1]). Thus it depends crucially on the dissipative term. However, miraculously, the flow of the Euler equations <u>is smooth</u> if one uses Lagrangian coordinates (Ebin-Marsden [1]).

We could use Lagrangian coordinates to prove our results for the Navier-Stokes equations as well, but it is simpler to use the present method.

Before proving smoothness we need a local existence theorem. Since this is readily available in the literature (Ladyzhenskaya [1]), we shall just sketch the method from a different point of view. (Cf. Soboleoskii [1].)

<u>Local Existence Theory</u>.

The basic method one can use derives from the use of integral equations as in the Picard method for ordinary differential equations. For partial differential equations this has been exploited by Segal [1], Kato-Jujita [1], Iooss [3,6],

Sattinger [2], etc. (See Carroll [1] for general background.)

The following result is a formulation of Weissler [1]. (See Section 9A for a discussion of Iooss' setup.)

First some notation: $\mathbb{E}_0$, $\mathbb{E}_1$, $\mathbb{E}_2$ will be three Banach spaces with norms $||\cdot||_0$, $||\cdot||_1$, $||\cdot||_2$, with $\mathbb{E}_2 \subset \mathbb{E}_1 \subset \mathbb{E}_0$, with the inclusions dense and continuous. (Some of the spaces may be equal.) Let e^{tA} be a C^0 linear semigroup on $\mathbb{E}_0$ which restricts to a contraction semigroup on $\mathbb{E}_2$. Assume, for $t > 0$, $e^{tA}: \mathbb{E}_1 \to \mathbb{E}_2$ is a bounded linear map and let its norm be denoted $\mu(t)$. Our first assumption is:

A1) For $T > 0$ assume $\int_0^T \mu(\tau)d\tau < \infty$.

For the Navier-Stokes equations, $A = \nu P\Delta$, and we can choose either

(i) $\mathbb{E}_0 = J_2$, $\mathbb{E}_1 = \tilde{H}^1$ and $\mathbb{E}_2 = \tilde{H}^2_0$

or (ii) $\mathbb{E}_0 = \mathbb{E}_1 = \tilde{H}^{-1/2}$ = completion of J with the norm $||(-\nu P\Delta)^{-1/4}u||_{L_2}$, $\mathbb{E}_2 = \tilde{H}^1$ = domain of $(-\nu P\Delta)^{1/2}$.

The case (ii) is that of Kato-Fujita [1].

The fact that A1) holds is due to the fact that A is a negative self adjoint operator on J with domain $\tilde{H}^2_0$ (Ladyzhenskaya [1]); generates an analytic semigroup, so the norm of e^{tA} from $\tilde{H}^2_0$ to J_2 is $\leq C/t$. (See Yosida [1].) However we have the Sobolev inequality (derived most easily from the general Sobolev-Nirenberg-Gagliardo inequality from Nirenberg [1]:

$$||D^j f||_{L_p} \leq C||D^m f||^a_{L^r}||f||^{1-a}_{L^q}$$

where

$$\frac{1}{p} = \frac{j}{n} + a\left(\frac{1}{r} - \frac{m}{n}\right) + (1-a)\frac{1}{q}, \quad \frac{j}{m} \le a \le 1$$

(if $m - j - \frac{n}{r}$ is an integer ≥ 1, $a = 1$ is not allowed)},

that

$$||f||_{H^1} \le ||f||_{H^2}^{1/2} ||f||_{L^2}^{1/2}$$

so we can choose $\mu(t) \le C/\sqrt{t}$, so A1) holds. Similarly in case (ii) one finds $\mu(t) = C/t^{3/4}$ (Kato-Fujita [1]).

As for the nonlinear terms, assume

A2) $J_t\colon \mathbb{E}_2 \to \mathbb{E}_1$ is a semi-Lipschitz map (i.e., Lipschitz on bounded sets), locally uniformly in t with $J_t(\phi)$ continuous in (t,ϕ). We can suppose $J_t(0) = 0$ for simplicity.

Consider the "formal" differential equation

$$\frac{d\phi}{dt} = A\phi + J_t(\phi) \tag{9.1}$$

in integral equation form (see e.g., Segal [1])

$$W(t,t_0)\phi = e^{(t-t_0)A}\phi + \int_{t_0}^{t} e^{(t-\tau)A} J_\tau(W(\tau,t_0)\phi)\,d\tau \tag{9.2}$$

where $t > t_0$. (Adding an inhomogeneous term causes no real difficulties.)

(9.4) <u>Theorem</u>. <u>Under assumptions</u> A1) <u>and</u> A2), <u>the equation</u> (9.2) <u>defines a unique local semiflow</u> (<u>i.e.</u>, <u>evolution system</u>) <u>on</u> $\mathbb{E}_2$ (<u>in the sense of Section 8A</u>) <u>with</u> $W(t,t_0)\colon \mathbb{E}_2 \to \mathbb{E}_2$ <u>locally uniformly Lipschitz</u>, <u>varying continuously in</u> t.

<u>Proof</u>. The proof proceeds by the usual contraction

mapping argument as for ordinary differential equations (see Lang [1]): Pick $0 < \alpha_0 < \alpha$, let $K_\alpha(t)$ be the Lipschitz constant of J_t from $\mathbb{E}_2$ to $\mathbb{E}_1$ on B_α the α ball about and pick T such that

$$\left(\int_0^{T-t_0} \mu(\tau)d\tau\right)\left(\sup_{\tau\in[t_0,T]} K_\alpha(\tau)\right) \le 1 - \frac{\alpha_0}{\alpha}. \qquad (9.3)$$

Now choose $\phi \in B_{\alpha_0}$ and let M be the complete metric space of C^0 maps Φ of $[0,T]$ to $\mathbf{E}_2$ with $\Phi(t_0) = \phi$, $\Phi(t) \in B_\alpha$ and metric

$$d(\Phi,\Psi) = \sup_{t\in[t_0,T]} ||\Phi(t)-\Psi(t)||_2$$

Define $\mathscr{F}: M \to M$ by

$$\mathscr{F}\Phi(t) = e^{(t-t_0)A}\phi + \int_{t_0}^{t} e^{(t-\tau)A}J_\tau(\Phi(\tau))d\tau .$$

From the definitions and (9.3) we have two key estimates: first

$$\begin{aligned} ||\mathscr{F}\Phi(t)||_2 &\le \alpha_0 + \int_{t_0}^{T} \mu(t-\tau)K_\alpha(\tau)\cdot\alpha\, d\tau \\ &\le \alpha_0 + \alpha\left(1 - \frac{\alpha_0}{\alpha}\right) = \alpha \end{aligned} \qquad (9.4)$$

(remember $J_\tau(0) = 0$ here), which shows $\mathscr{F}$ maps M to M and, in the same way

$$d(\mathscr{F}\Phi, \mathscr{F}\Psi) \le \left(1 - \frac{\alpha_0}{\alpha}\right)d(\Phi,\Psi) \qquad (9.5)$$

which shows $\mathscr{F}$ is a contraction.

The result now follows easily. □

(9.5) Exercises. 1. Show that $W(t,t_0)$ has $\mathbf{E}_2$ Lipschitz constant given by α/α_0. Verify that $W(t,s)W(s,t_0) = W(t,t_0)$.

2. If ϕ_t is a maximal solution of (9.2) on $[0,T)$ and $T < \infty$, show $\limsup_{t \to T} ||\phi_t||_2 = \infty$; i.e., verify the continuation assumption 8.2.

3. Use the Sobolev inequalities to verify that $J_t(u) = P((u \cdot \nabla)u)$ satisfies the hypotheses in case (i) above. For case (ii), see Kato-Fujita [1].

Next we want to see that we actually have a solution of the differential equation. Make

A3) Assume that the domain of A as an operator in $\mathbb{E}_0$ is exactly $\mathbb{E}_2$.

(9.6) Theorem. *If A1), A2) and A3) hold, then any solution of (9.2) solves (9.1) as an evolution system in $\mathbb{E}_0$ with domain $D = \mathbb{E}_2$; (in the terminology of Section 8A, $W(t,t_0)$ is the (time dependent) local flow of the operator $X(\phi) = A(\phi) + J_t(\phi)$, mapping $\mathbb{E}_2$ to $\mathbb{E}_0$). Solutions of (9.2) in $\mathbb{E}_2$ are unique.*

Proof. Let $\phi \in \mathbb{E}_2$ and $\phi(t) = W(t,t_0)\phi \in \mathbb{E}_2$ be the solution of (9.2), so taking $t_0 = 0$ for simplicity,

$$\phi(t) = e^{tA}\phi + \int_0^t e^{(t-\tau)A} J_\tau(\phi(\tau))d\tau.$$

It is easy to verify that

$$\begin{aligned}\frac{1}{h}\{\phi(t+h)-\phi(t)\} &= \frac{1}{h}\{e^{hA}\phi(t)-\phi(t)\} - J_t(\phi(t)) \\ &+ \frac{1}{h}\int_t^{t+h}\{e^{(t+h-\tau)A}J_\tau(\phi(\tau))-J_t(\phi(t))\}d\tau\end{aligned} \tag{9.6}$$

writing

$$e^{(t+h-\tau)A}J_\tau(\phi(\tau)) - J_t(\phi(t))$$

$$= e^{(t+h-\tau)A}[J_\tau(\phi(\tau))-J_t(\phi(t))] + e^{t+h-\tau}J_t(\phi(t))-J_t(\phi(t))$$

one sees that the last term of (9.6) $\to 0$ as $h \to 0$ in $\mathbb{E}_1$ and hence in $\mathbb{E}_0$. The first term of (9.6) tend to $A(\phi(t)) - J_t(\phi(t))$ in $\mathbb{E}_0$ as $h \to 0$ since $\phi(t) \in \mathbb{E}_2$, the domain of A. □

Thus we can conclude that the Navier-Stokes equations define a local semiflow on $\tilde{H}_0^2$ and that this semiflow extends to a local semiflow on $\tilde{H}^1$ (via the integral equation).

<u>Smoothness</u>.

(9.7) <u>Theorem</u>. <u>Let</u> A1), A2) <u>and</u> A3) <u>hold and assume</u>

A4) $J_t\colon \mathbb{E}_2 \to \mathbb{E}_1$ <u>is</u> C^∞ <u>with derivatives depending continuously on</u> t.

<u>Then the semiflow defined by equations</u> (9.1) <u>on</u> $\mathbb{E}_2$ <u>is a</u> C^∞ <u>semiflow on</u> $\mathbb{E}_2$; <u>i.e.</u>, <u>each</u> $W(t,t_0)$ <u>is</u> C^∞ <u>with derivatives varying continuously in</u> t <u>in the strong topology</u> (<u>see Section 8</u>).

<u>Proof</u>. We verify the hypotheses of (8A.31). Here we take $X = \mathbb{E}_0$ and $Y = \mathbb{E}_2$, with $D = Y$. Certainly a) holds by hypothesis. Since $Z(\phi_1,\phi_2)$ is the same type of operator as considered above, 9.4 shows b) holds. c') holds by the E_2 Lipschitzness of $W(t,t_0)$ proven in (9.4) and d) is clear.

Hence $W(t,t_0)\colon \mathbb{E}_2 \to \mathbb{E}_2$ is Gateaux differentiable.

The procedure can be iterated. The same argument applies to the semiflow

$$\tilde{W}(t,t_0)(\phi,\psi) = (W(t,t_0)\phi,\ DW(t,t_0)\phi\cdot\psi)$$

on $\mathbb{E}_2 \to \mathbb{E}_2$.

Hence W is C^1 (see (8A.32)), and by induction is C^∞. □

In the context of equation (9.1) the full power of the machinery in Section 8A; in particular the delicate results on time dependent evolution equations are not needed. One can in fact directly prove (9.7) by the same method as (8A.31). However it seems desirable to derive these types of results from a unified point of view.

(9.8) Problem. Assume, in addition that

A5) A generates an analytic semigroup.

Show that for $t > 0$ and $\phi \in \mathbb{E}_1$, $\phi(t)$ lies in the domain of every power of A and that $\phi(t)$ is a C^∞ function of t for $t > 0$. (Hint. Use (8A.33)). Also establish smoothness in ν if A is replaced by νA (see remarks following (8A.33)).

A more careful analysis actually shows that $W(t,t_0)$ are C^∞ maps on $\tilde{H}^1$ in the context of the Navier-Stokes equation (i.e., without assuming A3)). See Weissler [1].

Thus all the requisite smoothness is established for the Navier-Stokes equations, so the proof of (9.2) and hence the bifurcation theorems for those equations is established.

The Problem of Turbulence.

We have already seen how bifurcations can lead from stable fixed points to stable periodic orbits and then to stable 2-tori. Similarly we can go on to higher dimensional tori. Ruelle and Takens [1] have argued that in this or other situations, complicated ("strange") attractors can be expected and that this lies at the roots of the explanation of turbulence.

The particular case where tori of increasing dimension form, the model is a technical improvement over the idea of E. Hopf [4] wherein turbulence results from a loss of stability through successive branching. It seems however that strange attractors may form in other cases too, such as in the Lorenz equations (see Section 4B) [Strictly speaking, it has only a "strange" invariant set]. This is perfectly consistent with the general Ruelle-Takens picture, as are the closely related "snap through" ideas of Joseph and Sattinger [1].

In the branching process, stable solutions become unstable, as the Reynolds number is increased. Hence turbulence is supposed to be a necessary consequence of the equations and in fact of the "generic case" and just represents a complicated solution. For example in Couette flow as one increases the angular velocity Ω_1 of the inner cylinder one finds a shift from laminar flow to Taylor cells or related patterns at some bifurcation value of Ω_1. Eventually turbulence sets in. In this scheme, as has been realized for a long time, one first looks for a stability theorem and for when stability fails (Hopf [2], Chandresekar [1], Lin [1] etc.).

For example, if one stayed closed enough to laminar flow, one would expect the flow to remain approximately laminar. Serrin [2] has a theorem of this sort which we present as an illustration:

(9.9) Stability Theorem. *Let* $D \subset \mathbb{R}^3$ *be a bounded domain and suppose the flow* v_t^ν *is prescribed on* ∂D *(this corresponds to having a moving boundary, as in Couette flow). Let* $V = \max_{\substack{x \in D \\ t \geq 0}} \|v_t^\nu(x)\|$, d = *diameter of* D *and* ν *equal the viscosity. Then if the Reynolds number* $R = (Vd/\nu) \leq 5.71$, v_t^ν *is universally* L^2 *stable among solutions of the Navier-Stokes equations.*

Universally L^2 stable means that if $\bar{v}_t^\nu$ is *any* other solution to the equations and with the same boundary conditions, then the L^2 norm (or energy) of $\bar{v}_t^\nu - v_t^\nu$ goes to zero as $t \to 0$.

The proof is really very simple and we recommend reading Serrin [1,2] for the argument. In fact one has local stability in stronger topologies using Theorem 1.4 of Section 2A and the ideas of Section 8.

Chandresekar [1], Serrin [2], and Velte [3] have analyzed criteria of this sort in some detail for Couette flow.

As a special case, we recover something that we expect. Namely if $v_t^\nu = 0$ on ∂M is any solution for $\nu \to 0$ then $v_t^\nu \to 0$ as $t \to \infty$ in L^2 norm, since the zero solution is universally stable.

A traditional definition (as in Hopf [2], Landau-

Lifschitz [1]) says that turbulence develops when the vector field v_t can be described as $v_t(w_1, \ldots, w_n) = f(tw_1, \ldots, tw_n)$ where f is a quasi-periodic function, i.e., f is periodic in each coordinate, but the periods are not rationally related. For example, if the orbits of the v_t on the tori given by the Hopf theorem can be described by spirals with irrationally related angles, then v_t would such a flow.

Considering the above example a bit further, it should be clear there are many orbits that the v_t could follow which are qualitatively like the quasi-periodic ones but which fail themselves to be quasi-periodic. In fact a small neighborhood of a quasi-periodic function may fail to contain many other such functions. One might desire the functions describing turbulence to contain most functions and not only a sparse subset. More precisely, say a subset U of a topological space S is <u>generic</u> if it is a Baire set (i.e., the countable intersection of open dense subsets). It seems reasonable to expect that the functions describing turbulence should be generic, since turbulence is a common phenomena and the equations of flow are never exact. Thus we would want a theory of turbulence that would not be destroyed by adding on small perturbations to the equations of motion.

The above sort of reasoning lead Ruelle-Takens [1] to point out that since quasi-periodic functions are not generic, it is unlikely they "really" describe turbulence. In its place, they propose the use of "strange attractors." These exhibit much of the qualitative behavior one would expect from "turbulent" solutions to the Navier-Stokes equations and they are stable under perturbations of the equations; i.e., are "structurally stable".

For an example of a strange attractor, see Smale [1]. Usually strange attractors look like Cantor sets $\times$ manifolds, at least locally.

Ruelle-Takens [1] have shown that if we define a strange attractor A to be an attractor which is neither a closed orbit or a point, and disregarding non-generic possibilities such as a figure 8 then there are strange attractors on T^4 in the sense that a whole open neighborhood of vector fields has a strange attractor as well.

If the attracting set of the flow, in the space of vector fields which is generated by Navier-Stokes equations is strange, then a solution attracted to this set will clearly behave in a complicated, turbulent manner. While the whole set is stable, individual points in it are not. Thus (see Figure 9.1) an attracted orbit is constantly near unstable (nearly periodic) solutions and gets shifted about the attractor in an aimless manner. Thus we have the following reasonable definition of turbulence as proposed by Ruelle-Takens:

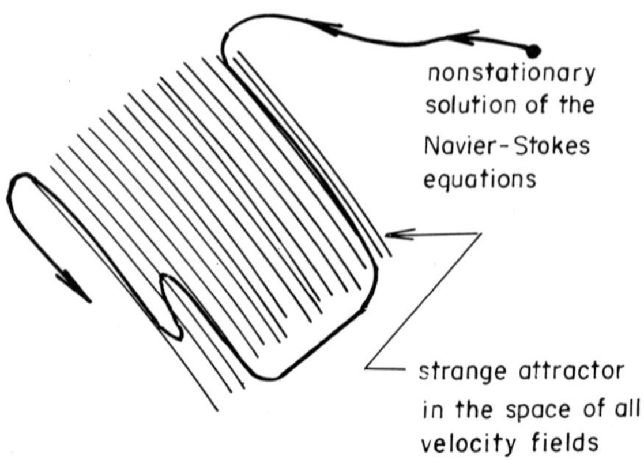

Figure 9.1

"...the motion of a fluid system is turbulent when this motion is described by an integral curve of a vector field X_μ which tends to a set A, and A is neither empty nor a fixed point nor a closed orbit."

One way that turbulent motion can occur on one of the tori T^k that occurs in the Hopf bifurcation. This takes place after a finite number of successive bifurcations have occurred. However as S. Smale and C. Simon pointed out to us, there may be an infinite number of other qualitative changes which occur during this onset of turbulence (such as stable and unstable manifolds intersecting in various ways etc.). However, it seems that turbulence can occur in other ways too. For example, in Example 4B.9 (the Lorenz equations), the Hopf bifurcation is subcritical and the strange attractor may suddenly appear as μ crosses the critical value without an oscillation developing first. See Section 12 for a description of the attractor which appears. See also McLaughlin and Martin [1,2], Guckenheimer and Yorke [1] and Lanford [2].

In summary then, this view of turbulence may be phrased as follows. Our solutions for small μ (= Reynolds number in many fluid problems) are stable and as μ increases, these solutions become unstable at certain critical values of μ and the solution falls to a more complicated stable solution; eventually, after a certain (finite) number of such bifurcations, the solution falls to a strange attractor (in the space of all time dependent solutions to the problem). Such a solution, which is wandering close to a strange attractor, is called turbulent.

The fall to a strange attractor may occur after a Hopf bifurcation to an oscillatory solution and then to invariant tori, or may appear by some other mechanism, such as in the Lorenz equations as explained above ("snap through turbulence").

Leray [3] has argued that the Navier-Stokes equations might break down and the solutions fail to be smooth when turbulence ensues. This idea was amplified when Hopf [3] in 1950 proved global existence (in time) of weak solutions to the equations, but not uniqueness. It was speculated that turbulence occurs when strong changes to weak and uniqueness is lost. However it is still unknown whether or not this really can happen (cf. Ladyzhenskaya [1,2].)

The Ruelle-Takens and Leray pictures are in conflict. Indeed, if strange attractors are the explanation, their attractiveness implies that solutions remain smooth for all t. Indeed, we know from our work on the Hopf bifucation that near the stable closed orbit solutions are defined and remain smooth and unique for all $t \geq 0$ (see Section 8 and also Sattinger [2]). This is already in the range of interesting Reynolds numbers where global smoothness is not implied by the classical estimates.

It is known that in two dimensions the solutions of the Euler and Navier-Stokes equations are global in t and remain smooth. In three dimensions it is unknown and is called the "global regularity" or "all time" problem.

Recent numberical evidence (see Temam et. al. [1]) suggests that the answer is negative for the Euler equations.

Theoretical investigations, including analysis of the

spectra have been inconclusive for the Navier-Stokes equations (see Marsden-Ebin-Fischer [1] and articles by Frisch and others in Temam et.al. [1]).

We wish to make two points in the way of conjectures:

1. In the Ruelle-Takens picture, global regularity for all initial data is not an a priori necessity; the basins of the attractors will determine which solutions are regular and will guarantee regularity for turbulent solutions (which is what most people now believe is the case).

2. Global regularity, if true in general, will probably never be proved by making estimates on the equations. One needs to examine in much more depth the attracting sets in the infinite dimensional dynamical system of the Navier-Stokes equations and to obtain the a priori estimates this way.

Two Major Open Problems:

(i) identify a strange attractor in a specific flow of the Navier-Stokes equation (e.g, pipe flow, flow behind a cylinder, etc.).

(ii) link up the ergodic theory on the strange attractor, (Bowen-Ruelle [1]) with the statistical theory of turbulence (the usual reference is Batchellor [1]; however, the theory is far from understood; some of Hopf's ideas [5] have been recently developed in work of Chorin, Foias and others).

SECTION 9A

ON A PAPER OF G. IOOSS

BY G. CHILDS

This paper [3] proves the existence of the Hopf bifurcation to a periodic solution from a stationary solution in certain problems of fluid dynamics. The results are similar to those already described. For instance, in the subcritical case, the periodic solution is shown to be unstable in the sense of Lyapunov when the real bifurcation parameter (Reynold's number) is less than the critical value where the bifurcation takes place; it is shown to be (exponentially) stable if this value is greater than the critical value in the supercritical case.

Iooss, in contrast to the main body of these notes, makes use of a linear space approach for almost all of what he does. Specifically, his periodic solution is a continuous function to elements of a Sobolev space on a fundamental domain Ω in $\mathbb{R}^3$. However, the implicit function theorem is extensively used. The three main theorems of the paper will be stated and the proofs will be briefly outlined to illustrate this method.

First, we formulate a statement of the problem. Let I be a closed interval of the real line. Let $V(I)$ be a neighborhood in $\mathbb{C}$ of this interval. For each $\lambda \in V(I)$, L_λ is a closed, linear operator on a Hilbert space H. The family $\{L_\lambda\}$ is holomorphic of type (A) in $V(I)$ (see Kato [3], p. 375). Also, each L_λ is m-sectorial with vertex-γ_λ. Finally, L_λ has a compact resolvent in H. Let $\mathcal{D}$ be the common domain of the L_λ. Assume K is a Hilbert space such that $\mathcal{D} \subseteq K \subseteq H$ with continuous injections and such that $\forall U \in K$, $||I(t)U||_{\mathcal{D}} \le ke^{\gamma_\lambda t}(1+t^{-\alpha})||U||_K$, $0 \le \alpha < 1$ where $I_\lambda(t)$ is the holomorphic semi-group generated by $-L_\lambda$. Let M be a continuous bilinear form: $\mathcal{D} \times \mathcal{D} \to K$. Now we can state the problem:

$$\begin{cases} \dfrac{\partial U}{\partial t} + L_\lambda U - M(U,U) = 0 \\ U \in C^0(0,\infty;\mathcal{D}) \cap C^1(0,\infty;H) \\ U(0) = U_0 \in \mathcal{D},\ U(0) = U(T) = U(2T) = \ldots \text{ for some } T > 0 \end{cases} \tag{9A.1}$$

Iooss shows that the equation of perturbation (from a stationary solution) for some Navier-Stokes configurations is of the above form (see also Iooss [5] for more details). In order to find a solution of (9A.1), it is necessary to make some additional hypotheses. Let $\xi_0(\lambda) = \sup_{\xi \in \sigma(-L_\lambda)} \{\mathrm{Re}\ \xi\}$. Then:

(H.1) $\exists\ \lambda_c \in \mathbb{R}$, <u>a left hand neighborhood</u> $V^-(\lambda_c)$ <u>and a right hand neighborhood</u> $V^+(\lambda_c)$ <u>such that</u> $\xi_0(\lambda_c) = 0$, $\lambda \in V^-(\lambda_c) - \{\lambda_c\} \Rightarrow \xi_0(\lambda) < 0$, $\lambda \in V^+(\lambda_c) - \{\lambda_c\} \Rightarrow \xi_0(\lambda) > 0$.

(H.2) <u>The operator</u> L_{λ_c} <u>admits as proper values pure imaginary numbers</u> $\zeta_0 = i\eta_0$ <u>and</u> $\overline{\zeta}_0$. <u>Also, these proper values are simple</u>. For $\lambda \in V(\lambda_c)$ there exist two analytic functions ζ_1 and $\overline{\zeta}_1 \in \sigma(-L_\lambda)$ such that $\zeta_1(\lambda_c) = \zeta_0$. The spectrum $\sigma(L_\lambda)$ separates into $\{-\zeta_1\} \cup \{-\overline{\zeta}_1\} \cup \tilde{\sigma}(L_\lambda)$. This separation gives a decomposition of H into invariant subspaces:

$$\forall\, U \in H, \quad U = X + Y, \quad X = E_\lambda U, \quad Y = P_\lambda U;$$

$E_\lambda = E(-\zeta_1) + E(-\overline{\zeta}_1)$, $(L_\lambda + \zeta_1)U_1(\lambda) = 0$, $(L_\lambda^* + \overline{\zeta}_1)W_1(\lambda) = 0$, $(U_1(\lambda), W_1(\lambda))_H = 1$, $(U_1(\lambda), W(0))_H = 1$. The eigenvectors $U_1(\lambda)$, $\overline{U_1(\lambda)}$ are a basis for $E_\lambda H$. For $\lambda \in V(\lambda_c)$, $L_\lambda U = L_{\lambda_c} U + \sum_{n=1}^{\infty} (\lambda-\lambda_c)^n L^{(n)} U$; $U_1(\lambda) = U^{(0)} + (\lambda-\lambda_c)U^{(1)} + \ldots$; $W_1(\lambda) = W^{(0)} + (\lambda-\lambda_c)W^{(1)} + \ldots$; $\zeta_1(\lambda) = \zeta_0 + (\lambda-\lambda_c)\zeta^{(1)} + \ldots$. In particular, we have $\zeta^{(1)} = -(L^{(1)}U^{(0)}, W^{(0)})_H$, $L_{\lambda_c} + \zeta_0 U^{(0)} = 0$, $L_{\lambda_c}^* + \overline{\zeta}_0 W^{(0)} = 0$, $(U^{(0)}, W^{(0)})_H = 1$. Now, we can make the hypothesis:

(H.3) $\mathrm{Re}(L^{(1)}U^{(0)}, W^{(0)})_H \neq 0$. By (H.1) this implies $\mathrm{Re}\,\zeta^{(1)} > 0$. These hypotheses are just the standard ones for the existence of the Hopf bifurcation. For the statement of the theorem we also need to know that $\gamma_0 = \gamma_{0_r} + i\gamma_{0_i} =$

$$-(M^{(0)}[U^{(0)}, L_{\lambda_c}^{-1} M^{(0)}(U^{(0)}, \overline{U^{(0)}})] +$$

$$M^{(0)}[\overline{U^{(0)}}, (L_{\lambda_c} + 2i\eta_0 1)^{-1} M(U^{(0)}, U^{(0)})],\ W^{(0)})_H \quad \text{where}$$

$$M^{(0)}(U,V) = M(U,V) + M(V,U).$$

We now state Iooss'

<u>Theorem 2</u>. <u>If the hypotheses</u> (H.1), (H.2), (H.3), <u>and</u>

$\gamma_{0_r} \neq 0$ are satisfied, there is a bifurcation to a nontrivial T-periodic solution of (9A.1) starting from λ_c. If $\gamma_{0_r} > 0$, the bifurcation takes place for $\lambda \in V^+(\lambda_c)$, whereas if $\gamma_{0_r} < 0$, it takes place for $\lambda \in V^-(\lambda_c)$. The solution $\mathscr{U} \in C^0(-\infty,\infty;\mathscr{D})$ is unique with the exception of Arg a corresponding to a translation in t. Finally, $\mathscr{U}(t)$ is analytic with respect to $\varepsilon = \sqrt{|\lambda-\lambda_c|}$, the period analytic with respect to $\lambda - \lambda_c$ and one can write $\mathscr{U}(t,\varepsilon) = \varepsilon\,\mathscr{U}^{(1)}(t) + \varepsilon^2\,\mathscr{U}^{(2)}(t) + \ldots$ where $\mathscr{U}^{(i)}(t)$ is T-periodic. Here Arg a is the phase of the $X(t)$ oscillations.

An outline of the proof will be given. Denote

$$\zeta_1(\lambda) = \xi(\lambda) + i\eta(\lambda)$$
$$N_\lambda = iE(-\zeta_1) - iE(-\bar{\zeta}_1).$$

Then the equations for the X and Y parts of $\mathscr{U}$ coming from (9A.1) are

$$\begin{cases} Y(t) = \tilde{B}_t(X+Y,X+Y;\lambda), \\ \dfrac{dX}{dt} - \eta N_\lambda X = \xi X + E_\lambda M(X+Y,X+Y) \equiv F(X,Y,;\lambda), \\ X(0) = X(T), \quad X \text{ and } Y \in C^0(-\infty,+\infty;\mathscr{D}), \end{cases} \tag{9A.2}$$

where $\tilde{B}_t(U,V;\lambda) = \int_{-\infty}^{t} I_\lambda(t-\tau)P_\lambda M[U(\tau),V(\tau)]d\tau$. By substituting $X(t) = A(t)U_1(\lambda) + \overline{A(t)U_1(\lambda)}$ in the right hand side of the equation for Y we obtain a right hand side which is analytic with respect to (λ,X,Y) in a neighborhood of $(\lambda_c,0,0)$ in $\mathbb{C} \times \{C^0(-\infty,+\infty;\mathscr{D})\}^2$. And the derivative with respect to Y is zero at $(\lambda_c,0,0)$. Then, denoting $X^{(0)}(t) =$

$A(t)U^{(0)} + \overline{A(t)U^{(0)}}$, and using the implicit function theorem, one has $Y(t) = \eta_t(X;\lambda) = \sum_{i,j\geq 2}^{\infty} (\lambda-\lambda_c)^i \eta_t^{(i,j)}(X^{(0)},\ldots,X^{(0)})$ where $\eta_t^{(i,j)}(\cdot,\ldots,\cdot)$ is a continuous functional, homogeneous of degree j. We must now solve $\frac{dX}{dt} = \eta N_\lambda X + F(X,\eta_t(X;\lambda);\lambda)$ with $X(0) = X(T)$, $X \in C^0(-\infty,\infty;\mathcal{D})$. The following form is assumed for X:

$$\begin{cases} X(t) = e^{\frac{2\pi t}{T}N_\lambda}\chi + \tilde{X}(t) \equiv \chi(t) + \tilde{X}(t) \\ \chi = \frac{1}{T}\int_0^T e^{-\frac{2\pi t}{T}N_\lambda} X(t)\,dt = aU_1(\lambda) + \overline{aU_1(\lambda)} \end{cases} \tag{9A.3}$$

Decomposing the equation for $X(t)$ according to (9A.3), one can solve for $\tilde{X}(t)$ using the implicit function theorem

$$\tilde{X}(t) = \mathcal{X}_t(\chi,\lambda,T) = \sum_{i,j,k\geq 2}^{\infty} (\lambda-\lambda_c)^i (T-T_0)^j \mathcal{X}^{(i,j,b)}(\chi;t), \qquad t \in [0,T],$$

where $\mathcal{X}^{(i,j,k)}(\chi;t)$ is homogeneous of degree k with respect to χ. $\tilde{X}(t)$ is now replaced with $\mathcal{X}_t(\chi,\lambda,T)$ in the other equation (the one for χ). Splitting the result into real and imaginary parts one obtains

$$\xi + f(|a|^2,\lambda,T) = 0$$

$$\eta = \frac{2\pi}{T} + g(|a|^2,\lambda,T) = 0$$

with $f(0,\lambda,T) = g(0,\lambda,T) = 0$. The development in Taylor series about the point $(0,\lambda_c,T_0)$ has first term $-\gamma_0|a|^2$. It is in this way that a non-zero value of γ_{0_r} allows a solution for $|a|^2$ and T by the implicit function theorem. This completes the determination of $X(t)$, $Y(t)$ and therefore $\mathcal{U}(t,\varepsilon)$.

Now that we have the periodic solution it is desired to exhibit its stability properties. We consider a nearby solution $U(t)$ and set $U(t) = \mathscr{U}(t+\delta,\varepsilon) + U'(t)$. Then $U'(t)$ satisfies

$$\frac{\partial U'}{\partial t} = A_\varepsilon(t+\delta)U' + M(U',U')$$

$$U'(0) = U_0 = \mathscr{U}(\delta,\varepsilon) \in \mathscr{D}$$

$$U' \in C^0(0,\infty;\mathscr{D}) \cap C'(0,\infty;H)$$

where

$$A_\varepsilon(t+\delta) = -L_\lambda + M^{(0)}[\mathscr{U}(t+\delta,\varepsilon),\cdot], \quad \lambda = \lambda_c + \varepsilon^2 \operatorname{sgn}(\lambda-\lambda_c).$$

Therefore, in order to study stability one examines the properties of the solution of the linearized equation:

$$\frac{\partial V}{\partial t} = A_\varepsilon(t+\delta)V, \quad V \in C^0(0,T_1;\mathscr{D}) \cap C^1(0,T_1;H)$$

$$V(0) = V_0 \in \mathscr{D}, \; T_1 < \infty$$

The solution of this equation is:

$$V(t) = I_\lambda(t)V_0 + \int_0^t I_\lambda(t-\tau)M^{(0)}[\mathscr{U}(t+\delta,\varepsilon),V(\tau)]d\tau.$$

We denote this solution as

$$V(t) = G_\varepsilon(t,\delta)V_0.$$

The stability properties will come from the properties of the spectrum of $G_\varepsilon(T,\delta)$, which plays the role of the Poincaré map.

We can now state Iooss'

<u>Theorem 3.</u> <u>The hypotheses of Theorem 2 being satisfied and the operator</u> $G_\varepsilon(T,\delta)$ <u>being defined by the equation above</u>, <u>the spectrum of</u> $G_\varepsilon(T,\delta)$ <u>is independent of</u> $\delta \in \mathbb{R}$. <u>It is constituted on the one hand</u>, <u>by two real simple proper</u>

values in a neighborhood of 1, which are 1 and $1 - 8\pi\xi^{(1)}(\lambda-\lambda_c) + o(\lambda-\lambda_c)$. On the other hand, the remainder of the spectrum is formed from a denumerable infinity of proper values of finite multiplicities, the only point of accumulation being 0, there remaining in the interior of a disk of radius $\zeta < 1$ independent of $\varepsilon \in \mathscr{V}(0)$.

The following is then a direct result of Lemma 5 of Judovich [10].

Corollary. The hypotheses of Theorem 2 being satisfied, if further $\gamma_{0_r} < 0$, the bifurcation takes place for $\lambda \in \mathscr{V}^-(\lambda_c)$ and the secondary solution is unstable in the sense of Lyapunov.

Now the proof of Theorem 3 is outlined. The operator $G_\varepsilon(T,\delta)$ is compact in $\mathscr{D}$. Hence, its spectrum is discrete. Let $\sigma \in$ spectrum of $G_\varepsilon(T,\delta)$. Then there exists $V \neq 0$ such that $\sigma V = G_\varepsilon(T,\delta)V$. Then, if $W = G_\varepsilon(nT-\delta,\delta)V$ with $n \in \mathbb{N}$ such that $\delta < nT$, $\sigma W = G_c(T,0)W$. Hence, spectrum $(G_c(T,\delta)) \subset$ spectrum $(G_\varepsilon(T,0))$. Similarly, the reverse containment holds. To establish $1 \in$ spectrum of $G_\varepsilon(T,\delta)$, it suffices to show $G_\varepsilon(T,0)\frac{\partial\mathscr{U}}{\partial t}(0,\varepsilon) = \frac{\partial\mathscr{U}}{\partial t}(0,\varepsilon)$. Note that $I_{\lambda_c}(T_0) = \lim_{\varepsilon\to 0} G_\varepsilon(T,0)$.

It can be shown that 1 is a semi-simple (multiplicity 2) proper value of $I_{\lambda_c}(T_0)$. With all proper values ζ_n of $I_{\lambda_c}(T_0)$ correspond a finite number of proper values of L_{λ_c} of the form $-T_0^{-1}(\mathrm{Log}\,|\zeta_n| + 2k\pi i)$. If one removes $i\eta_0$ from the spectrum of L_{λ_c}, the proper values ζ_i are remaining such that $\mathrm{Re}\,\zeta_i > \tilde{\xi} > 0 \Rightarrow \mathrm{Log}\,|\zeta_n| < -\tilde{\xi}T_0 < 0$. Thus, the remainder of the spectrum of $I_{\lambda_c}(T_0)$ other than 1 is contained in a disk of radius strictly less than 1. Because of the continuity of the discrete spectrum the same is true of $G_\varepsilon(T,0)$ for $\varepsilon \in \mathscr{V}(0)$. The other proper value of $G_\varepsilon(T,0)$ or $G_\varepsilon(T,\delta)$ is found by studying the degenerate operator

$$\varepsilon^{-1}[E(\varepsilon)G_\varepsilon E(\varepsilon) - E(\varepsilon)] = \tilde{G}(\varepsilon)$$

where $E(\varepsilon)$ is $\frac{1}{2\pi i}\int_\Gamma [\zeta 1 - G_\varepsilon]^{-1} d\zeta$ with Γ being a circle of sufficiently small radius about 1 and $G_\varepsilon \equiv G_\varepsilon(T(\varepsilon),0)$. One uses the expansion $G_\varepsilon = I_{\lambda_c}(T_0) + \varepsilon G^{(1)} + \varepsilon\hat{G}(\varepsilon)$ where $\hat{G}(\varepsilon) = o(1)$ and methods of the theory of perturbations to obtain $\hat{G}(\varepsilon) = \varepsilon\tilde{G}^{(1)} + o(\varepsilon)$,

$$\tilde{G}^{(1)} = -T_0\begin{pmatrix} |a^{(1)}|^2\gamma_0 & \left(a^{(1)}\right)^2\gamma_0 \\ \left(\bar{a}^{(1)}\right)^2\gamma_0 & |a^{(1)}|^2\bar{\gamma}_0 \end{pmatrix}$$

(in the basis $\{U^{(0)},\overline{U^{(0)}}\}$). The bijection $\sigma \to \varepsilon^{-1}(\sigma^{-1})$ makes a correspondence between the proper values of $G(\varepsilon)$ and $\tilde{G}(\varepsilon)$. The proper values of $\tilde{G}^{(1)}$ are 0 and $-4T_0\gamma_{0_r}|a^{(1)}|^2 = -8\pi\xi^{(1)}\operatorname{sgn}\gamma_{0_r}$. This gives us 1 and $1 - 8\pi\xi^{(1)}(\lambda-\lambda_c) + o(\varepsilon^2)$ as proper values of $G_\varepsilon(T,\delta)$.

We now state

<u>Theorem 4</u>. <u>If the hypotheses of Theorem 2 are satisfied and</u> $\gamma_{0_r} > 0$, <u>the bifurcation takes place for</u> $\lambda \in \mathscr{V}^+(\lambda_c)$. <u>There exist</u> $\mu > 0$ <u>and a right hand neighborhood</u> $\mathscr{V}^+(\lambda_c)$ <u>such that if</u> $\lambda \in \mathscr{V}^+(\lambda_c)$, <u>and if one can find</u> $\delta_0 \in [0,T]$ <u>such that the initial condition</u> U_0 <u>satisfies</u>

$$||U_0 - \mathscr{U}(\delta_0,\varepsilon)|| \le \mu\varepsilon^2 \quad (\varepsilon = \sqrt{\lambda-\lambda_c}),$$

then there exists $\delta_\ell \in [0,T]$ such that $||U(t)-\mathcal{U}(t+\delta_\ell,\varepsilon)||_{\mathcal{D}} \to 0$ exponentially when $t \to \infty$, $U(t)$ being the solution of (9A.1) satisfying $U(0) = U_0$.

This case is not so simple since there is the proper value, 1. The theorem results from the following lemma:

Lemma 9. Given $V_0 \in V(\delta)$, that is to say, satisfying $E_\delta V_0 = \mathcal{G}(P_\delta V_0,\varepsilon,\delta)$ and $||P_\delta V_0||_{\mathcal{D}} \leq \mu_1\varepsilon^2$, then the equation,

$$\frac{\partial V}{\partial t} = A_\varepsilon(t+\delta)V + M(V,V), \quad V(0) = V_0,$$

admits a unique solution in $C^0(0,\infty;\mathcal{D}) \cap C^1(0,\infty;H)$ such that $||V(t)||_{\mathcal{D}} \leq \mu_2\varepsilon^2 e^{-\sigma/2\ t}$, $\forall t \geq 0$.

We must identify some of the terminology. First, E_δ is the projection onto a vector collinear to $\frac{\partial\mathcal{U}}{\partial t}(\delta,\varepsilon)$, and $P_\delta = 1 - E_\delta$. We have $\mathcal{G}(P_\delta V_0,\varepsilon,\delta) \doteq E_\delta \check{\mathcal{B}}_0\{\eta_\tau[W_0(P_\delta V_0,\varepsilon,\delta),\varepsilon,\delta]$, $\eta_t[\cdots];\varepsilon,\delta\}$. The notation on the right hand side is associated with the following problem:

$$V(t) = G_\varepsilon(t,\delta)W_0 + \hat{\mathcal{B}}_t(V,V;\varepsilon,\delta) + \check{\mathcal{B}}_t(V,V;\varepsilon,\delta)$$

with

$$\hat{\mathcal{B}}_t(U,V;\varepsilon,\delta) = \int_0^t G_\varepsilon(t-\tau,\tau+\delta)P_{\delta+\tau}M[U(\tau),V(\tau)]d\tau,$$

$$\check{\mathcal{B}}_t(U,V;\varepsilon,\delta) = -\int_t^\infty G_\varepsilon(t-\tau,\tau+\delta)E_{\delta+\tau}M[U(\tau),V(\tau)]d\tau,$$

where W_0 is such that $E_\delta W_0 = 0$, and where one searches for V in the Banach space $\mathcal{B}_\beta = \{V: t \to e^{\beta t}V(t) \in C^0(0,\infty;\mathcal{D})\}$, provided with the norm $|V|_\beta = \sup_{t\in(0,\infty)} ||e^{\beta t}V(t)||_{\mathcal{D}}$, with $\beta = \sigma/2$.

These estimates can be shown:

$$|G_\varepsilon(t,\delta)W_0|_\beta \le M_1||W_0||_{\mathscr{D}},$$

$$|\hat{\mathscr{B}}_t(U,V;\varepsilon,\delta)|_\beta \le M_2\gamma\sigma^{-1}|U|_\beta|V|_\beta,$$

$$|\check{\mathscr{B}}_t(U,V;\varepsilon,\delta)|_\beta \le M_2\gamma\sigma^{-1}|U|_\beta|V|_\beta$$

where γ is a bound for the bilinear form M. It results that there exists a μ_0 independent of ε such that for $||W_0|| \le \mu_0\varepsilon^2$, there exists a unique V in $\mathscr{B}$ satisfying the above problem. The solution is denoted $V(t) = \eta_t(W_0,\varepsilon,\delta)$. Now, $V(0) = \eta_0(W_0,\varepsilon,\delta) = W_0 + \check{\mathscr{B}}_0[\eta_\tau(W_0,\varepsilon,\delta),\ \eta_\tau(W_0,\varepsilon,\delta);\varepsilon,\delta]$ which gives after decomposition:

$$E_\delta V_0 = E_\delta\,\check{\mathscr{B}}_0[\eta_\tau(W_0,\varepsilon,\delta),\eta_\tau(W_0,\varepsilon,\delta);\varepsilon,\delta],$$

$$P_\delta V_0 = W_0 + P_\delta\,\check{\mathscr{B}}_0[\eta_\tau(W_0,\varepsilon,\delta),\eta_\tau(W_0,\varepsilon,\delta);\varepsilon,\delta].$$

After having remarked that $\frac{\partial}{\partial W_0}[\eta_\tau(W_0,\varepsilon,\delta)]_{W_0=0} = G_\varepsilon(t,\delta)$, it is easy to find μ_1 independent of ε such that if $||P_\delta V_0|| \le \mu_1\varepsilon^2$, then the second of the above equations is solvable with respect to W_0 by the implicit function theorem, and $W_0 = W_0(P_\delta V_0,\varepsilon,\delta)$ satisfies $||W_0||_{\mathscr{D}} \le \mu_0\varepsilon^2$. The notation is completely explained. The proof from here on is not too difficult. The uniqueness comes from the uniqueness of the solutions to (9A.1) on a bounded interval for ε sufficiently small. The lemma will be demonstrated as soon as it is shown that the solution of our above problem is the same as the solution of the equation in the lemma. But this follows immediately utilizing $\frac{\partial}{\partial t}G_\varepsilon(t-\tau,\tau+\delta) = A_\varepsilon(\tau+\delta)G_\varepsilon(t-\tau,\tau+\delta)$ in evaluating $\frac{\partial}{\partial t}\hat{\mathscr{B}}_t(V,V;\varepsilon,\delta)$ and $\frac{\partial}{\partial t}\check{\mathscr{B}}_t(V,V;\varepsilon,\delta)$. (It is seen that the manifold $V(\delta)$ is nothing more than the set of V_0 such that W_0 can be found with $||W_0|| \leq \mu_0 \varepsilon^2$ and satisfying the equations for $P_\delta V_0$ and $E_\delta V_0$.)

The techniques of Iooss are very similar to those of Judovich [1-12] (see also Bruslinskaya [2,3]). These methods are somewhat different from those of Hopf which were generalized to the context of nonlinear partial differential equations by Joseph and Sattinger [1].

Either of these methods is, nevertheless, basically functional-analytic in spirit. The approach used in these notes attempts to be more geometrical; each step is guided by some geometrical intuition such as invariant manifolds, Poincaré maps etc. The approach of Iooss, on the other hand, has the advantage of presenting results in more "concrete" form, as, for example, $\mathcal{U}(t,\varepsilon)$ in a Taylor series in ε. This is also true of Hopf's method. However, stability calculations (see Sections 4A, 5A) are no easier using this method.

Finally, it should be remarked that Iooss [6] presents analogous results to this paper for the case of the invariant torus (see Section 6).

SECTION 9B

ON A PAPER OF KIRCHGÄSSNER AND KIELHÖFFER

BY O. RUIZ

The purpose of this section is to present the general idea that Kirchgässner and Kielhöffer [1] follow in resolving some problems of stability and bifurcation.

The Taylor Model.

This model consists of two coaxial cylinders of infinite length with radii r_1' and r_2' $(r_1' < r_2')$ rotating with constant angular velocities ω_1 and ω_2. Due to the viscosity an incompressible fluid rotates in the gap between the cylinders. If λ is the Reynolds number $\lambda = \dfrac{r_1'\omega_1(r_2'-r_1')}{\nu}$ (ν the viscosity), we have for small values of λ a solution independent of λ, called the Couette flow. As λ increases, several types of fluid motions are observed, the simplest of which is independent of ϕ and periodic in z, when we consider cylindrical coordinates. If we restrict our considerations to these kinds of flows and require that the solution v be invariant under the groups of translations T_1 generated by $z \to z + 2\pi/\sigma$ and $\phi \to \phi + 2\pi$, $\sigma > 0$ and we consider the "basic" flow $V = (V_r, V_\phi, V_z), P$ in cylindrical coordinates, (we assume V, P are given) we may write the N-S equation in the form

$$\begin{aligned}
&\text{(a)} \quad D_t u - \tilde{\Delta} u + \lambda L(V)u + \lambda \Delta q = -\lambda N(u), \\
&\text{(b)} \quad \nabla \cdot u = 0, \quad u|_{r=r_1,r_2} = 0, \; u(Tx,t) = u(x,t), \\
&\qquad\qquad T \in T_1, \\
&\text{(c)} \quad u|_{t=0} = u^0
\end{aligned} \tag{9B.1}$$

where

$$v = V + u, \quad p = P + q, \quad D_t = \frac{\partial}{\partial t},$$

$$\nabla = \left(\frac{\partial}{\partial r}, 0, \frac{\partial}{\partial z}\right), \quad \text{(gradient)}$$

$$\Delta = \frac{\partial}{\partial r^2} + \frac{1}{r}\frac{\partial}{\partial r} + \frac{\partial^2}{\partial z^2}, \text{ Laplacian, } \tilde{\Delta}_{ik} = (\Delta - (1-\delta_{i3})/r^2)\delta_{ik}$$

$$L^0_{ik}(V) = -2V_\phi \delta_{i1}\delta_{2k}/r + (V_r \delta_{2k} + V_\phi \delta_{1k})\delta_{i2}/r,$$

$$L(V)u = L^0(V)u + (V \cdot \nabla)u + (u \cdot \nabla)V,$$

$$Q(u)_i = -u_\phi^2 \delta_{i1}/r + u_\phi u_r \delta_{i2}/r,$$

$$N(u) = (u \cdot \nabla)u + Q(u).$$

<u>The Bénard model</u> consists of a viscous fluid in the strip between 2 horizontal planes which moves under the influence of viscosity and the buoyance force, where the latter is caused by heating the lower plane. If the temperature of the upper plane is T_1, and the temperature of the lower plane is T_0 $(T_0 > T_1)$, the gravity force generates a pressure distribution which for small values of $T_1 - T_0$ is balanced by

the viscous stress resulting in a linear temperature distribution. If however $T_1 - T_0$ is above a critical point of value, a convection motion is observed. Let α, h, g, ν, ρ, k denote the coefficient of volume expansion, the thickness of the layer, the gravity, the kinematic viscosity, the density and the coefficient of thermodynamic conductivity respectively. We use Cartesian coordinates where the x_3-axis points opposite to the force of gravity, $\tilde{\theta}$ denotes the temperature and p the pressure. By the N-S equations we have the following initial-value problem for an arbitrary reference flow V, T, P. If $\omega = (u,\theta)$, $v = V + u$, $p = P + q$, $\tilde{\theta} = T + \theta$, we have

$$\begin{aligned} &(a) \quad D_t\omega - \tilde{\Delta}\omega + \lambda L(V)\omega + \nabla q = -N(\omega), \\ &(b) \quad \nabla\cdot\omega = 0, \quad \omega|_{x_3=0,1} = 0, \qquad (9B.2) \\ &(c) \quad \omega|_{t=0} = \omega^0 \end{aligned}$$

where $\lambda = \alpha g(T_0 - T_1)h^3/\nu^2$ (Reynolds number or Grashoff number)

$$\nabla = (\partial/\partial x_1, \partial/\partial x_2, \partial/\partial x_3, 0),$$

$$\tilde{\Delta}_{ik} = \left(\frac{\partial^2}{\partial x_1^2} + \frac{\partial^2}{\partial x_2^2} + \frac{\partial^2}{\partial x_3^2}\right)\left(\delta_{ik} + \frac{1}{Pr}\,\delta_{i4}\right), \quad Pr = k/\nu$$

$$L^0_{ik} = -\delta_{i3}\delta_{k4} - \frac{1}{Pr}\,\delta_{i4}\delta_{k3},$$

$$L(V)\omega = L^0\omega + (V\cdot\nabla)\omega + (u\cdot\nabla)V,$$

$$N(\omega) = (u\cdot\nabla)\omega.$$

Since experimental evidence shows that the convection

takes place in a regular pattern of closed cells having the form of rolls, we are going to consider the class of solutions such that

$$\omega(Tx,t) = \omega(x,t), \quad q(Tx,t) = q(x,t)$$

where $T \in T_1$, and T_1

is the group generated by the translations

$$x_1 \to x_1 + 2\pi/\alpha, \quad x_2 \to x_2 + 2\pi/\beta, \quad \alpha^2 + \beta^2 \neq 0.$$

$$u(Tx,t) = Tu(x,t)$$

$$q(Tx,t) = q(x,t) \qquad T \in T_2$$

$$\theta(Tx,t) = \theta(x,t)$$

where T_2 is the group of rotations generated by

$$T_\alpha = \begin{pmatrix} \cos\alpha & -\sin\alpha & 0 \\ \sin\alpha & \cos\alpha & 0 \\ 0 & 0 & 1 \end{pmatrix}$$

Note. It is possible to show that all differential operators in the differential equation preserve invariance under T_1 and T_2. An interesting fact is that a necessary condition for the existence of nontrivial solutions is $\alpha = 2\pi/n$, $n \in \{1,2,3,4,6\}$, and that there are only 7 possible combinations of n,α,β, which give different cell patterns (no cell structure, rolls, rectangles, hexagons, squares, triangles).

Functional-Analytic Approach.

The analogy between (9B.1) and (9B.2) in the Taylor's and

Bénard's models suggests to the authors an abstract formulation of the bifurcation and stability problem. In this part I am going to sketch the idea that the authors follow to convert the differential equations (9B.1) and (9B.2) in a suitable evolution equation in some Hilbert space.

We may consider D an open subset of R^3 with boundary ∂D which is supposed to be a two dimensional C^2-manifold. T_1 denotes a group of translations and Ω its fundamental region of periodicity which we may suppose is bounded. Assume

$$D = \bigcup_{T \in T_1} T\Omega .$$

Now we consider the following sets ($c\ell$ = closure)

$$C^{T,\infty}(\bar{D}) = \{\omega | \omega: c\ell(D) \to R^n, \text{ infinitely often differentiable in } c\ell(D),\ \omega(Tx) = \omega(x),\ T \in T_1\},$$

$$C_0^{T,\infty}(D) = \{\omega | \omega \in C^{T,\infty}(\bar{D}),\ \text{supp } \omega \subset D\},$$

$$C_{0,\infty}^{T,\infty}(D) = \{\omega | \omega \in C_0^{T,\infty}(D),\ \nabla \cdot u(x) = 0,\ \omega = (u,v),\ u \in R^3\}.$$

Defining

$$(v,\omega)_m = \sum_{|\gamma| \leq m} (D^\gamma v, D^\gamma \omega), \quad |v|_m = \{(v,v)_m\}^{1/2},$$

where

$$(D^\gamma v, D^\gamma \omega)_2 = \int_\omega (D^\gamma v(x) \cdot D^\gamma \omega(x))dx$$

and γ is a multiindex of length 3; one obtains the following Hilbert spaces

$$L_2^T = c\ell_{||_0} C_0^{T,\infty}(D), \quad \mathfrak{J}^T = c\ell_{||_0} C_{0,\sigma}^{T,\infty}(D),$$

$$\overset{\circ}{H}{}^T_{1,\sigma} = c\ell_{||_1} C_{0,\sigma}^{T,\infty}(D), \quad H_m^T = c\ell_{||_m} C^{T,\infty}(\bar{D}),$$

For the Taylor problem we have

$$n = 3,\ D = (r_1, r_2) \times [0, 2\pi)$$

$$\Omega = (r_1, r_2) \times [0, 2\pi) \times [0, 2\pi/\sigma)$$

and for the Bénard problem

$$n = 4, \quad D = \mathbb{R}^2 \times (0,1), \quad \mathbb{R} = \text{the real numbers}$$

$$\Omega = [0, 2\pi/\alpha) \times [0, 2\pi/\beta) \times (0,1) .$$

We may consider that the differential equations of the form (9B.1) or (9B.2) are written with operators in L_2^T.

From H. Weyl's lemma it is possible to consider $L_2^T = \overset{\circ}{\mathcal{J}}{}^T \oplus G^T$, where G^T contains the set of ∇_q such that $q \in H_1^T$. We may use the orthogonal projection $P: L_2^T \to \overset{\circ}{\mathcal{J}}{}^T$ for removing the differential equation $D_t\omega - \tilde{\Delta}\omega + \lambda L(V)\omega + \lambda\nabla q = -\lambda N(\omega)$ with the additional conditions of boundary and periodicity, to a differential equation in $\overset{\circ}{\mathcal{J}}{}^T$. If we consider $q \in H_1^T$, $P\nabla q \equiv 0$ and since we look for solutions on $\overset{\circ}{J}{}^T$, we have $Pu = u$, and we may write the new equation in $\overset{\circ}{J}{}^T$ as

$$\frac{d\omega}{dt} + P\tilde{\Delta}u + \lambda PL(V) = -\lambda PN(\omega) \tag{9B.3}$$

with initial condition $\omega_{t=0} = \omega^0$.

The authors write (9B.3) in the form

$$\frac{d\omega}{dt} + \tilde{A}(\lambda)\omega + h(\lambda)R(\omega) = 0, \quad \omega|_{t=0} = \omega^0,$$

where $\tilde{A}(\lambda) = P\tilde{\Delta} + \lambda PL(V)$, $R(\omega) = PN(\omega)$ and where $h(\lambda) = \lambda$ for Taylor's model and $h(\lambda) = 1$ for Benard's model.

Also, they show using a result of Kato-Fujita, that it is possible to define fractional powers of $\tilde{A}(\lambda)$ by

$$\tilde{A}(\lambda)^{-\beta} = \frac{1}{\Gamma(\beta)} \int_0^{\infty} \exp(-\tilde{A}(\lambda)t)t^{\beta-1}dt, \quad \beta > 0,$$

and that this operator is invertible.

The above fact is useful in resolving some bifurcation problems in the stationary case. If $A = P\tilde{\Delta}$, $M(V) = PL(V)$ the stationary equation form (9B.3) is

$$\text{(SP)} \qquad A\omega + \lambda M(V)\omega + h(\lambda)R(\omega) = 0$$

where V is any known stationary solution with $V \in$ Domain of A.

If we consider the substitution $A^{3/4}\omega = v$

$$K(V) = A^{-1/4}M(V)A^{-3/4} \qquad T(v) = A^{-1/4}R(A^{-3/4}v)$$

we may write (SP) in the form

$$v + \lambda K(V) + h(\lambda)T(v) = 0.$$

If one uses a theorem of Krasnoselskii, it is possible to show the following theorem.

<u>Theorem 4.1.</u> <u>Let be</u> $\lambda_j \in R$, $\lambda_j \neq 0$ <u>and</u> $(-\lambda_j)^{-1}$ <u>be an eigenvalue of</u> K <u>of odd multiplicity</u>, <u>then</u>

i) <u>in every neighborhood of</u> $(\lambda_j, 0)$ <u>in</u> $R \times \overset{0}{J}{}^{T}$ <u>there exists</u> (λ,ω), $0 \neq \omega \in D(A)$ <u>such that</u> ω <u>solves the stationary equation</u> (SP).

ii) <u>if</u> $(-\lambda_j)^{-1}$ <u>is a simple eigenvalue</u>, <u>then there exists a unique curve</u> $(\lambda(\alpha),\omega(\alpha))$ <u>such that</u> $\omega(\alpha) \neq 0$ <u>for</u> $\alpha \neq 0$ <u>which solves</u> (SP), <u>moreover</u> $(\lambda(0),\omega(0)) = (\lambda_j, 0)$.

Now if we assume that $V \in C(\overline{\Omega})$ <u>and ∂D is a C^{∞} - manifold</u> which is satisfied for the Taylor and Bénard Problem, and we consider solutions of (SP) on the space $H_{m+2} \cap H_{1,\sigma}$, $m > 3/2$, we may write the (SP) equation in the form

$$\text{(SP)}' \qquad A_m \omega + \lambda M \omega + h(\lambda) R(\omega) = 0$$

where A_m is an operator whose domain is $D(A_m) = H_{m+2} \cap H_{1,\sigma} \subset PH_m$, $A_m(\omega) = A\omega$, $\omega \in D(A)$. If besides we consider that in $R(\omega) = PN(\omega)$, $N(\omega)$ is an arbitrary polynomial operator including differentiation operators up to the order 2, and $K_m = MA_m^{-1}$ we may obtain the following theorem.

Theorem 4.2. *Let $V \in C(\bar{D})$, ∂D be a C^∞-manifold; let M such that there exists constants C_1 and C_2 such that $|M\omega|_{m+1} \leq C_1 |\omega|_{m+2}$ and $|MA_m^{-1}\omega|_{m+1} \leq C_2 |\omega|_m$. Then for every eigenvalue $(-\lambda_j)^{-1}$ of K_m, $\lambda_j \neq 0$, of odd multiplicity, $(\lambda_j, 0)$ is a bifurcation point of (SP)'. The solutions ω are in $C(\bar{D})$ and fulfill the boundary condition $\omega|_{\partial D} = 0$.*

We may note that this theorem shows that we may obtain a strong regularity for the branching solutions. Besides, we note that in theorems 4.1 and 4.2 the existence of non-trivial solutions of (SP) is reduced to the investigation of the spectrum of K or K_m.

Now, we are going to apply these theorems to the Taylor's and Bénard models.

Taylor Model. For this model we take $K = A^{-1/4} M(v^0) A^{-3/4}$ where v^0 is the Couette flow. In cylindrical coordinates the solution is given by $v^0 = (0, v_\phi^0, 0)$ where $v_\phi^0 = ar + b/r$

$$a = \frac{1}{r_1\omega_1}\frac{\omega_2 r_2^2-\omega_1 r_1^2}{r_2^2-r_1^2} \qquad b = \frac{1}{r_1\omega_1}\frac{(\omega_1-\omega_2)r_1^2 r_2^2}{r_2^2-r_1^2}$$

When $a \geq 0$, $v_\phi^0 \geq 0$, Synge shows that the Couette flow is locally stable.

For $a < 0$, $v_\phi^0(r) > 0$, _Velte and Judovich_ proved the following theorem for the K operator.

Theorem 4.3. _Let be_ $a < 0$, $v_\phi^0(r) > 0$ _for_ $r \in (r_1,r_2)$, T_1 _the group of translations generated by_ $z \to z + 2\pi/\sigma$; $\phi \to \phi + 2\pi$, $\sigma > 0$. _Then for all_ $\sigma > 0$, _except at most a countable number of positive numbers, there exists a countably many sets of real simple eigenvalues_ $(-\lambda_i)^{-1}$ _of_ K. _Every point_ $(\lambda_i,0) \in R \times D(A)$ _is a bifurcation point of the stationary problem where exactly one nontrivial solution branch_ $(\lambda(\alpha),\omega(\alpha))$ _emanates. These solutions are Taylor vortices._

Strong experimental evidence suggest that all solutions branching off $(\lambda_i,0)$ where $\lambda_i \neq \lambda_1$ are unstable; however, no proof is known.

Benard model. In this model if $\lambda = \alpha g(T_0-T_1)h^3/\nu^2$, $\sigma = (\alpha^2+\beta^2)^{1/2}$, α, β like on page 318, it is known that for some $\lambda_1(\sigma)$, $\lambda \in [0,\lambda_1]$, $\omega = 0$ is the only solution of the stationary problem.

For this model the bifurcation picture is determined by the spectrum of $K = A^{-1/4}M(v^0)A^{-3/4}$, where v^0 is given in Cartesian coordinates by

$$v_0 = 0, \quad P_0(x_3) = -\frac{gh^3}{\nu^2}(x_3+\alpha(T_0-T_1)), \quad \theta_0(x_3) = -x_3.$$

In order to obtain simple eigenvalues, Judovich introduces even solutions $u(-x) = (-u_1(x), -u_2(x), u_3(x))$, $\theta(-x) = \theta(x)$, $q(-x) = -q(x)$, and he shows in his articles, On the origin of convection (Judovich [6]) and Free convection and bifurcation (1967) the following theorem.

Theorem 4.6. i) The Bénard problem possesses for apprximately all α and β countably many simple positive characteristic values λ_i. Furthermore $(\lambda_i, 0) \in \mathbb{R} \times D(A)$ is a bifurcation point.

ii) If n, α, β are chosen according to the note on pg. 318, the branches emanating from $(\lambda_i, 0)$ are doubly periodic, rolls, hexagons, rectangles, and triangles.

iii) If λ_1 denotes the smallest characteristic value, then the nontrivial solution branches to the right of λ_1 and permits the parametrization $\omega(\lambda) = \pm(\lambda - \lambda_1)^{1/2} F(\lambda)$ where $F: R \to D(A)$ is holomorphic in $(\lambda - \lambda_1)^{1/2}$.

It is interesting to observe that since the characteristic values are determined only by σ, we may consider differentials α, β with the same value of σ, and to note that we have solutions of every possible cell structure emanating from each bifurcation point.

Stability.

About this topic I am going to give a short description of the principal results.

It is known that the basic solution loses stability for some $\lambda_c \in (0, \lambda_1]$, λ_1 as in the past section. Under the assumption that $\lambda_c = \lambda_1$, and λ_1 simple, the nontrivial solution branch emanating from $(\lambda_1, 0)$ gains stability for $\lambda > \lambda_1$ and is unstable for $\lambda < \lambda_1$. This result can be derived using Leray-Schauder degree or by analytic perturbation methods.

Precisely if we take $V = v_0 + \omega^+$ where ω^+ is a stationary solution of (SP) _V is called stable_ if for $\tilde{A}(\lambda) = A + \lambda M(V)$, $\omega = 0$ is stable in sense with respect to strict solutions in $D(A^\beta)$, $3/4 < \beta < 1$. (Strict solution in the sense of Kato-Fujita of the article), and _V is called unstable_ if it is not stable in $\mathcal{J}^{\sigma T}$. If we consider that the nontrivial solution branch $(\lambda(\alpha), V(\alpha))$ in $R \times \mathcal{J}^{\sigma T}$, $|\alpha| \leq 1$ $(\lambda(0), V(0)) = (\lambda_1 v_0)$ can be written in the form

$$\begin{aligned} \alpha(\lambda) &= \pm c_1 |\lambda - \lambda_1|^{1/r} \\ & \qquad\qquad\qquad\qquad r \in N \\ V(\lambda) &= v_0 + |\lambda - \lambda_1|^{1/r} F(\lambda) \end{aligned}$$

where $F\colon R \to D(A)$ is analytic in $(\lambda - \lambda_1)^{1/r}$ and $F(\lambda_1) \neq 0$ (valid conditions for Bénard's and Taylor's models (Theorem 4.6 and Corollary 4.4), we have the following Theorem or Lemma 5.6.

Lemma 5.6. Assume $\lambda_c = \lambda_1$ _and_ $\operatorname{Re} \mu \geq \alpha > 0$ _for all nonvanishing_ μ _in the spectrum of_ $\sigma(\tilde{A}(\lambda_1, v_0))$. _Let_ λ_1 _be a simple characteristic value of_ $K(v_0)$, 0 _a simple eigenvalue of_ $\tilde{A}(\lambda_1, v_0)$.

Then, if λ _is restricted to a suitable neighborhood of_ λ_1

i) v_0 _is stable for_ $\lambda < \lambda_1$ _and unstable for_ $\lambda > \lambda_1$

ii) $V(\lambda)$ _is stable for_ $\lambda > \lambda_1$ _and unstable for_

$\lambda < \lambda_1$.

For the Bénard's model, the assumptions of Lemma 5.6 are satisfied for fixed n, α, β; λ_1 is a simple characteristic value of $K(v_0)$ by Theorem 4.6 and $\lambda_c = \lambda_1$ follows from Lemma 4.5 of the article.

For the Taylor's model only the simplicity of λ_1 as a characteristic value of $K(v_0)$ is known. The simplicity of $\mu = 0$ in $\sigma(\tilde{A}\lambda_1, v_0)$ is an open problem.

However, we may give the following theorem.

Theorem 5.7. i) For the Bénard's problem, every solution with a given cell pattern (fixed n, α, β) exists in some right neighborhood of λ_1 and is asymptotically stable in $D(A^\beta)$, $\beta \in (3/4,1)$. The basic solution v_0 is asymptotically stable for $\lambda < \lambda_1$ and unstable for $\lambda > \lambda_1$.

ii) For the Taylor's problem, let the assumptions of Lemma 5.6 on the spectrum of $\tilde{A}(\lambda; v_0)$ be valid. Then for every period (σ fixed) $V(\lambda)$ is asymptotically stable if it exists for $\lambda > \lambda_1$, and is unstable if it exists for $\lambda < \lambda_1$.

Finally, we remark that these results can also be obtained using the invariant manifold approach. (See, for example, Exercise 4.3). That this is possible was noted already by Ruelle-Takens [1] in their elegant and simple proof of Velte's theorem. We also note that Prodi's basic results relating the spectral and stability properties of the Navier-Stokes equations are contained in the smoothness properties of the flow from Section 9 and the results of Section 2A.

SECTION 10

BIFURCATION PHENOMENA IN POPULATION MODELS

BY

G. OSTER AND J. GUCKENHEIMER

1. Introduction: The Role of Bifurcations in Population Models.

Biological systems tend to be considerably more complex than those studied in physics or chemistry. In analyzing models, one is frequently presented with two alternatives: either resorting to brute force computer simulation or to reducing the model further via such drastic approximations as to render it biologically uninteresting. Neither alternative is attractive. Indeed, the former alternative is hardly viable for most situations in ecology since sufficient data is rarely available to quantitatively validate a model. This contrasts starkly with the physical sciences where small differences can often discriminate between competing theories. The situation is such that many ecologists seriously question whether mathematics can play any useful role in biology. Some claim that there has not yet been a single fundamental advance in biology attributable to mathematical theory.* Where complex systems are concerned, they assert that the appropriate language is English, not mathematical. A typical attitude among biologists is that models are useful only insofar as they explain the unknown or suggest new experiments. Such models are hard to come by.

In the face of such cynicism, perhaps mathematicians who would dabble in biology should set themselves more modest goals. Rather than presenting the biological community with an exhaustive analysis of an interesting model, it might be better to produce a "softer" analysis of a meaningful model. From this viewpoint, the role of mathematics is not to generate proofs, but to act as a guide to one's intuition in perceiving what nature is up to. This is no excuse for avoiding hard analysis where it can be done, but as models mimic nature more closely it becomes harder to prove theorems.

In this spirit, we shall discuss several instances where some concepts of bifurcation theory have proved useful in ecological modelling. We shall discuss (briefly) bifurcation phenomena in three kinds of population models: (i) discrete generation populations modelled by difference equations, (ii) continuously breeding populations modelled by ordinary differential equations, and (iii) populations with age structure which require partial or functional differential equations.

---

*Perhaps excluding the Hardy-Weinberg law--which is trivial mathematically.

These represent three successive stages of increasing biological realism as well as mathematical intractibility. Thus we shall proceed from less realistic models based upon solid mathematical foundations to more realistic models based upon mathematical intuition. In all cases, however, the usefulness of bifurcation theory transcends our ability to cite theorems. By furnishing a qualitative modelling mechanism, it provides a conceptual framework within which we can view a number of important ecological processes.

2. Populations with Discrete Generations.

(2.1) Consider an insect population which breeds once a year. A plot of the total number of individuals as a function of time might look like Figure 10.1a:

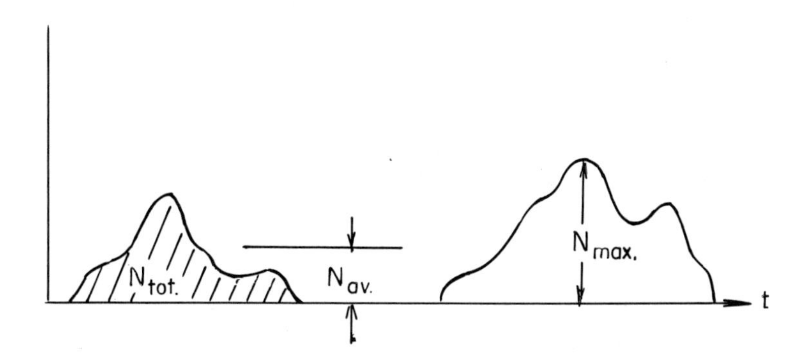

Figure 10.1a

If we are only interested in either the mean, total or average number each year, we might consider an approximate difference equation model as shown in Figure 10.1b:

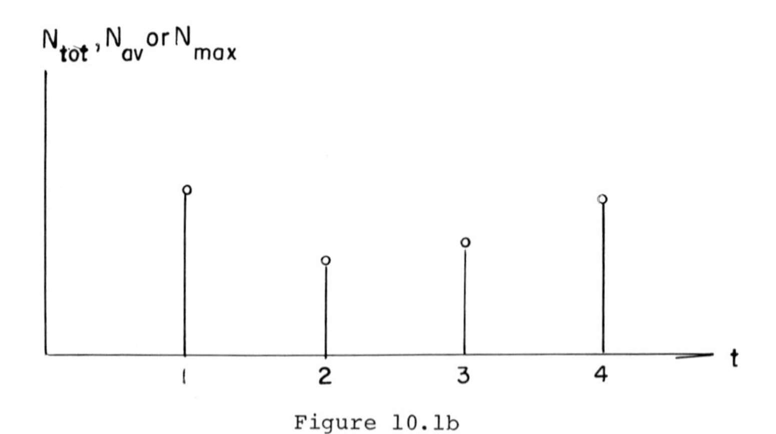

Figure 10.1b

i.e., an equation of the form:

$$N_{t+1} = F(N_t). \tag{2.1}$$

Models of this kind are commonly employed in entomology (Hassell & May, [1]; Varley, Gradwell & Hassell, [1]). In general, $F(\cdot)$ will have the shape shown in Figure 10.2:

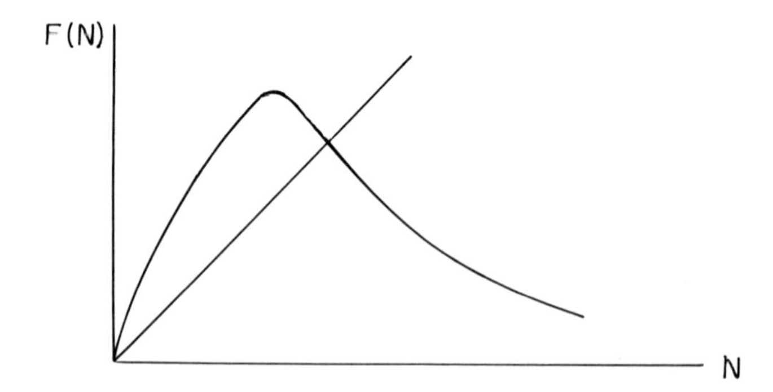

Figure 10.2

The reason for this is that as the population density increases crowding effects, such as competition for food, tend to increase deathrates and decrease birthrates.

Typical functional forms that have been employed in modelling insect populations are:

$$N_{t+1} = N_t e^{r(1-N_t/K)} \tag{2.2a}$$

$$N_{t+1} = N_t \frac{\lambda}{(1+N_t)^b} \tag{2.2b}$$

$$N_{t+1} = \frac{N_t}{1+e^{-a(1-N_t/K)}} \tag{2.2c}$$

$$N_{t+1} = \begin{cases} \lambda N_t, & N_t < 1 \\ \lambda N_t^{1-b}, & N_t > 1 \end{cases} \tag{2.2d}$$

Each of these models has the origin as a fixed point and have the "1-hump" characteristic of Figure 10.2, i.e., single critical point less than the positive fixed point. Beyond this, however, they are largely empirical, generated ad hoc by regression of one generation on the next.* By and large, however, such simple-minded models have been surprisingly effective in reproducing the generation-to-generation variations in population levels. (Auslander, Oster, Huffaker, [1]; Varley, et. al., op. cit.)

---

*Note that, for $r << 1$, equation (2.2a) is

$$N_{t+1} - N_t = rN_t(1 - \frac{N_t}{K}) \tag{2.2a*}$$

which is just the forward difference equation corresponding to the familiar logistic equation for population growth: $\frac{dN}{dt} = rN(1 - \frac{N}{K})$. Thus, r in (2.2) can be interpreted as the net generation-to-generation reproductive rate. Although Equation (2.2a) can exhibit bifurcations, (2.2a*) cannot. (i.e., for large r, (2.2a) has a critical point, so it cannot be a finite version of (2.2a*)).

(2.2) Whether or not the population, as modelled by any of equations (2.2), settles down to a steady generation-to-generation level depends on the stability of the fixed point, $F(N) = N$, $N > 0$, and perhaps the initial condition. This, in turn, depends on the particular parameter values, such as r in (2.2a). Let us consider (2.2a) as the prototype for our discussion. The eigenvalue of $F(\cdot)$ at the fixed point $\overline{N} = K$ is $\lambda(r) = F'(\overline{N}) = 1-r$. As the reproduction rate increases past 2, $\lambda(r)$ moves across the unit circle and $\overline{N}$ ceases to be an attractor. However, if F i has a critical point--as we have supposed in the models (2.2) --then the composition of F with itself will have at least 3 critical points and we can look at the period-2 fixed points:

$$F^2(\overline{N}_2) \equiv F \circ F(\overline{N}_2) = \overline{N}_2 \tag{2.3}$$

and the eigenvalues of F^2 at these points:

$$\lambda_2(r) = DF^2(\overline{N}_2). \tag{2.4}$$

The stability of the pair of period-2 points, which have split off from the original fixed point as r crosses 2, is determined by the eigenvalues, $\lambda_2(r)$. Initially stable, these period-2 points bifurcate when $|\lambda_2(r)| \geq 1$. In this case, the nature of the bifurcation depends on whether it occurs at $+1$ or -1. At $\lambda < -1$ each period 2 point bifurcates into a pair of stable points with period 4. This process continues as r increases: bifurcations from $\lambda_k(r) = -1$ giving rise to pairs of attracting points of period $2k$ while bifurcations from $\lambda_k(r) = +1$ either create or destroy periodic points. (c.f., Figure 10.3.)

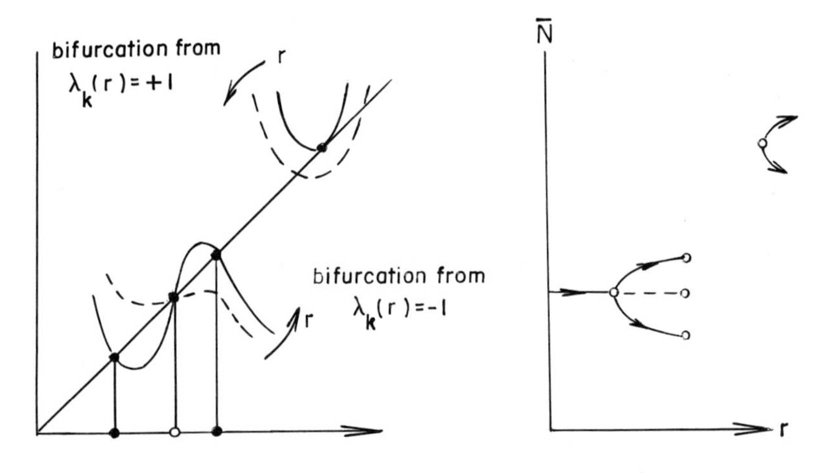

Figure 10.3

The orbit generated by $F(\cdot)$ becomes successively more complicated with each bifurcation--the initially stable fixed point splitting to orbits of successively higher periods. This can continue indefinitely, with bifurcation points occuring closer and closer together. As r is increased past 2, exciting successively stable higher periodic orbits, there can occur a limit point, r_c, beyond which completely aperiodic points appear. That is, orbits are generated -which do not tend asymptotically to a periodic orbit. Sufficient conditions for such aperiodic orbits to exist has been given by Li and Yorke [1]. If $F(\cdot)$ folds some interval onto itself as shown in Figure 10.4a, then there exist aperiodic points (i.e., initial conditions which do not lie in the domain of attraction of any stable fixed point).

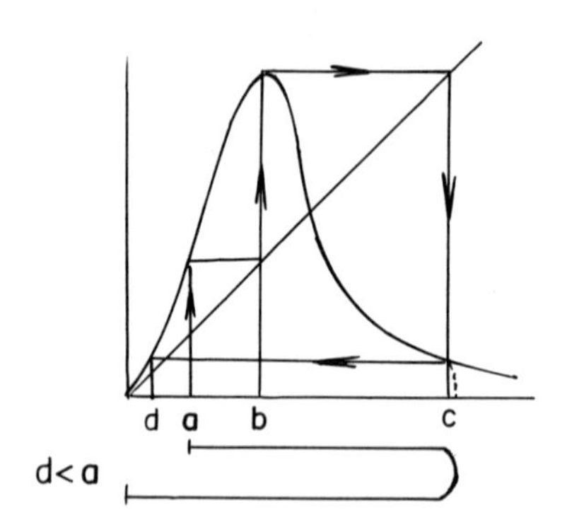

Figure 10.4a

Alternatively, if the population exhibits a "3-point cycle" wherein the population rises 2 years in succession and then crashes past the original level, then non-periodic motion will ensue, (c.f. Figure 10.4b):

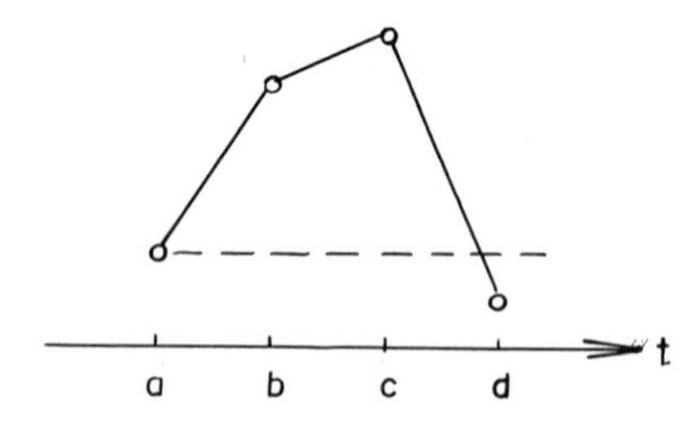

Figure 10.4b

The consequences of this phenomenon for ecological modelling are profound. We can have confidence in a model only if it is subject to experimental validation. If a series of yearly censes are collected of some population, and they appear chaotic, exhibiting no perceivable regularities, then we can conclude one of three things: (a) the system is truly stochastic--dominated by random influences; (b) experimental error is of such magnitude that all regularities are obscured; (c) a very simple deterministic mechanism is operating, but is obscured by the phenomenon described above. As an extreme case, the orbit generated by the simple map shown in Figure 10.5 is indistinguishable from a sequence of Bernoulli trials!

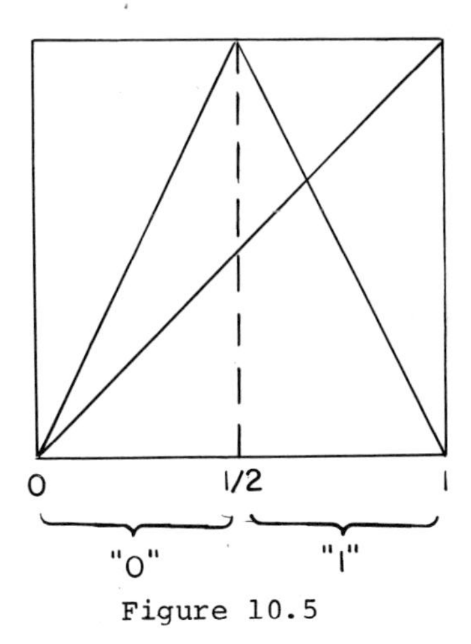

Figure 10.5

For systems of 2 interacting populations (e.g., predator-prey, parasite-host, etc.) the situation is even more delicate and little is known about the transition to aperiodic motion. However, May [2] has simulated some 2-population difference equation models. He found that the population trajectories exhibited chaotic behavior for quite reasonable parameter ranges.

3. Populations with Overlapping Generations.

(3.1) If a population breeds continuously, so that the

generations overlap, then the appropriate model is an ordinary differential equation of the form

$$\frac{dN}{dt} = Nf(N) = N\left[\begin{pmatrix}\text{per capita}\\ \text{birth rate}\end{pmatrix} - \begin{pmatrix}\text{per capita}\\ \text{death rate}\end{pmatrix}\right] \qquad (3.1)$$

The number of such models in the literature is legion, and we shall comment only briefly on certain aspects pertaining to their bifurcation behavior.

(3.2) A recurrent theme in ecology is the phenomenon of population oscillations. The earliest prototype was the predator-prey equations of Volterra and Lotka (see, for example, May, [1]):

$$\begin{aligned} \dot{N}_1 &= N_1 f_1(N_1,N_2) = N_1[c_1 - c_2N_2] \\ \dot{N}_2 &= N_2 f_2(N_1,N_2) = N_2[-c_3 + c_4N_1]. \end{aligned} \qquad (3.2)$$

The solutions to (3.2) are indeed periodic (Hirsch and Smale, [1]), but are neutrally stable, the amplitude of the oscillations depending on the initial conditions.

Recently, May [1,2] has shown that virtually all of the models for predator-prey systems possess either a stable equilibrium or a stable limit cycle (in the first quadrant). His demonstration hinges on showing that most models fall within the purview of a theorem by Kolmogorov [1], which is essentially an application of the Poincaré-Bendixson Theorem to systems of the form (1). [Essentially, any population model such that 1) there is a single unstable singularity in the first quadrant and 2) the axes are invariant (e.g.,

$\dot{\underset{\sim}{N}} = \underset{\sim}{N}\underset{\sim}{f}(\underset{\sim}{N}))$ will have a limit cycle since large radius orbits must be directed inward due to the finite population limitation that must be imposed on any realistic model.]

A typical predator-prey system which exhibits limit cycle behavior is: (Rosenzweig, [1]; May, [1]):

$$\begin{aligned} \dot{N}_1 &= N_1[r(1 - \frac{N_1}{K})] - kN_2(1-e^{-cN_1}) \\ \dot{N}_2 &= N_2[-b + \beta(1-e^{-fN_1})] \quad . \end{aligned} \tag{3.3}$$

The interpretation is that the prey, N_1, in the absence of the predator, grows logistically and the predator, in the absence of prey, dies out exponentially. The second term in the first equation models a predator population whose capacity to capture prey gradually satiates.

The equilibrium point of equation (3.3) can be computed explicitly:

$$\begin{aligned} \bar{N}_1 &= \ln(1 + \frac{b}{\beta})^{-1/f} \\ \bar{N}_2 &= r\bar{N}_1(1 - \frac{\bar{N}_1}{K})/k[1 - (1+ \frac{b}{\beta})^{c/f}] . \end{aligned}$$

Then, computing the Jacobian at the equilibrium it is easy to check that the signs of the determinant and trace depend on the magnitude of the parameters $\{r,K,k,c,b,\beta,f\} \equiv \underset{\sim}{\pi}$. Thus, there exists a family of curves, parametrized by some combination of members of $\underset{\sim}{\pi}$, carrying the eigenvalues into the RHP (c.f. Figure 10.6). Since large radius orbits move inward, the limit cycles are indeed generated by the Hopf mechanism. We also note that, since the eigenvalue trajectories are controlled by more than one parameter, the limit cycle can appear at finite, rather than zero amplitude.

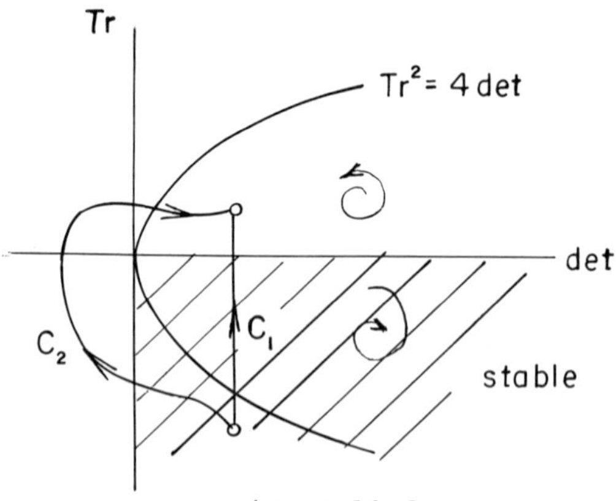

Figure 10.6

(3.3) Predator-prey type equations have been used by Bell [1] to model "populations" of antibody and antigen in the immune response. Using Friedrichs' [1] version of the bifurcation theorem, Pimbly [1] demonstrated that Bell's equations exhibit periodic behavior which can be interpreted biologically in terms of the mechanism controlling infection.

(3.4) For systems of 3 or more species the possibility of higher order bifurcations raises the same operational problems as we encountered for difference equation models. Successive bifurcations beyond the first occur when the eigenvalues of the Poincaré map passes outside the unit circle (Hirsch and Smale, [1]), thus higher periods, (and aperiodic behavior) of this difference equation will produce quite chaotic-looking population records. This phenomenon is quite well known in Hamiltonian systems (Arnold and Avez, [1]). Since it is generally much more difficult to obtain a reliable experimental record for population systems, the existence of such "strange attractors" would imply that the model may well not be experimentally verifiable.

Thus we find that bifurcation in model equations are a mixed blessing, explaining some phenomena and obscuring others.

4. Age-Structured Populations.

(4.1) By using ordinary differential and difference equation models we have taken a naive view of population dynamics by assuming that the state of the population is specified by total population number alone. A moments reflection shows that, in order to predict the growth of a population, account must be taken of internal variables such as age and size distributions. Clearly a thousand individuals past breeding age, or all of one sex do not constitute a viable population. In this section we shall illustrate some of the consequences of including the population age structure as a state variable.

(4.2) The equation of motion for an age-structured population is easy to write down. Let $n(a,t)$ = population age density function, i.e., $N(t) = \int_0^\infty n(a,t)da$ = total population. Then a conservation equation can be written for n:

$$\frac{\partial n}{\partial t} + \text{div } J_n = \text{loss by deaths.} \tag{4.1}$$

Since the flux of individuals, J_n, through age-time is just vn, where $v = \frac{da}{dt} = 1$, we can write

$$\frac{\partial n}{\partial t} + \frac{\partial n}{\partial a} = -\mu n \tag{4.2}$$

where $\mu(a,t,\cdot)$ is the age-specific deathrate. The special feature of the equation is the boundary condition giving the birthrate,

$$n(0,t) = \int_0^\infty b(a,t,\cdot)n(a,t)da \tag{4.3}$$

where $b(a,t,\cdot)$ is the age-specific birthrate. As we have indicated, the birth and death rates are functions of other variables as well. For example, population density frequently affects mortality and fecundity, so that

$$\mu = \mu(a,t,N) \tag{4.4a}$$

$$b = b(a,t,N) \tag{4.4b}$$

where

$$N = \int_0^\infty n(a,t,N)da. \tag{4.5}$$

is the total population. With appropriate smoothness and boundedness assumptions, Gurtin and MacCamy [1] proved existence and uniqueness for the system (4.2) - (4.5). We note that in engineering terms the age equations constitute a "distributed parameter positive feedback system." This easily implies that, as birthrates increase and/or deathrates decrease, the system will pass from a stable to an unstable regime. In the next subsection we examine the bifurcation behavior of a single population feeding off a single resource. Then we model a host-parasite system by coupling two age systems together. In both cases the existence of bifurcations must be inferred from qualitative and numerical arguments, since direct verification is unavailable. Nevertheless, we shall gain significant insights into some interesting ecological phenomena via our models.

(4.3) In one of the best known experiments in ecology, the Australian entomologist A. J. Nicholson maintained a population of sheep blowflies on a diet of chopped liver and sugar for several years. In Figure 10.7 we have reproduced a portion of his data. The biological explanation for the violent oscillations is straightforward: Nicholson deliberately kept the food supply to the adult flies below the level required to sustain a population the size of one of the peaks. At moderate population levels competition prevents any individual from obtaining enough protein. Protein starvation, in turn, reduces the fecundity of each adult fly so that the next generation is much smaller. For this smaller generation the food supply is adequate and the fecundity rebounds to its maximum level.

A model for this situation must include some accounting for the nutritional state of the adult flies since this governs the rate of egg laying. Accordingly, we shall define a variable, ξ, which measures the nutritional state (e.g., mass, "health") (Oster and Auslander, [1]). A conservation equation in (t,a,ξ) coordinates takes the form

$$\frac{\partial n}{\partial t} + \frac{\partial n}{\partial a} + \frac{\partial}{\partial \xi}(gn) = -\mu n \tag{4.5}$$

where

$$\frac{d\xi}{dt} = g(t,a,\xi,f) \tag{4.6}$$

is the growthrate of ξ, which depends on the food supply, $f(t)$. The birthrate is then

$$n(0,t,\xi) = \iint da\, d\xi'\, nb(t,a,\xi',\xi). \tag{4.7}$$

The equation for the food abundance is

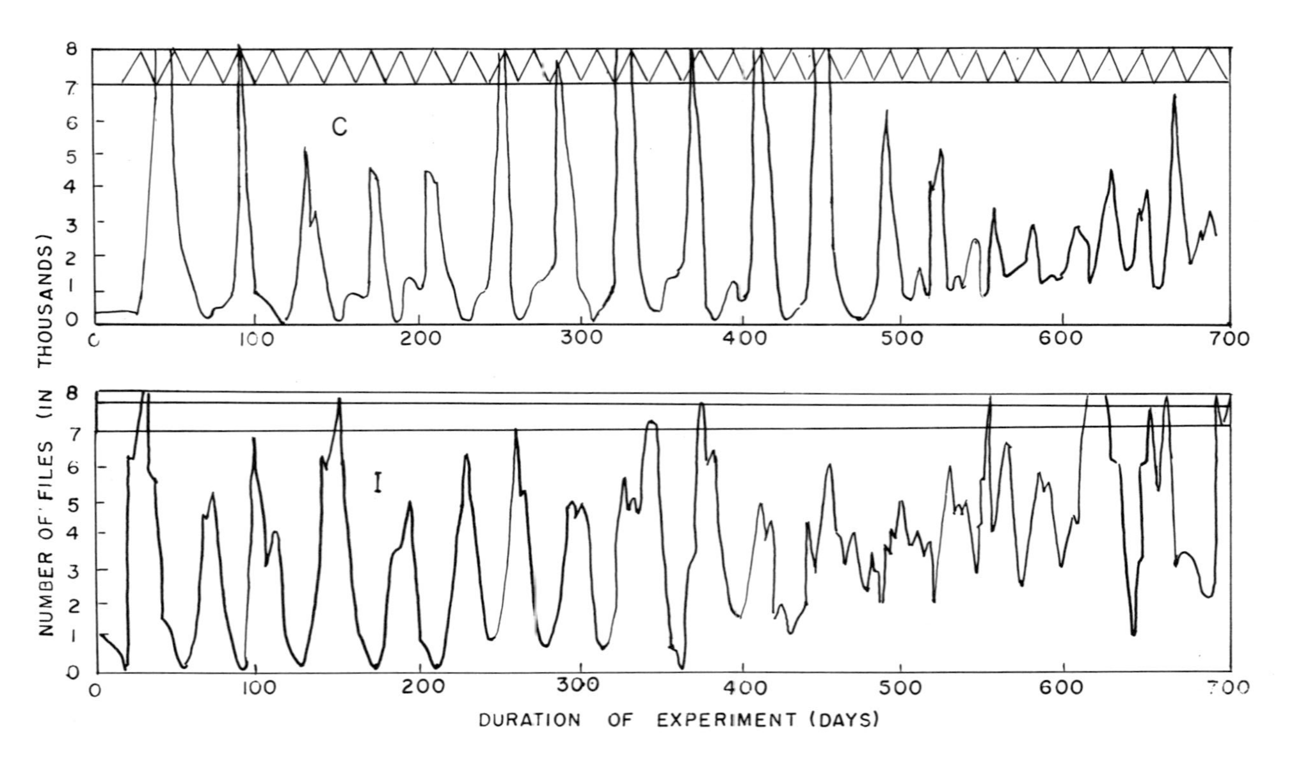

Figure 10.7. Top: periodically forced system Bottom: constant food supply

$$\frac{df}{dt} = u(t) - C(t,a,n) \tag{4.8}$$

where $u(t)$ is the rate food is supplied to the population and $C(\cdot)$ is the consumption rate by the adult flies. Reasonable empirical forms for the functions $b(\cdot)$, $\mu(\cdot)$, C are shown in Figure 10.8. A careful numerical simulation of this

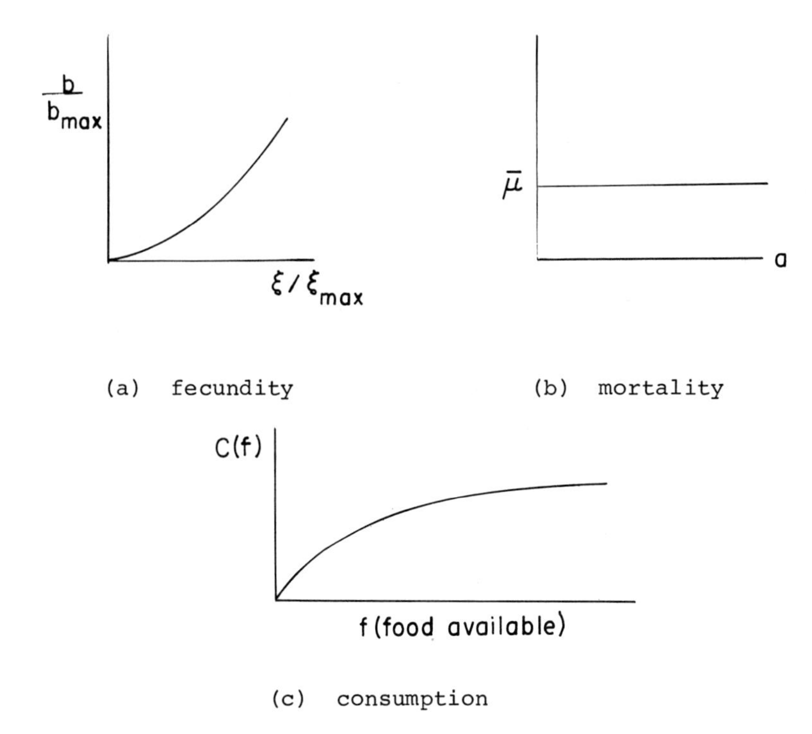

(a) fecundity (b) mortality

(c) consumption

Figure 10.8. Constitutive Relations

model shows reasonable agreement with experiment, Figure 10.9 (Oster and Auslander, [1]). However, we would like to see how the model generates these oscillations; naturally, the mechanism of bifurcations suggests itself. In order to examine this mechanism let us consider the simpler population-resource system

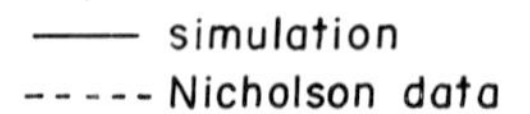

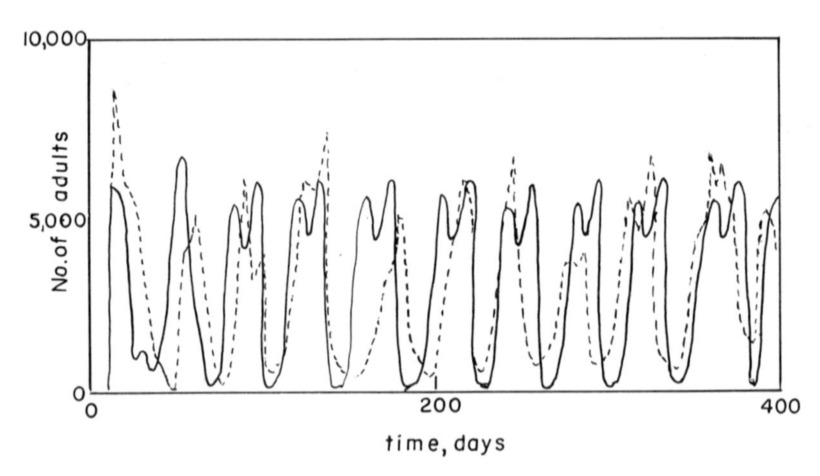

Figure 10.9

Comparison of simulated results with experimental data.

$$\frac{\partial n}{\partial t} + \frac{\partial n}{\partial a} = -\mu n \tag{4.9a}$$

$$n(0,t) = \int b(a,R)n\, da \tag{4.9b}$$

$$\frac{dR}{dt} = F(R,n). \tag{4.9c}$$

Regardless of the form of the functions the equations, linearized about an equilibrium state, will take the form (Oster and Takahashi, [1])

$$\frac{\partial x}{\partial t} + \frac{\partial x}{\partial a} = -\bar{\mu}x \tag{4.10a}$$

$$x(0,t) = gR(t) + \bar{b}\int_{\alpha}^{\alpha+\gamma} x(t,a)da \tag{4.10b}$$

$$\frac{dR}{dt} = -AR(t) - By(t) + Cu(t) \tag{4.10c}$$

where g, A, B, C, $\bar{\mu}$ and $\bar{b}$ are linearization constants and $y(t) = \int_0^\infty x\,da$. One way to obtain the response of system (4.10) to various food supply schedules, $u(t)$, is to compute the "transfer function" (Takahashi, Rabins, Auslander, [1]). That is, equation (4.10a) can be written as $\frac{\partial x}{\partial t} = \underset{\sim}{L}x$ where $\underset{\sim}{L}$ is a linear operator. Note that the initial conditions for the linearized system are zero, thus taking the Laplace Transform with respect to time is equivalent to the eigenvalue equation $Lx(a,s) = sx(a,s)$, where $s \in \mathbb{C}$. Therefore, we can compute the spectrum of the system (4.10) as follows. Taking the Laplace Transform of system (4.10) we obtain

$$X(a,s) = e^{-(s+\bar{\mu})a}G(s)R(s) \tag{4.11}$$

$$R(s) = \frac{C}{S+A+\frac{B}{s+\bar{\mu}}G(s)}U(s) \tag{4.12}$$

where $X(a,s)$, $R(s)$ and $U(s)$ are the transformed variables and

$$G(s) = g/1 - \frac{\bar{b}}{s+\bar{\mu}}\{e^{-(s+\bar{\mu})\alpha} - e^{-(s+\bar{\mu})(\alpha+\gamma)}\}\,. \tag{4.13}$$

Thus, the response, or "output", $X(s,a)$ can be expressed in terms of the input, $U(s)$, as

$$X(a,s) = g(s)U(s) \tag{4.14}$$

where $g(s) = e^{-(s+\bar{\mu})a}G(s)$ is the "transfer function". The response of the total population to food supply can be written

$$Y(s) = \frac{1}{s+\bar{\mu}}G(s)R(s) \tag{4.15}$$

where $Y(s) = \int_0^\infty X(a,s)da$. The characteristic equation for the system is given by the denominator of equation (4.1):

$$1 + \frac{Bg}{(s+A)\{s+\bar{\mu} - \bar{b}e^{-(s+\bar{\mu})\alpha}(1-e^{-(s+\bar{\mu})\gamma})\}} = 0. \quad (4.16)$$

The roots of (4.16) yield the system eigenvalues, as can be verified directly by substitution into (4.10). The product Bg can be interpreted as measuring (effect of population on food level) × (effect of food level on birthrate). Thus, by varying the "gain", Bg, each of the infinite number of eigenvalues traces out a path on the complex plane. Clearly the system is asymptotically stable for $u(t) = 0$, for the population will eventually starve. For $u(t) > 0$, the dynamics are controlled by, the parameter (Bg). We can get some idea of the effect of varying (Bg) on the system eigenvalues by examining the special case of $b(a) = b^*\delta(a-\alpha)$, i.e., all births occur at age α. The characteristic equation then becomes:

$$1 + \frac{Bg}{(s+A)(s+\bar{\mu})(1-b^*e^{-(s+\bar{\mu})\alpha})} = 0. \quad (4.17)$$

At $Bg = 0$, (4.17) has roots as $s = -A$, $s = -\bar{\mu}$, and those satisfying

$$e^{(s+\bar{\mu})\alpha} = b^*. \quad (4.18)$$

Setting $s = \sigma + i\omega$, the roots of (4.18) are seen to be at $\sigma = \rho$, $\omega = \frac{2n\pi}{\alpha}$, $n = 1,2,\ldots$, where $s = \rho$ is the real root of (4.18). In Figure 10.10 we sketch the "root locus" (Takahashi, Rabins and Auslander, [1]) for equation (4.17) as (Bg) is varied from 0 to ∞ (Oster and Takahashi, [1]).

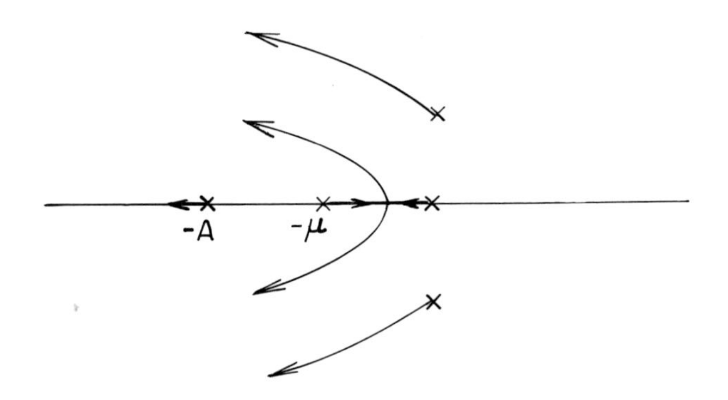

Figure 10.10

The branches start at $Bg = 0$ (denoted by x) and, as $Bg \to \infty$, approach the asymptotes $\omega = \pm \frac{2n\pi}{\alpha}$, $n = 0,1,\ldots$. What is apparent from Figure 10.9 is that there is some range of parameter values for which the linearized model passes from stability to instability. That is, as the interaction parameter (Bg) is varied in the appropriate range the leading pair of eigenvalues cross the imaginary axis. At this point the linearized system begins to exhibit small amplitude oscillations, which grow as the parameter is further increased. (Of course, sooner or later other root pairs cross to the RHP; these are associated with secondary frequencies, and will not concern us here.)

Simulation studies of the model system (4.5 - 4.8) indicate that the oscillations do not grow from zero amplitude, but bifurcate to finite amplitude oscillations. This suggests that the bifurcation is controlled by 2 parameters rather than 1. (c.f. Takens [1]) as shown in Figure 10.11.

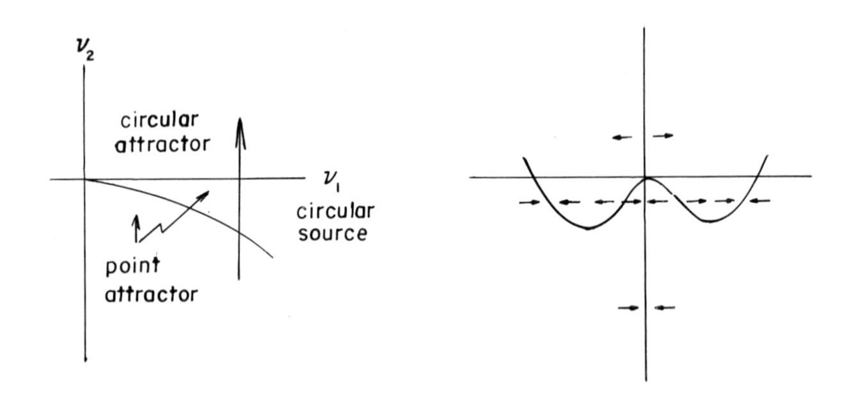

Figure 10.11

(4.4) We can use the age model to answer a puzzling question in the ecological literature. Over a period of several years Professor C. B. Huffaker maintained an experimental ecosystem containing a parasitic wasp which lays its eggs in the larvae of a certain moth. He noticed that, very quickly after initiation, the populations settled into stable oscillations. These oscillations were characterized by age structures which were practically discrete generations. Conventional predator-prey models do not suffice to explain these oscillations since phase plane trajectories cross--the explanation lies in the age structure dynamics. We can couple two conservation equations like (4.2) by an age specific interaction that models the searching behavior of the parasite. The resulting model looks like:

$$\frac{\partial p}{\partial t} + \frac{\partial p}{\partial a} = -\mu_p(a,t)p \qquad (4.19)$$

$$p(0,t) = \int_{\gamma_p}^{\alpha_p+\gamma_p} b_p(a',t,H_0)p(a',t)da' \tag{4.20}$$

$$\frac{\partial h}{\partial t} + \frac{\partial h}{\partial a} = -\mu_h(a,t,H,H_0,P_1)h \tag{4.21}$$

$$h(0,t) = \int_{\alpha_h}^{\alpha_h+\gamma_h} b_h(t,a',H_1(t-\tau))h(a',t)da' \tag{4.22}$$

where

$$H_0(t) = \int_{\beta}^{\beta+\delta} h(a',t)da' = \text{no. host larvae} \tag{4.23}$$

$$H_1(t) = \int_{\alpha_h}^{\alpha_h+\gamma_h} h(a',t)da' = \text{no. host adults} \tag{4.24}$$

$$H(t) = \int_0^{\infty} h(a',t)da' = \text{total no. hosts} \tag{4.25}$$

$$P_1(t) = \int_{\alpha_p}^{\alpha_p+\gamma_p} p(a',t)da' = \text{no. parasite adults.} \tag{4.26}$$

The form of the interaction between the populations can be derived by assuming a random search by each parasite for host larvae and employing a mean "area of discovery," A, for each. If the hosts are distributed randomly (Poisson) in a plane, the inter-arrival times are distributed exponentially. Thus the interaction takes the form: (Auslander, Oster, Huffaker, op. cit.)

$$[\text{no. hosts parasitized}](a) = bh(a)(1-e^{-A(s)P_1}) \tag{4.27}$$

This is added to the natural mortality (assumed constant) to obtain the total host mortality. As indicated in equation (4.22) the host birthrate includes a delayed effect that depends on the nutritional history of the host. This is because fecundity is a function of adult size, which depends on the available food.

The above model was simulated numerically using Huffaker's data (Auslander, Oster and Huffaker, op. cit.), and some of the results are shown in Figure 10.12. First of all, as the strength of the interaction is increased (e.g. by increasing the area of discovery, A) the system undergoes a transition from a state wherein all age classes are represented in both populations to one wherein only a few age classes are represented. That is, the age profiles of both species condense into "travelling waves," which propagate through the age structure in such a fashion that in a "stroboscopic photograph," the generations appear virtually discrete. The phase relationship of the population waves in each population determine the extent to which the populations can coexist. If the parameters are adjusted so that all age classes are represented, then the populations do not coexist: the parasite eliminates the host and then dies out itself.

Following the same procedure outlined for the single population model, we can linearize equations (4.19) - (4.26), Laplace transform and examine the roots of the characteristic equation as the coupling parameter is increased. Clearly, at zero coupling the parasite system is stable about the zero solution while the host population approaches a stable age distribution. Simulation indicates that a stationary age profile also exists with all age classes represented in both populations (continuous generations). Furthermore, at sufficiently high coupling strength the system is stable at zero. Thus, a root locus study, which reveals a leading root pair crossing the imaginary axis as the coupling is increased,

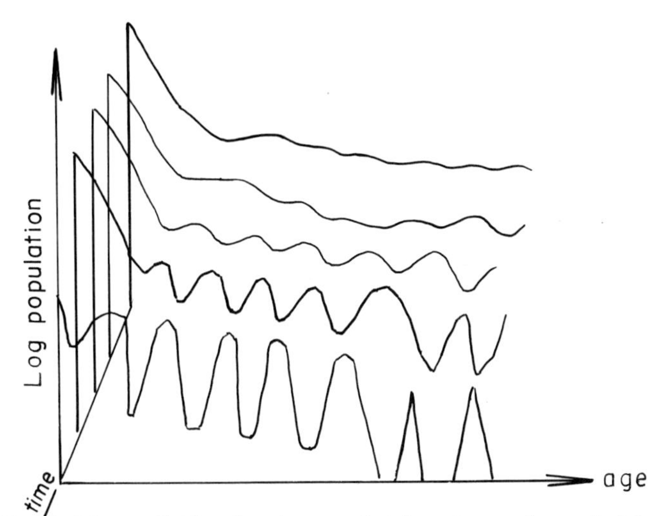

Evolution of the host population toward a stable age distribution in the absence of the parasite. The initial waves were induced by the periodic addition of adult females.

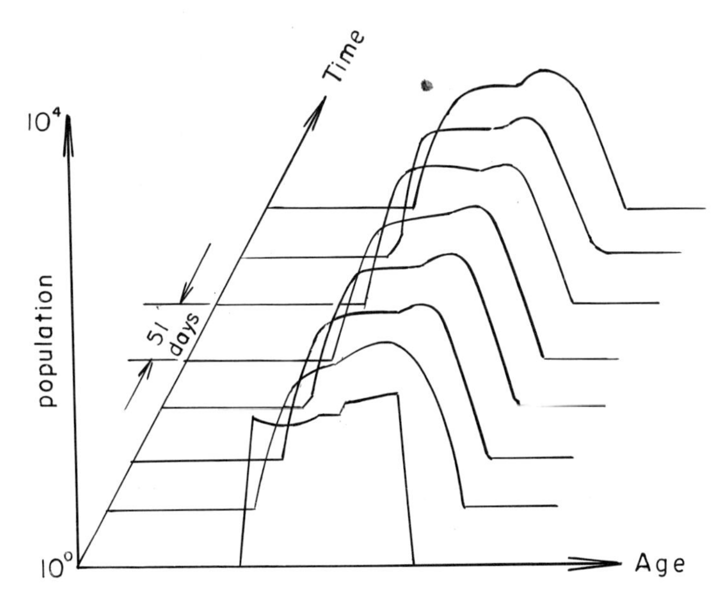

Stroboscopic shot at one generation-time intervals (~51 days) of a "pulse" of hosts which evolves to a stable periodic solution.

Figure 10.12. Simulation of Host-Parasite System

leads us to conclude that an intuitive explanation for the population waves is a bifurcation phenomenon. Moreover, it is this mechanism which gives us a satisfying explanation for how the two populations are able to coexist in a homogeneous environment--the bifurcation phenomenon creates a "phase niche" within which the host can escape total annihilation by the parasite.

(4.5) Two other phenomena involving the age structure deserve comment. First, an examination of Figure 10.7b shows that if a periodic signal (in this case a periodic food supply) is applied to the population system, the forcing frequency interacts with the natural resonant frequency to produce a "beat" frequency with a wavelength longer than either component (Oster and Auslander, [1]). This suggests a possible explanation for certain population periodicities observed in nature which do not appear to track any apparent environmental cycle. Secondly, there appear to be component frequencies higher than that of the major resonance. This suggests that secondary bifurcations from the basic cycle may play a role in the dynamics. If the age system is discretized along the characteristics, the resulting set of difference equations corresponds to the Leslie model well known to demographers. Beddington and Free [1] simulated such a discrete age class model and found that, as certain parameters are varied, transitions to chaotic behavior occurred reminiscent of the aperiodic orbits discussed in Section 2 for single difference equations. Thus it appears that the bifurcation phenomenon can supply a satisfying mathematical mechanism for explaining not only cyclic regularities in population dynamics, but perhaps some of the irregularities as well.

SECTION 11

A MATHEMATICAL MODEL OF TWO CELLS VIA TURING'S EQUATION

BY

S. SMALE

(11.1) Here we describe a mathematical model in the field of cellular biology. It is a model for two similar cells which interact via diffusion past a membrane. Each cell by itself is inert or dead in the sense that the concentrations of its enzymes achieve a constant equilibrium. In interaction however, the cellular system pulses (or expressed perhaps over dramatically, becomes alive!) in the sense that the concentrations of the enzymes in each cell will oscillate indefinitely. Of course we are using an extremely simplified picture of actual cells.

The model is an example of Turing's equations of cellular biology [1] which are described in the next section. I would like to thank H. Hartman for bringing to my attention the importance of these equations and for showing me Turing's paper.

The general idea of our model is to first give abstractly an example of a dynamical system for the chemical kinetics of four chemicals (or enzymes). This dynamics represents the reaction of these chemicals with each other and has the property that every solution tends to one unique stationary point or equilibrium in the space of concentrations as time goes to ∞. This is the sense in which the cell is dead, where the cell consists of these four chemicals. After a period of transition, the chemical system stays at equilibrium. We emphasize that our reaction process is an abstract mathematical one and that we have not tried to find four chemicals with this kind of chemical kinetics.

The next step is to give four positive diffusion constants for the membrane which could describe the diffusion of the four chemicals past the membrane. The cellular system consisting of the two cells separated by the membrane will be described by differential equations according to Turing. With our choice of the chemical kinetics and diffusion constants this new dynamical system will have a nontrivial periodic solution and essentially every solution will tend to this periodic solution. Thus no matter what the initial conditions, the interacting system will tend toward an oscillation (with fixed period). After an interval of transition, it will oscillate.

Both the equilibrium of the isolated cell and the oscillating solution of the interacting system described above are stable (or are attractors) and even stable in a global way. But more than this, the equations themselves are stable so that any equations near ours have the same properties. Our dynamical systems are "structurally stable." This gives them at least a physical possibility of occurring.

In Turing's original paper some examples of Turing's equations are given with oscillation. However, these are linear and it is impossible to have an oscillation in any structurally stable linear dynamical system. Linear analysis can be used primarily to understand the neighborhood of an equilibrium solution. Development of linear Turing theory has been carried very far in the very pretty paper of Othmer and Scriven [1].

Our example has reasonable boundary conditions, as one or more of the concentrations goes to 0 or to ∞. Also, a complete phase portrait of the differential equation in eight dimensions for the cellular system is obtained.

This example and Turing's equations as well go beyond biology. The model here shows how the linear coupling of two different kinds of processes, each process in itself stationary, can produce an oscillation. This is the coupling of transport processes (in this case diffusion across a membrane) and transformation processes (in this case chemical reactions). In ecology, Turing's equations have another interpretation; see, e.g., Levin [1]. Also S. Boorman's Harvard Thesis has a related interpretation and analysis.

We finally remark that our results could equally well be interpreted as putting a single cell into an environment which could start it pulsating.

(11.2) We give a brief description of Turing's equations [1]. These are sometimes called Rashevsky-Turing equations because of earlier work of Rashevsky on this subject.

One starts from a cell-complex, in either the biological or mathematical sense of the word, e.g., as given in Figure 11.1.

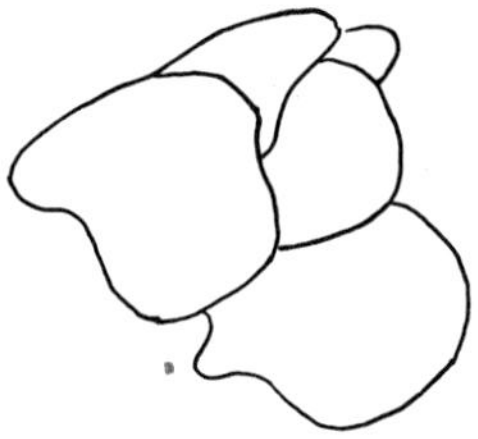

Figure 11.1

From the mathematical point of view this system is a cell-complex structure on a two- or three-dimensional manifold (e.g., an open set of R^2 or R^3). Suppose there are N cells and they are numbered $1,\dots,N$.

It is supposed that the cells contain enzymes (or chemicals, or "morphogens" in the terminology of Turing) which react with each other. Suppose there are m of these chemicals. Then the state space for each cell is the space

$$P = \{x \in R^m \mid x = (x^1,\dots,x^m),\ \ x^i \geq 0,\ \text{each}\ \ i\},$$

where x^i denotes the concentration of the ith chemical.

The state space for the system under discussion is the Cartesian product $P \times \cdots \times P$ (N times) or $(P)^N$. Thus a state for this cellular system is a point, $x \in (P)^N$,

$x = (x_1,\ldots,x_n)$ with each $x_i \in P$ giving all the concentrations for the ith cell, $i = 1,\ldots,N$. The dynamics for the typical cell by itself is given by an ordinary differential equation on P; this can be described by a map $R: P \to R^m$ and $dx/dt = R(x)$. This R describes how the chemicals react with each other in that cell; the subject of chemical kinetics deals with the nature of R. Most typically, the dynamics of $dx/dt = R(x)$ on P is described by the existence of a single equilibrium $\bar{x} \in P$ such that every solution tends to $\bar{x}$; at least this will be the case if conservation laws have been taken into account one way or another (as in the situation in Turing [1] or Othmer and Scriven [1]).

A natural boundary condition on this equation is that if $x \in P$, $x = (x^1,\ldots,x^m)$, with $x^k = 0$, then the kth component $R^k(x)$ of $R(x)$ is positive.

So far we have discussed each cell in some kind of hypothetical isolation. The cells are separated from each other by a membrane which allows for diffusion from one cell to adjoining cells. In the simplest case of diffusion, if a certain chemical has a bigger concentration in the rth cell than an adjoining well, then the concentration of that chemical decreases in the rth cell, at a rate proportional to the difference. This gives some motivation to Turing's equations which add this diffusion term to give an interaction between the cells.

$$\text{(T)} \quad \frac{dx_k}{dt} = R(x_k) + \sum_{\substack{i \in \text{set of cells} \\ \text{adjoining } k\text{th cell}}} \mu_{ik}(x_i - x_k), \quad k = 1,\ldots,N.$$

Let us explain (T) in detail. The first term above,

$R(x_k)$ gives the chemical kinetics in the kth cell. The 2nd term above describes the diffusion processes between cells. Thus $x_i - x_k \in R^m$ represents the difference of the concentrations of all the chemicals between the ith and k^{th} cells. Here μ_{ik} is a linear transformation from R^m to R^m or an $m \times m$ matrix. In the most natural simple case, and the case we develop here, μ_{ik} is a positive diagonal matrix. Also the chemical kinetics for each cell is considered the same. (T) is a 1st order system of ordinary differential equations on the state space $(P)^N$ of the biological system, and will tell how a state moves in time.

We specialize to a case of 2 cells adjoined along a membrane which is the example pursued in the rest of the paper.

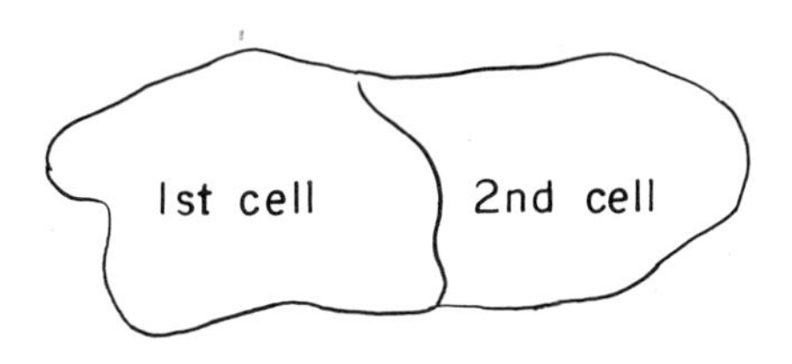

Figure 11.2

$$(T_2) \qquad \begin{aligned} dz_1/dt &= R(z_1) + \mu(z_2 - z_1), \\ dz_2/dt &= R(z_2) + \mu(z_1 - z_2). \end{aligned}$$

This is an equation on $P \times P$ with $(z_1, z_2) \in P \times P$. Here μ will be of the form

$$\begin{pmatrix} \mu_1 & & 0 \\ & \ddots & \\ 0 & & \mu_m \end{pmatrix}$$

with each $\mu_i > 0$. This is the most simple case. We may also write (T_2) as given by a vector field X on $P \times P$ where

$$X(z_1,z_2) = (R(z_1) + \mu(z_2-z_1),\ R(z_2) + \mu(z_1-z_2)).$$

(11.3) Here we state our results.

Main Theorem. Let $P = \{z \in R^4,\ z = (z^1,z^2,z^3,z^4),\ z^i \geq 0\}$. There exists a smooth (C^∞) map $R: P \to R^4$, and $\mu_1,\ldots,\mu_4 > 0$ with the following properties (1), (2), (3) below:

(1) The differential equation $dz/dt = R(z)$ on P is globally asymptotically stable and is structurally stable.

In other words there is a unique equilibrium $\bar{z} \in P$ of the differential equation and every solution tends to $\bar{z}$ as $t \to \infty$. That R is structurally stable means that the equation $dz/dt = R_0(z)$ has the same structural properties as $dz/dt = R(z)$ if R_0 is a C^1 perturbation of R. (See [1] for details on these kinds of stability and background on ordering differential equations.)

(2) On $P \times P$ with $z = (z_1,z_2) \in P \times P$ the differential system,

$$(T)\qquad \begin{aligned} dz_1/dt &= R(z_1) + \mu(z_2-z_1), \\ dz_2/dt &= R(z_2) + \mu(z_1-z_2), \end{aligned} \qquad \mu = \begin{pmatrix} \mu_1 & & 0 \\ & \ddots & \\ 0 & & \mu_4 \end{pmatrix},$$

is a "global oscillator" and is structurally stable.

More precisely, a global oscillator on $P \times P$ is a dynamical system which has a nontrivial attracting periodic solution γ and except for a closed set Σ of measure 0, *every solution tends to* γ, *as* $t \to \infty$. *In fact, here* Σ *is a six-dimensional smoothly imbedded cell and on* Σ *every solution tends to the unique equilibrium* $(\bar{z},\bar{z})$ *of* (T).

(3) *The boundary conditions are reasonable in the following way. Let* $z_0 = (z_0^1, z_0^2, z_0^3, z_0^4) \in P$ *satisfy* $z^k = 0$ *for some* k *between one and four. Then the* k^{th} *component* $R^k(z_0)$ *of* $R(z_0)$ *is positive*.

The above theorem gives mathematical precision to the statements made in Section 1 via the interpretation of (T) as in Section 2.

The example is related to the phenomena of Hopf bifurcation; see Section 3. Our analysis is more global however, and we have a complete description of the phase portrait.

(11.4) In this section we show how to construct the differential equations of the previous section.

Towards obtaining the vector field $Q: P \to R^4$ of the main theorem of Section 3 we will find a C^∞ vector field $Q: R^4 \to R^4$ on R^4 and

$$\mu = \begin{pmatrix} \mu_1 & & 0 \\ & \ddots & \\ 0 & & \mu_4 \end{pmatrix}$$

with these properties:

(1) Q has the origin, 0, as a global attractor for the equation $dz/dt = Q(x)$ on R^4.

(2) There is a $K > 0$ such that if $z \in R^4$, $||z|| \geq K$, then $Q(z) = -z$.

(3) On $R^4 \times R^4$ the vector field

$$(z_1,z_2) \to (2Q(z_1) + \mu(z_2-z_1),\ 2Q(z_2) + \mu(z_1-z_2))$$

is a global oscillator.

Once such a Q has been found we finish as follows: Choose some $\bar{z} \in P$ so that $z-\bar{z} \in P$ for all z with $||z|| \leq K$. Let $R(z) = 2Q(z-\bar{z})$. Then R will have the properties of Section 3.

This changes the question from P to R^4 where we can use the linear structure systematically (even though our equations are not linear).

To find a Q as needed above we first ignore property (2), or behaviour of Q at ∞ and concentrate on (1) and (3). In fact we shall find $S: R^4 \to R^4$ satisfying (1) and (3) with S replacing Q and after that S is modified to satisfy (2).

Toward constructing this S, observe that the set $\Delta = \{(z_1,z_2) \in R^4 \times R^4 \mid z_1 = z_2\}$ has the property that the vector field

$$(*) \qquad (2S(z_1) + \mu(z_2-z_1),\ 2S(z_2) + \mu(z_1-z_2))$$

is tangent to Δ. That is, Δ is invariant under the flow and on Δ the flow is contracting to the origin.

Now suppose S satisfies $S(-z) = -S(z)$ or that S is odd. Then on

$$\Delta^{\perp} = \{(z_1,z_2) \in R^4 \times R^4 | z_1 = -z_2\}$$

the vector field (*) is invariant and has the form

$$z \to S(z) - \mu(z)$$

(up to a factor of 2). From these considerations we are motivated to seek an odd map $S: R^4 \to R^4$ which satisfies the following:

(S1) S has 0 has a global attractor and $S - \mu$ is a global oscillator on R^4.

(S2) $\Delta^{\perp}$ is an attractor for (*) on $R^4 \times R^4$.

(S3) Boundary conditions can be made good.

The heart of the matter lies in (1); we consider that next. First consider the matrix

$$\overline{\mu} = \begin{pmatrix} a & 0 & \gamma a & 0 \\ 0 & a & 0 & \gamma a \\ -\gamma a & 0 & -2a & 0 \\ 0 & -\gamma a & 0 & -2a \end{pmatrix}$$

where $a < -1$ and $\sqrt{2} < \gamma < 3/2$, in linear coordinates $y = (y^1,\ldots,y^4)$ on R^4.

Note that $\overline{\mu}$ has real positive eigenvalues say μ_1, μ_2,μ_3,μ_4. This can be checked easily since $\overline{\mu}$ resembles a 2×2 matrix. Also there is a linear change of coordinates which changes the matrix

$$\mu = \begin{pmatrix} \mu_1 & & 0 \\ & \ddots & \\ 0 & & \mu_4 \end{pmatrix} \quad \text{into } \overline{\mu}.$$

The coordinates of chemical concentrations are those in which $\overline{\mu}$ has the diagonal form. Thus the y_i do not represent concentrations. However the y coordinates are much easier to work with.

We give now S as the sum of a linear map $S_1: R^4 \to R^4$ and a cubic map S_3 in terms of the y-coordinates.

Thus let $S = S_1 + S_3$ with

$$S_1 = \begin{pmatrix} 1+a & 1 & \gamma a & 0 \\ -1 & a & 0 & \gamma a \\ -\gamma a & 0 & 2a & 0 \\ 0 & -\gamma a & 0 & 2a \end{pmatrix}$$

$$S_3(y) = (-(y^1)^3, 0, 0, 0).$$

One notes now that S is odd and since the inner product $\langle Sy, y\rangle < 0$ if $y \neq 0$ (an easy check) it follows that the origin of R^4 is a global attractor for S.

The next step is to check that $S - \mu$ takes the form

$$\begin{pmatrix} \begin{pmatrix} 1 & 1 \\ -1 & 0 \end{pmatrix} & 0 \\ 0 & \begin{pmatrix} 4a & 0 \\ 0 & 4a \end{pmatrix} \end{pmatrix} + S_3.$$

Thus the (y^1, y^2) 2-dimensional subspace is a contracting invariant subspace in R^4 for $S - \mu$, since $a < 0$; on this subspace, the equations for $S - \mu$ take the form

$$dy'/dt = y^2 - ((y')^3 - y^1), \quad dy^2/dt = -y^1.$$

This is Van der Pol's equation (see Hirsch-Smale [1]), which we know is a global oscillator.

Thus $S - \mu$ is a global oscillator on R^4.

The next step is to show that (S2) is true. This can be proved along the following lines.

The vector field X on $R^4 \times R^4$ given by

$$X(z) = (2S(z_1) + \mu(z_2 - z_1),\ 2S(z_2) + \mu(z_1 - z_2))$$

can be written in the form $Y_1(z) + Y_2(z)$ where $Y_1(z) \in \Delta$, $Y_2(z) \in \Delta^{\perp}$. Then

$$Y_1(z) = (S(z_1) + S(z_2), S(z_1) + S(z_2)),$$

$$Y_2(z) = (S(z_1) - S(z_2) + \mu(z_2-z_1),$$
$$-(S(z_1) - S(z_2) + \mu(z_2-z_1))).$$

That X points toward $\Delta^{\perp}$ follows from the lemma.

<u>Lemma</u>. $\langle S(z_1) + S(z_2), z_1 + z_2 \rangle \leq 0$.

<u>Proof of Lemma</u>. Write $S = S_1 + S_2$. We already know

$$\langle S_1(z_1) + S_1(z_2), z_1+z_2 \rangle = \langle S_1(z_1+z_2), z_1+z_2 \rangle \leq 0.$$

But

$$\langle S_3(z_1) + S_3(z_2), z_1+z_2 \rangle = [-(y_1^1)^3 - (y_2^1)^3] \cdot (y_1^1 + y_2^1)$$

and that is ≤ 0 since, for any real numbers a and b, $(a^3+b^3)(a+b) \geq 0$.

Finally one "straightens out" the flow of S outside some large ball. One uses a smooth function $\phi: R^+ \to R^+$, $0 \leq \phi \leq 1$, $\phi \equiv 0$ in a neighborhood of 0, and $\phi(r) \equiv 1$ for large enough r; then

$$Q(z) \equiv (1-\phi(||z||))S(z) - \phi(||z||)z.$$

It can be shown that Q satisfies (1), (2), (3) with suitable constants in the definition of ϕ.

(11.5) We end this note with some discussion of our results.

Various forms of Turing's equations, or reaction-diffusion equations have appeared in one form or another in many works and fields. However, any sort of systematic understanding or analysis seems far away. Before one can expect any general understanding, many examples will have to be thought through, both on the mathematical side and on the experimental side. This is one reason why I have worked out this model.

Moreover, the work here poses a sharp problem, namely to "axiomatize" the properties necessary to bring about oscillation via diffusion. In the 2-cell case, just what properties does the pair (R,μ) need to possess (where R is "dead") to make the Turing interacting system oscillate? In the many-cell case, how does the topology contribute?

We have not hesitated to make simplifying assumptions here, because we were not making an analysis, but producing an example. Because of the structural stability properties of this example, one can use it to obtain more complicated examples, e.g., with as many cells (more than one) as one wants, as many chemicals (more than three) as one wants and complicated diffusion matrices. But it is more difficult to reduce the number of chemicals to two or even three. Also it is a problem to construct a model with three cells and two or three chemicals.

There is a paradoxical aspect to the example. One has two dead (mathematically dead) cells interacting by a diffusion process which has a tendency in itself to equalize the concentrations. Yet in interaction, a state continues to pulse indefinitely.

Several chemists have pointed out to me that interpreting the reaction R to be an "open system" makes the model more acceptable.

There is quite a history of numerical work on related systems which I will not try to cover here.

Finally, there is a partial differential equation analogue to the version of Turing's equations studied here. This can be found in Turing's paper [1]. In this P.D.E. context the recent work of L. Howard and N. Kopell on the Zhabotinsky oscillation bears strong analogies to the present work.

SECTION 12

A STRANGE, STRANGE ATTRACTOR

BY

JOHN GUCKENHEIMER

Examples have been given by Abraham-Smale [1], Shub [1], and Newhouse [2] of diffeomorphisms on a compact manifold which are not in the closure of diffeomorphisms satisfying Smale's Axiom A or in the closure of the set of Ω-stable diffeomorphisms (Smale [1]). The suspension construction (Smale [1]) allows one to give analogous examples for vector fields on compact manifolds.

This note gives another example of a vector field on a compact manifold which does not lie in the closure of Ω-stable or Axiom A vector fields. The interest of this example is that the violation of Axiom A' occurs differently than in the examples previously given. This example has additional instability properties not verified for the previous

---

Research partially supported by the National Science Foundation.

examples. A vector field X is said to be topologically Ω stable if nearby vector fields (in the C^1 topology on the space of vector fields) have nonwandering sets homeomorphic to the nonwandering set of X. Our example is not topologically Ω stable. Moreover, it provides another negative answer to the following question about dynamical systems: is it generically true that the singularities of a vector field are isolated in its nonwandering set? Previous examples of Newhouse have nonisolated singularities in non-attractive parts of the nonwandering set.

The example is based upon numerical studies of a system of differential equations introduced by Lorenz [1]. The system studied by Lorenz seems to have the dynamical behavior of our example, but we do not attempt to make the estimates necessary to prove this statement. I would like to acknowledge the assistance of Alan Perelson in doing the numerical work which underlies this note and conversations with R. Bowen, C. Pugh, S. Smale, and J. Yorke. Finally, we mention the explicit equations of Lorenz which display such marvelous dynamics (see Example 4B.8, p. 141):

$$\dot{x} = -10x + 10y, \quad \dot{y} = -xz + 28x - y, \quad \dot{z} = xy - 8/3\, z.$$

We define a C^∞ vector field X in a bounded region of $\mathbb{R}^3$. Inside the region there will be a compact invariant set A which is an attractor in the sense that A has a fundamental system of neighborhoods, each of which is forward invariant under the flow of X. The set A is two dimensional. To describe the construction of X, we use coordinates (x,y,z) in $\mathbb{R}^3$.

The vector field X is to have three singular points. The first, $p = (0,0,0)$, is a saddle with a two dimensional stable manifold $W^s(p)$. The rectangle $\{(x,y,z) | x = 0, -1 \leq y \leq 1, 0 \leq z \leq 1\}$ is to be contained in $W^s(p)$. The stable eigenvectors of X at p are $\frac{\partial}{\partial y}$ with an eigenvalue of large absolute value and $\frac{\partial}{\partial z}$ with an eigenvalue of small absolute value. The unstable manifold $W^u(p)$ contains the segment from $(-1,0,0)$ to $(1,0,0)$ and has an eigenvalue of intermediate absolute value. Other conditions on $W^u(p)$ are imposed below.

The other two singular points of X are $q_\pm = (\pm 1, \pm 1/2, 1)$. These are saddle points with one dimensional stable manifolds $W^s(q_\pm)$. The segments from $(\pm 1, -1, 1)$ to $(\pm 1, 1, 1)$ are contained in $W^s(q_\pm)$. The negative eigenvalues of X at $q_\pm$ have large absolute values. The remaining eigenvalues of $q_\pm$ are complex with eigenspaces spanned by $\frac{\partial}{\partial y}$ and $\frac{\partial}{\partial z}$. The real parts of these eigenvalues are small.

Consider the square $R = \{(x,y,z) | -1 \leq x \leq 1, -1 \leq y \leq 1, z = 1\}$ and its Poincaré return map θ. The map θ is not defined when X is ± 1 or 0 since these points lie in the stable manifold of one of the singular points. The orbits in R for $X = \pm 1$ never leave R while those for $x = 0$ never return. At all other points of R, θ is defined. Let R_+ be the set $R \cap \{(x,y,z) \mid 0 < x < 1\}$ and R_- be the set $R \cap \{(x,y,z) \mid -1 < x < 0\}$. Define $\theta_\pm$ to be θ restricted to $R_\pm$. We assume that there are functions $f_\pm$, $g_\pm$ and a number $\alpha > 1$ with the properties that

$\theta_\pm(x,y) = (f_\pm(x), g_\pm(x,y))$, $0 < \partial g_\pm/\partial y < 1/2$, and $df_\pm/dx > \alpha$. The numbers $\lim_{x \to 0} f_\pm(x)$, denoted $\rho_\pm$, are assumed to have the properties $\rho_+ < 0$, $\rho_- > 0$, $\theta_-(\rho_+) < 0$, and $\theta_+(\rho_-) > 0$. The first intersections of $W^u(p)$ with R occur at the points with $x = \rho_\pm$. Finally, it is assumed that the images of $g_\pm$ are contained in the intervals $[\pm 1/4, \pm 3/4]$. Figure 12.1 illustrates these essential features of the flow X.

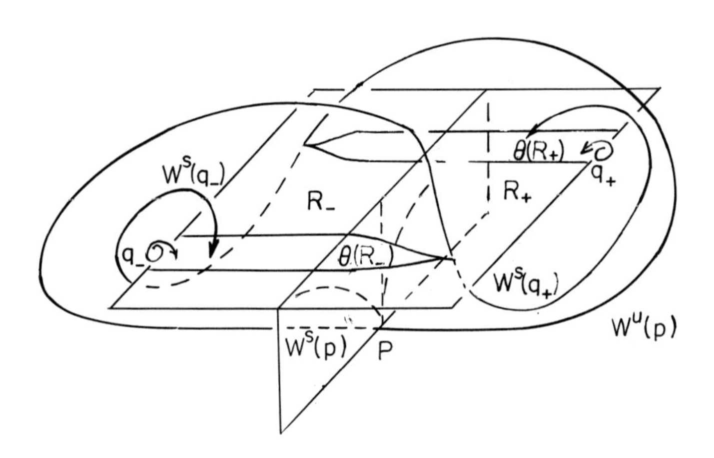

Figure 12.1

We remark that the conditions imposed on the eigenvalues of X at p imply that $\lim_{x \to 0} \partial g_\pm(x,y)/\partial y = 0$ and $\lim_{x \to 0} df_\pm/dx = \infty$. The reason for this behavior is given by solving a linear system of differential equations near a saddle point. The return maps $\theta_\pm$ acquire singularities like a power of x because the trajectories of $R_\pm$ come arbitrarily close to p.

In the theorems which we now state, we assume that the

vector field X is extended to a vector field on a compact three manifold M. We continue to denote the extended vector field X. Note that the only properties used in defining X which do not remain after perturbation are the existence of the functions $f_{\pm}$ and $g_{\pm}$. These functions are introduced to simplify the discussion and are not essential properties of X.

(12.1) Theorem. There is a neighborhood $\mathcal{U}$ of X in the space of C^r vector fields on M $(r \geq 1)$ and a set $\mathcal{V}$ of second category in $\mathcal{U}$ such that if $Y \in \mathcal{V}$, then Y has a singular point which is not isolated in its nonwandering set.

(12.2) Theorem. The vector field X has a neighborhood $\mathcal{U}$ in the space of C^r vector fields on M $(r \geq 1)$ with the property that if $\mathcal{V} \subset \mathcal{U}$ is an open set in the space of C^r vector fields, then there are vector fields in $\mathcal{V}$ whose nonwandering sets are not homeomorphic to each other.

Theorem (12.2) states that X is not in the closure of the set of topologically Ω-stable vector fields.

We attack the proofs of both of these theorems by giving a description of the nonwandering set of X. This description is given largely in terms of "symbolic dynamics" (Smale [4]).

Consider the return map θ of R. We pick out four subsets of $\theta(R)$ which will be used in analyzing the symbolic dynamics of the nonwandering set of X. Denote

$$R_1 = \theta(R_+) \cap \{\rho_+ < x < 0\}$$

$$R_2 = \theta(R_+) \cap \{0 < x < f_+(\rho_-)\}$$

$$R_3 = \theta(R_-) \cap \{f_-(\rho_+) < x < 0\}$$

$$R_4 = \theta(R_-) \cap \{0 < x < \rho_-\}.$$

Figure 12.2 shows these sets. The image of R_1 under θ extends horizontally across R_3 and R_4. $\theta(R_2)$ extends horizontally across R_1. Similarly, $\theta(R_3)$ extends across R_4, and $\theta(R_4)$ extends across R_1 and R_2.

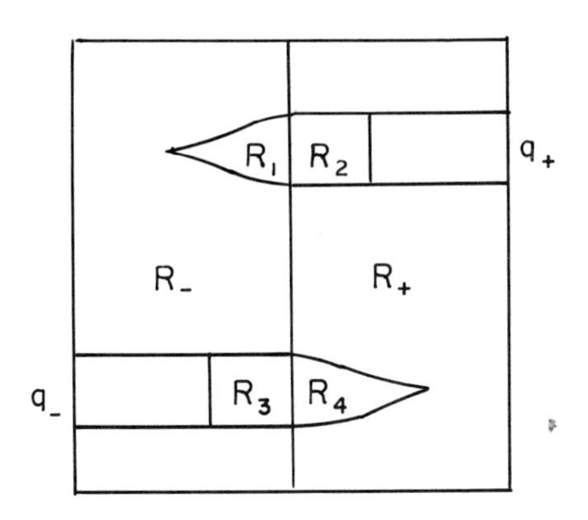

Figure 12.2

Now consider sequences $\{a_k\}_{k=0}^{\infty}$ of the integers 1, 2, 3, and 4 such that, for each k, (a_k, a_{k+1}) is one of the pairs (3,1), (4,1), (1,2), (4,3), (1,4), or (2,4). The set of such sequences forms the underlying space Σ of a "sub-shift of finite type" with transition matrix

$$\begin{pmatrix} 0 & 1 & 0 & 1 \\ 0 & 0 & 0 & 1 \\ 1 & 0 & 0 & 0 \\ 1 & 0 & 1 & 0 \end{pmatrix}.$$

Corresponding to each finite sequence $\{a_0,\ldots,a_n\}$ constructed from "admissible" pairs listed above, the intersection $\bigcap_{k=0}^{n} \theta^k(R_{a_k})$ contains a component which extends horizontally across R_{a_0}. For example, if $a_0 = 1$, then the images of R_2 and R_4 extend across R_1. If $a_1 = 2$, then only the image of R_4 need extend across R_2. Hence $a_2 = 4$, $\theta(R_4)$ extends across R_{a_2}, and $\theta^2(R_4)$ extends across R_1. As n increases, the vertical height of these strips decreases exponentially. If $\{a_k\} \in \Sigma$, then $\bigcap_{k=0}^{\infty} \theta^k(R_{a_k})$ contains an arc crossing R_{a_0} horizontally. There are an uncountable number of sequences in Σ, hence $S = \bigcap_{k=0}^{\infty} \theta^k(\bigcup_{i=1}^{4} R_i)$ contains an uncountable number of arcs extending across each R_i.

We want to investigate whether S is contained in the nonwandering set of θ. If each arc contained in S has an image under some iterate of θ which extends across each R_i, then S will be contained in the nonwandering set of θ. In these circumstances, we prove that 0 is not isolated in the nonwandering set of X. Whether or not every arc in S has an image extending across the set R_i depends only on the functions $f_\pm$ acting on the intervals $(\rho_+,0)$ and $(0,\rho_-)$. Denote by f the discontinuous map $f\colon (\rho_+,\rho_-) \to (\rho_+,\rho_-)$ determined by $f_\pm$ (with, say, $f(0) = 0$.) Consider a subinterval $\gamma \subset (\rho_+,\rho_-)$. Since $df_\pm/dx > \alpha > 1$, the sum of the lengths of the components of $f^k(\gamma)$ is at least $c\alpha^k$. Therefore, some image of γ has more than one component. The only point of discontinuity for f is $x = 0$, so there is a $k > 0$ and an $x \in \gamma$ with $f^k(x) = 0$.

The map θ has a periodic point of period 2 in R_1 because $\theta^2(R_1)$ crosses R_1 horizontally. Therefore, f has a point r of period 2. Any neighborhood of r has an image which eventually covers (ρ_+,ρ_-). Now assume that there is an open set $U \subset (\rho_+,\rho_-)$, none of whose images cover (ρ_+,ρ_-). Then no image of U contains p. It follows that if U_1 and U_2 are two open sets, none of whose images cover (ρ_+,ρ_-), then $U_1 \cup U_2$ also has this property (because r is in none of its images.) Thus there is a largest open set $U \subset (\rho_+,\rho_-)$ with the property that none of its images cover (ρ_+,ρ_-). It follows that $f^{-1}(U) = U = f(U)$.

We observed above that any interval contains a point which is eventually mapped to 0 by the iterates of f. Thus U contains a neighborhood of 0 and, hence, neighborhoods of $\rho_\pm$. This implies that U contains a neighborhood of each point which eventually maps to 0. Since these points are dense, U is a dense subset of (ρ_+,ρ_-). Notice that the property $f^{-1}(U) \subset U$ implies that the components of U must map <u>onto</u> the components of U. Let (ξ_-,ξ_+) be the component of U containing 0. Some image of $(\xi_-,0)$ contains 0, and hence (ξ_-,ξ_+). (Since $f_-(0) = \rho_-$, the images of 0 are endpoints of components of U.) The first time an image of $(\xi_+,0)$ contains 0, that power of f is continuous on $(\xi_-,0)$. Since f is orientation preserving, it follows that

ξ_- is mapped by this power of f to ξ_-. Therefore ξ_- is a periodic point of f. We conclude that $\rho_\pm$ have images for some power of f which are periodic points of f.

For the return map θ of R, this implies that the images of the vertical lines $x = \rho_\pm$ each remain within a finite set of vertical lines. Because θ contracts in the vertical direction, the intersections of R with $W^u(p)$ have θ-trajectories which tend asymptotically to periodic orbits of of θ. These periodic θ trajectories lie on periodic orbits γ_1, γ_2 for the flow X. Because θ is uniformly hyperbolic (apart from its discontinuity), these periodic orbits are hyperbolic with two dimensional stable and unstable manifolds. Applying the Kupka-Smale Theorem (Smale [1]), we note that it is a generic property of vector fields that the stable manifold of a hyperbolic periodic trajectory intersect the unstable manifold of a singular point transversally. This is not the case here. Thus we conclude that in the open set of vector fields which we have described, those vector fields for which any arc of S eventually extends across each R_i form a set of second category. I do not know whether there is an open set of vector fields with this property.

Proof of Theorem (12.1): Let us assume now that X is chosen so that θ has the property that some image of every arc in S eventually extends across each R_i. If $w \in S$ and U is a rectangular neighborhood of w in R, then $\theta^k(U)$ extends across each R_i for k sufficiently large. Also $\theta^{-k}(U)$ extends vertically across R for k sufficiently large because θ contracts the vertical direction. It follows that $\theta^{-k}(U) \cap \theta^{k}(U) \neq \emptyset$ for k very large. Thus $\theta^{2k}(U) \cap U \neq \emptyset$ and w is nonwandering. We conclude that S is contained in the nonwandering set of θ. Since S intersects $W^s(p)$, p is in the nonwandering set of X. This proves Theorem (12.1). □

The nonwandering sets of the vector fields satisfying Theorem (12.1) have a two dimensional attractor Λ which contains the origin. The intersection of Λ with R contains S. We want to go further in describing the structure of Λ. This can be done most completely when p is a homoclinic point with $W^u(p) \subset W^s(p)$. This happens when there are powers of f which map ρ_+ and ρ_- to 0.

For purposes of definiteness, we shall describe Λ in the case that $f^2(\rho_\pm) = 0$. Afterwards we indicate the modifications which are necessary when higher powers of f map ρ_+ and ρ_- to 0. Now $R \cap \Lambda = \overline{S}$. If $f^2(\rho_\pm) = 0$, then $\theta(R_1) \subset \overline{R_3 \cup R_4}$, $\theta(R_2) \subset R_1$, $\theta(R_3) \subset R_4$, and $\theta(R_4) \subset \overline{R_1 \cup R_2}$. Consequently, if $\{a_k\}_{k=0}^{\infty}$ is a sequence with $a_i \in \{1,2,3,4\}$, then $\bigcap_{k=0}^{\infty} \theta^k(R_{a_k}) \neq \emptyset$ if and only if $\{a_k\} \in \Sigma$. If $\{a_k\} \in \Sigma$, then there is a segment extending across R_{a_0} which lies in S and hence in Λ. This presents the following picture for Λ. There is a Cantor set of arcs, corresponding to points of Σ, each of which extends across some of the R R_i's. These are joined at their ends by $W^u(p)$. See Figure 12.3. Note that points of $\Lambda - W^u(p)$ have neighborhoods which are homeomorphic to a 2-disk x Cantor set.

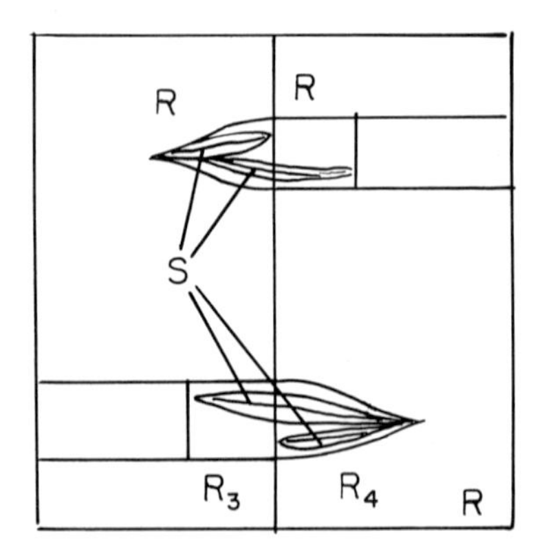

Figure 12.3

If higher powers of f map ρ_+ and ρ_- to 0, then we construct another subshift of finite type as follows. Cut the image of $\theta(R)$ along vertical lines passing through each point in the orbit. $\theta(\rho_+)$ and $\theta(\rho_-)$. This will divide $\theta(R)$ into a number of components, say $R_1,\ldots,R_n$. Define the $n \times n$ matrix T by

$$T_{ij} = \begin{cases} 1 & \text{if } \theta(R_j) \cap R_i \neq \emptyset \\ 0 & \text{if } \theta(R_j) \cap R_i = \emptyset. \end{cases}$$

Let Σ be the (one-sided) subshift of finite type with transition matrix T. Corresponding to each sequence in Σ, there will be exactly one arc crossing R_i which lies in the attractor Λ. The closure of these segments will be $\Lambda \cap R$ as before, because $\bigcap_{k=0}^{\infty} \theta^k(R_{a_k}) = \emptyset$ if $\{a_k\} \notin \Sigma$. Finally, we remark that if θ does not preserve vertical segments in R, then R is to be cut along components of $W^s(p) \cap R$ which also contain points of $W^u(p)$.

Proof of Theorem (12.2): We prove Theorem (12.2) in two steps. In the first step, we consider two flows, X and $\tilde{X}$, of the general sort considered in this paper such that, for the flow X, $W^u(p) \subset W^s(p)$, and for the flow $\tilde{X}$, $W^u(p) \cap W^s(p) = \{p\}$. We prove that X and $\tilde{X}$ have nonwandering sets which are not homeomorphic. The second step demonstrates that vector fields of each of these two classes are dense in some open set in the space of C^r vector fields.

We have described above the attractor $\Lambda(X)$ of a vector field X for which $W^u(p) \subset W^s(p)$. In this case, Λ is path connected and $\Lambda - W^u(p)$ is locally homeomorphic to the product of a 2-disk and a Cantor set. Furthermore, $W^u(p)$ is homeomorphic to the wedge product of two circles, a "figure eight."

Now consider the attractor $\Lambda(\tilde{X})$ of a vector field $\tilde{X}$ for which $W^u(p) \cap W^s(p) = \{p\}$ and Λ is a two dimensional set containing p. If $\Lambda(\tilde{X})$ is to be homeomorphic to $\Lambda(X)$, then $\Lambda(\tilde{X})$ must be path connected. Consider the set C of points $w \in \Lambda(\tilde{X})$ such that no neighborhood of w is homeomorphic to a 2-disk x Cantor set. It is easily seen that $W^u(p) \subset C$ since there are no points of $\Lambda \cap R$ to the left of the line $\rho_- = x$ or to the right of the line $\rho_+ = x$. If $\Lambda(X)$ is homeomorphic to $\Lambda(\tilde{X})$, then C is homeomorphic to the wedge product of two spheres. Since $W^u(p) \not\subset W^s(p)$ for $\tilde{X}$ and $W^u(p) \subset C$, there must be two points of $C - \{p\}$ which are the ω-limit sets of the two trajectories in $W^u(p) - \{p\}$. A single point which is the ω-limit set of a trajectory must be a singular point. There are no singular points of $\tilde{X}$ in $\Lambda(\tilde{X})$ other than p, so we conclude that C

is not homeomorphic to the wedge product of two spheres. Hence $\Lambda(X)$ and $\Lambda(\tilde{X})$ are not homeomorphic. This concludes the first step of the proof.

We now prove that the sets of vector fields X, $\tilde{X}$ of the sort considered above are each dense in some open set. The Kupka-Smale Theorem implies that vector fields like $\tilde{X}$ in that $W^s(p) \cap W^u(p) = \{p\}$ form a set of second category. Since the set of vector fields with $p \in \Lambda$ and Λ two dimensional is a second category subset of an open set, there is a dense set of vector fields of the form of $\tilde{X}$ in some open set of vector fields.

The only thing remaining to prove is that there is a dense subset of an open set of vector fields for which $W^u(p) \subset W^s(p)$. Consider the effect on $W^u(p)$ of a perturbation Y of $\tilde{X}$ parallel to the x-axis which has the effect of decreasing ρ_- and increasing ρ_+. See Figure 12.4.

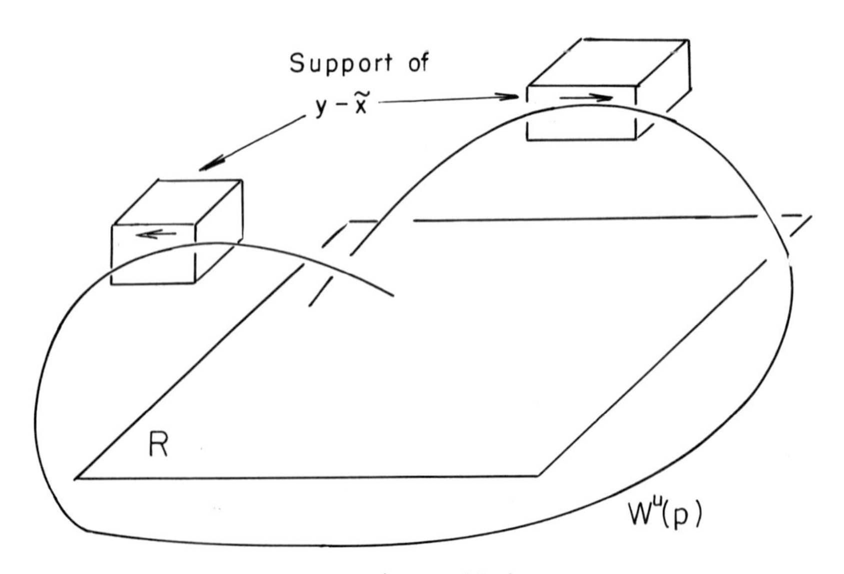

Figure 12.4

We examine successive intersections of $W^u(p)$ with R for the vector fields Y and $\tilde{X}$. The functions f_+ and f_- are orientation preserving. Consequently, as long as the corresponding, successive intersections for the two vector fields lie on the same side of the line $x = 0$ in R, the effect of the perturbation is push the intersections following ρ_- along $W^u(p)$ to the left and to push the intersections following ρ_+ to the right. Furthermore, since the map θ expands in the x direction, the distance between the corresponding, successive points of intersection grows exponentially. The distance cannot grow indefinitely, so after sometime, the corresponding points of intersection lie on opposite sides of the line $x = 0$. Thus, for some perturbation intermediate between Y and $\tilde{X}$, there are points of intersection of $W^u(p)$ with R which lie on the line $x = 0$ (in both directions along $W^u(p)$.) This means that $W^u(p) \subset W^s(p)$ for the intermediate perturbation. We conclude that there is a dense set of vector fields in some open set of the space of vector fields for which $W^u(p) \subset W^s(p)$ to finish the proof of Theorem (12.2).

As is traditional in dynamical systems, we end with a question. The vector fields described here are very pathological from the point of view of topological dynamics. Yet they seem to preserve as much hyperbolicity as they possibly could without satisfying Axiom A. There is now a well developed "statistical mechanics" for attractors satisfying Axiom A (Bowen-Ruelle [1]). How much of this statistical theory can be extended to apply to the vector fields described here?

REFERENCES

Abraham, R. [1] Introduction to Morphology, Publ. Dept. Math. Lyon 4 (1972), 38-114.

[2] Macrodynamics and Morphogenesis, in "Evolution in the Human World", E. Jantsch and C. Waddington, eds. (to appear).

Abraham, R. and Marsden, J. [1] "Foundations of Mechanics", Benjamin (1967).

Abraham, R. and Robbin, J. [1] "Transversal Mappings and Flows", Benjamin (1968).

Abraham R. and Smale, S. [1] Nongenericity of Ω-Stability, Proceedings of Symposia in Pure Mathematics, vol. XIV, (1970), 5-8.

Alexander, J. C. and York, J. A. [1] Global Bifurcation of periodic orbits. Ann. of Math. (to appear).

Andreichikov, I. P. and Yudovich, V. I. [1] On auto-oscillatory regimes branching out from the Poisuille flow in a plane channel, Sov. Math. Dokl., 13 (1972), 791-794.

Andronov, A. A. and Pontriagin [1] Coarse Systems, Dokl. Akad. Nauk SSSR, 14 (1937), 247-251.

Andronov, A. A. and Chaikin, C. E. [1] "Theory of Oscillations", Princeton University Press (1949). Translation of 1937 Russian edition. Second edition (1959), in Russian by Andronov, Witt and Chaikin, English Translation, Pergamon Press, New York (1966).

Andronov, A. A., Leontovich, E. A., Gordon, I. I. and Maier, A. G. [1] "Qualitative Theory of Second Order Dynamic Systems", Halsted Press, New York (1973).

[2] "Theory of Bifurcations of Dynamic Systems on a Plane", Halsted Press, New York (1973).

Andronov, A. A. and Witt, A. [1] Sur la theórie mathematiques des autooscillations, C. R. Acad. Sci. Paris 190 (1930), 256-258.

Antman, P. and Keller, J. [1] "Bifurcation Theory and Nonlinear Eigenvalue Problems", Benjamin (1969).

Arnold, V. I. [1] "Ordinary Differential Equations", M.I.T. Press (1973).

[2] Lectures on bifurcation and versal families, Russian Math. Surveys 27 (1972), 54-123.

[3] Singularities of Smooth Mappings, Russian Math. Surveys, 23 (1968), 1-43.

[4] Remarks on the Stationary Phase Method and Coxete Numbers, Russian Math. Surveys, 29 (1974), 19-48.

[5] Sur la geometrié differentielle des groupes de Lie de dimension infinie et ses applications à l'hydrodynamique des fluids parfaits, Ann. Inst. Grenoble 16 (1) (1966), 319-361.

Arnold, V. I. and Avez, A. [1] "Ergodic Problems in Classical Mechanics", Benjamin (1970).

Aronson, D. G. [1] Stability for nonlinear parabolic equations via the Maximum Principle (to appear).

[2] A Comparison Method for Stability Analysis of Nonlinear Parabolic Equations (to appear).

Aronson, D. G. and Thames, H. D. Jr. [1] Stability and Oscillation in Biochemical Systems with Localized Enzymes, Arch. Rat. Mech. and Anal. (to appear).

Aronson, D. G. and Weinberger, H. [1] Nonlinear diffusion in population genetics, combustion and nerve propagation, Proc. Tulane Program in Partial Differential Equations, Springer Lecture Notes (to appear).

Auslander, D. G. and Huffaker, C. [1] Dynamics of interacting populations, J. Frank. Inst. (1974), 297 (5).

Ball, J. M. [1] Continuity properties of nonlinear semigroups, J. Funct. An. 17 (1974), 91-103.

[2] Measurability and Continuity Conditions for Nonlinear Evolutionary Processes, Proc. A.M.S. (to appear).

[3] Saddle Point Analysis for an Ordinary Differential Equation in a Banach Space, and an application to dynamic buckling of a beam, in "Nonlinear Elasticity", Dixon, ed. Academic Press, New York (1974).

Baserga, R. (ed.) [1] "The Cell Cycle and Cancer", Marcel Dekker, New York (1971).

Batchelor, G. K. [1] "The Theory of Homogeneous Turbulence", Cambridge University Press (1953).

Bautin, N. N. [1] On the number of limit cycles which appear with the variation of coefficients from an equilibrium of focus or center type, Transl. A.M.S. (1954), #100.

Beddington, J. C. and Free, A. [1] Age structure, density dependence and limit cycles in predator-prey interactions, J. Theoret, Popul. Biol. (to appear).

Bell, G. I. [1] Predator-prey equations simulating an immune response, Math. Biosci. 16 (1973), 291-314.

Bellman, R. [1] "Stability Theory of Differential Equations, McGraw Hill, New York (1953), Dover (1969).

Berger, M. [1] On Von Karman's Equations and the Buckling of a Thin Elastic Plate, Comm. Pure and Appl. Math., 28 (1967), 687-719, 21 (1968), 227-241.

Birkhoff, G. [1] "Hydrodynamics; a Study in Logic, Fact and Similitude", Princeton University Press (1950).

Birkhoff, G. O. [1] "Dynamical Systems", A.M.S. Colloquim Publ. (1927).

Bochner, S. and Montgomery, D. [1] Groups of differentiable and real or complex analytic transformations, Ann. of Math. 46 (1945), 685-694.

Bonic, R. and Frampton, J. [1] Smooth functions on Banach manifolds, J. Math. Mech. 16 (1966), 877-898.

Bogoliouboff, N. and Mitropolski, Y. A. [1] "Les methodes Asymptotiques en Théorie des Oscillations non Linéares", Paris: Gunthier-Villars (1962).

Bourbaki, N. [1] "Elements of Mathematics, General Topology Part 2", Hermann, Paris and Addison-Wesley, Reading, Mass. (1966).

Bourguignon, J. P. and Brezis, H. [1] Remarks on the Euler Equation, J. Funct. An. 15 (1974), 341-363.

Bowen, R. and Ruelle, D. [1] The ergodic theory of axiom A flows, Inv. Math. (to appear).

Brauer, F. and Nohel, J. A. [1] "Qualitative Theory of Ordinary Differential Equations", Benjamin, New York (1969).

Bruslinskaya, N. N. [1] Qualitative integration of a system of n differential equations in a region containing a singular point and a limit cycle, Dokl. Akad. Nauk. SSSR 139 (1961), 9-12, Sov. Math. Dokl. 2 (1961), 9-12.

[2] On the behavior of solutions of the equations of hydrodynamics when the Reynolds number passes through a critical value, Dokl. Akad. Nauk. SSSR, 162 (1965), 731-734.

[3] The origin of cells and rings at near-critical Reynolds numbers, Uspekhi Mat. Nauk. 20 (1965), 259-260, (Russian).

Cantor, M. [1] Perfect Fluid Flows over R^n with Asymptotic Conditions, J. Funct. An. 18 (1975), 73-84.

Carroll, R. [1] "Abstract Methods in Partial Differential Equations", Harper and Row (1969).

Cesari, L. [1] "Asymptotic Behavior and Stability Problems in Ordinary Differential Equations", Springer-Verlag, New York (1973).

Chafee, N. [1] The bifurcation of one or more closed orbit from an equilibrium point of an autonomous differential system, J. Diff. Eq. 4 (1968), 661-679.

[2] A bifurcation problem for a functional differential equation of finitely retarded type, J. Math. An. and Appl. 35 (1971), 312-348.

Chandresekar, S. [1] "Hydrodynamic and Hydromagnetic Stability", Oxford University Press (1961).

Chen, T. S. and Joseph, D. D. [1] Subcritical bifurcation of plane Poiseuille flow, J. Fluid Mech. 58 (1973), 337-351.

Chernoff, P. [1] A note on continuity of semigroups of maps (to appear).

Chernoff, P. and Marsden, J. [1] "Properties of Infinite Dimensional Hamiltonian Systems", Springer Lecture Notes #425, (1974).

[2] On continuity and smoothness of group actions, Bull. Am. Math. Soc. 76 (1970), 1044.

Choquet, G. [1] "Lectures on Analysis", (3 vols.), W. A. Benjamin (1969).

Chorin, A. J. [1] Numerical Study of Slightly Viscous Flow, Jour. Fluid Mech. 57 (1973), 785-796.

[2] "Lectures on Turbulence Theory", Publish or Perish, Boston, Mass. (1975).

Chow, S. N. and Mallet-Paret, J. [1] Integral Averaging and Bifurcation, to appear, Journal of Differential Equations.

Cohen, D. [1] Multiple Solutions and Periodic Oscillations in Nonlinear Diffusion Processes, SIAM J. on Appl. Math. 25 (1973), 640-654.

Cohen, D. and Keener, J. P. [1] Oscillatory Processes in the Theory of Particulate Formation in Supersaturator Chemical Solutions, SIAM J. Appl. Math. (to appear).

Cole, J. D. [1] "Perturbation Methods in Applied Mathematics", Blaisdell (1968).

Coles, D. [1] Transition in circular Couette flow, J. Fluid Mech. 21 (1965), 385-425.

Cooke, K. L. and Yorke, J. A. [1] Some equations modelling growth processes and gonorrhea epidemics, Math. Biosci. 16 (1973), 75-101.

Crandall, M. and Rabinowitz, P. [1] Bifurcation from simple eigenvalues, J. Funct. An. 8 (1971), 321-340.

[2] Bifurcation, perturbation of simple eigenvalues and linearized stability, Arch. Rat. Mech. An. 52 (1974), 161-180.

Cronin, J. [1] One-side bifurcation points, J. Diff. Eq. 9 (1971), 1-12.

Dajeckii, Ju. L. and Krein, M. G. [1] "Stability of Solutions of Differential Equations in Banach Spaces", A.M.S. Translations of Math. Monographs, 43 (1974).

Davey, A., DiPrima, R. C. and Stuart, J. T. [1] On the instability of Taylor vortices, J. Fluid Mech. 31 (1968), 17-52.

Desoer, C. and Kuh, E. [1] "Basic Circuit Theory", McGraw-Hill, New York (1969).

Dorroh, J. R. [1] Semi-groups of maps in a locally compact space, Canadian J. Math. 19 (1967), 688-696.

Dorroh, J. R. and Marsden, J. E. [1] Smoothness of nonlinear semigroups, (to appear).

Dunford, N. and Schwartz, J. [1] "Linear Operators", Vol. I, Interscience (1958).

[2] "Linear Operators", Vol. II, New York, Wiley (1963).

Durand, G. [1] Application de la theorie de la bifurcation et de la stabilité aux equations du champ magnetique terrestre, Lecture Notes #128, Univ. de Paris XI, Orsay (1975).

Ebin, D. G. [1] Espace des metriques Riemanniennes et mouvement des fluides via les varietes d'applications, Lecture Notes, Ecole Polytechnique, et Université de Páris VII (1972).

Ebin, D. G. and Marsden, J. [1] Groups of diffeomorphisms and the motion of an incompressible fluid, Ann. of Math. 92 (1970), 102-163.
(see also Bull. Am. Math. Soc. 75 (1969), 962-967.)

Fenichel, N. [1] The Orbit Structure of the Hopf Bifurcation Problem, J. Diff. Eqns. 17 (1975), 308-328.

[2] Center manifolds and the Hopf bifurcation problem (preprint).

Fife, P. and Joseph D. [1] Existence of convective solutions of generalized Bénard problems, Arch. Rat. Mech. An. 33 (1969), 116-138.

Freedman, H. I. [1] On a bifurcation theorem of Hopf and Friedrichs (to appear).

Friedman, A. [1] "Partial Differential Equations", Holt, New York (1969).

Friedrichs, K. O. [1] "Advanced Ordinary Differential Equations", New York, Gordon and Breach (1965).

[2] "Advanced Ordinary Differential Equations", Springer (1971).

Fujita, H. and Kato, T. [1] On the Navier-Stokes Initial Value Problem, Arch. Rat. Mech. Anal. 16 (1964), 269-315.

Gavalas, G. R. [1] "Nonlinear Differential Equations of Chemically Reacting Systems", New York, Springer (1968).

Glansdoff, P. and Prigogine, I. [1] "Thermodynamic Theory of Structure, Stability and Fluctuations", New York, Wiley-Interscience (1971).

Glass, L. and Kauffman, S. A. [1] The logical analysis of continuous, nonlinear biochemical control networks, J. Theor. Biol. 39 (1973), 103-129.

Goodwin, B. C. [1] Oscillatory Behavior in Enzymatic Control Processes, in "Advances in Enzyme Regulation", Vol. 3, G. Weber (ed.), Pergamon (1965), 425-438.

Greenberg, J. M. and Hoppensteadt, F. [1] Asymptotic Behavior of Solutions to a Population Equation, SIAM J. Appl. Math. 28 (1975), 662-674.

Guckenheimer, J. [1] Review of "Stabilité Structurelle et Morphogénèse" by R. Thom., Bull. A.M.S. 79 (1973), 878-890.

Guckenheimer, J., Oster, G., and Ipaktchi, A. [1] The Dynamics of Density Dependent Population Models, (preprint).

Gurel, O. [1] Dynamics of Cancerous Cells, Cancer 23 (1969), 497-505.

Gurel, O. [2] Qualitative Study of Unstable Behavior of Cancerous Cells, Cancer 24 (1969), 945-947.

[3] Limit Cycles with Stability of the First Level, J. Franklin Inst. 288 (1969), 235-238.

[4] Biomolecular Topology and Cancer, Physiol. Chem. & Physics 3 (1971), 371-388.

[5] Global Studies of Orbit Structures of Biomolecular Dynamical Systems, Collective Phenomena 1 (1972), 1-4.

[6] Bifurcation Theory in Biochemical Dynamics, Analysis and Simulation - North-Holland - Amsterdam (1972).

[7] Bifurcation Models of Mitosis, Physiol. Chem. & Physics 4 (1972), 139-152.

[8] A Classification of the Singularities of (X, f), Mathematical Systems Theory, 7 (1973), 154-163.

[9] Topological Dynamics in Neurobiology, Intern. J. Neuroscience 6 (1973), 165-179.

[10] Bifurcations in Nerve Membrane Dynamics, Intern. J. Neuroscience 5 (1973), 281-286.

[11] Peeling and Nestling of A Striated Singular Point, Collective Phenomena 2 (1974), 000-000.

[12] Limit cycles and Bifurcations in Biochemical Dynamics, Biol. Systems (to appear).

[13] Partial Peeling, Proc. Lefschetz Conf., Brown University (1974) (to appear).

Gurel, O. and Lapidus, L. [1] Liapunov Stability Analysis of Systems with Limit Cycle, Chem. Eng. Symp. Ser. 61, #55 (1965), 78-87.

[2] A Guide to The Generation of Liapunov Functions, Industrial and Engineering Chemistry, March (1969), 30-41.

Gurtin, M. and MacCamy, R. [1] Non-linear age-dependent population dynamics, Arch. Rat. Mech. An. (1974), 281-300.

Haag, J. [1] "Les mouvements vibratoires", Paris, Presses Univ. de France I, II (1955).

Haag, J. and Chaleant, R. [1] "Problèmes de Théorie Générale des Oscillations et de Chronométrie", Paris, Gunter-Villars (1960).

Hale, J. K. [1] Integral manifolds of perturbed differential systems, Ann. Math. 73 (1961), 496-531.

[2] "Oscillations in Nonlinear Systems", McGraw-Hill, (1963).

[3] "Ordinary Differential Equations", New York, Wiley-Interscience (1969).

Hale, J. K. and LaSalle, J. P. (eds.) [1] "Differential Equations and Dynamical Systems", Academic Press (1967).

Hall, W. S. [1] The bifurcation of solutions in Banach spaces, Trans. Am. Math. Soc. 161 (1971), 207-218.

Hanusse, P. [1] De l'existence d'un cycle limite dans l'evolution des systemes chimiques ouverts, C. R. Acad. Sci. Paris 274, Series C (1972), 1245-1247.

Hartman, P. [1] "Ordinary Differential Equations", Second Edition, P. Hartman (1973).

Hassell, M. and May, R. [1] Stability in insect host-parasite models, J. Animal Ecology 42 (1973), 693-726.

Hastings, S. P. [1] Some Mathematical Problems from Neurobiology, Am. Math. Monthly 82 (1975), 881-894.

Hastings, S. P. and Murray, J. D. [1] The existence of oscillatory solutions in the Field-Noyes Model for the Belousob-Zhabotinskii Reaction, SIAM J. Appl. Math. 28 (1975), 678-688.

Hastings, S. P., Tyson, J. and Webster, D. [1] Existence of Periodic Solutions for Negative Feedback Cellular Control Systems (preprint).

Hearon, J. Z. [1] The kinetics of linear systems with special reference to periodic reactions, Bull. Math. Biophysics 15 (1953), 121-141.

Henrard, J. [1] Liapunov's center theorem for a resonant equilibrium, J. Diff. Eq. 14 (1973), 431-441.

Henry, D. [1] "Geometric Theory of Semilinear Parabolic Equations", Mimeographed, Univ. of Kentucky (1974).

Higgins, J. [1] A Chemical Mechanism for Oscillations of Glycolytic Intermediates in Yeast Cells, Proc. N.A.S., 51 (1964), 988-994.

[2] The Theory of Oscillatory Reactions, Ind. & Eng. Chem. 59 (1967), 19-62.

Hille, E. and Phillips, R. [1] "Functional Analysis and Semigroups", Vol. 31 AMS Colloq. Publ. (1957).

Hirsch, M. and Pugh, C. [1] Stable manifolds and hyperbolic sets, Proc. Symp. Pure Math. XIV, Am. Math. Soc. (1970), 133-163.

Hirsch, M. and Pugh, C. and Schub, M. [1] "Invariant Manifolds", Springer Lecture Notes (to appear).

Hirsch, M. and Smale, S. [1] "Differential Equations, Dynamical Systems and Linear Algebra", New York, Academic Press (1974).

Hodgkin, A. L. and Huxley, A. F. [1] A Quantitative Description of Membrane Current and its Application to Conduction and Excitation in Nerve, J. Physiol. 117 (1952), 500-544.

Hopf, E. [1] Abzweigung einer periodischen Losung von einer stationaren Losung eines Differentialsystems, Ber. Math-Phys. Sachsische Adademie der Wissenschaften Leipzig 94 (1942), 1-22.

[2] A mathematical example displaying the features of turbulence, Comm. Pure Appl. Math. 1 (1948), 303-322.

[3] Uber die Anfanswert-aufgabe fur die hydrodynamische Grundgleichungen, Math. Nachr. 4 (1951), 213-231.

[4] The partial differential equation $u_t+uu_x = \mu u_{xx}$, Comm. Pure. Appl. Math. 3 (1950), 201-230.

[5] Repeated branching through loss of stability, an example, Proc. Conf. on Diff. Equations, Univ. of Maryland (1955).

[6] Remarks on the functional-analytic approach to turbulence, Proc. Symps. Appl. Math., XIII, Amer. Math. Soc. (1962), 157-163.

[7] On the right weak solution of the Cauchy problem for a quasi-linear equation of first order, J. Math. Mech. 19 (1969/1970), 483-487.

Hsü, I. and N. D. Kazarinoff [1] An applicable Hopf bifurcation formula and instability of small periodic solutions of the Field-Noyes model, J. for Math. Anal. and Appl. (to appear).

[2] Existence and stability of periodic solutions of a third order nonlinear autonomous system simulating immune response in animals (preprint).

Hughes, T. and Marsden, J. [1] "A Short Course in Fluid Mechanics", Publish or Perish (1976).

Iooss, G. [1] Contribution a la theorie nonlineare de la stabilite des ecoulements laminaires, These, Faculte des Sciences, Paris VI (1971).

[2] Theorie non linearire de la stabilite des ecoulements laminaires dans le cas de <<l'echange des stabilites >>, Arch. Rat. Mech. An. 40 (1971), 166-208.

[3] Existence de stabilite de la solution periodique secondaire intervenant dans les problemes d'evolution du type Navier-Stokes, Arch. Rat. Mech. An. 49 (1972), 301-329.

[4] Bifurcation d'une solution T-periodique vers une solution nT-periodique, pour certains problèmes d'evolution du type Navier-Stokes, C. R. Acad. Sc. Paris 275 (1972), 935-938.

[5] "Bifurcation et Stabilite", Lecture Notes, Université Paris XI (1973).

[6] Bifurcation of a periodic solution into an invariant torus for Navier-Stokes equations and their respective stabilities, Arch. Rat. Mech. An. (to appear).

Irwin, M. C. [1] On the stable manifold theorem, Bull. London Math. Soc. 2 (1970), 196-198.

Ize, J. [1] Thesis, New York Univeristy (1975).

Joseph, D. D. [1] Stability of convection in containers of arbitrary shape, J. Fluid Mech. 47 (1971), 257-282.

[2] Response curves for plane Pouseuille flow. In Advances in Applied Mechanics, Vol. XIV, (ed. C. S. Yih), New York, Academic Press (1974).

[3] Repeated supercritical branching of solutions arising in the variational theory of turbulence, Arch. Rat. Mech. Anal. 53 (1974), 101-130.

[4] "Global Stability of Fluid Motions", Springer (to appear).

Joseph, D. D. and Chen, T. S. [1] Friction factors in the theory of bifurcating flow through annular ducts. J. Fluid Mech. 66 (1974), 189-207.

Joseph, D. D. and Nield, D. A. [1] Stability of Bifurcating Time-periodic and Steady Solutions of Arbitrary Amplitude, Archive for Rat. Mech. and An. (to appear).

Joseph, D. D. and Sattinger, D. H. [1] Bifurcating time periodic solutions and their stability, Arch. Rat. Mech. An 45 (1972), 79-109.

Jost, R. and Zehnder, E. [1] A generalization of the Hopf bifurcation theorem, Helv. Phys. Acta 45 (1972), 258-276.

Judovich, V. I. [1] Periodic motions of a viscous incompressible fluid, Sov. Math. Dokl. 1 (1960), 168-172.

[2] Nonstationary flows of an ideal incompressible fluid, Z. Vycis l. Mat. i, Fiz. 3 (1963), 1032-1066.

[3] Two-dimensional nonstationary problem of the flow of an ideal incompressible fluid through a given region, Mat. Sb. N. S. 64 (1964), 562-588.

[4] Example of the generation of a secondary stationary or periodic flow when there is a loss of stability of the laminar flow of a viscous incompressible fluid, Prikl. Math. Mek. 29 (1965), 453-467.

[5] Stability of stress flows of viscous incompressible fluids, Dokl. Akad. Nauk. SSSR 16 (1965), 1037-1040 (Russian), Soviet Phys. Dokl. 104 (1965), 293-295 (English).

[6] On the origin of convection, Prikl. Mat. Mech. (J. Appl. Math. Mech.) 30 no. 6 (1966), 1193-1199.

[7] Secondary flows and fluid instability between rotating cylinders, Prikl. Mat. Mech. (J. Appl. Math. Mech.) 30 (1966A), 688-698.

[8] On the stability of forced oscillations of fluid, Soviet Math. Dokl. 11 (1970), 1473-1477.

[9] On the stability of self-oscillations of a liquid, Soviet Math. Dokl. 11 (1970), 1543-1546.

[10] On the stability of oscillations of a fluid, Dokl. Akad. Nauk. SSSR 195 (1970), 292-295 (Russian).

[11] The birth of proper oscillations in a fluid, Prikl. Math. Mek. 35 (1971), 638-655.

Kato, T. [1] Integration of the equations of evolution in a Banach space, J. Math. Soc. Japan 5 (1953), 208-234.

[2] Abstract evolution equations of parabolic type in Banach and Hilbert space, Nagoya Math. J. 19 (1961), 93-125.

[3] "Perturbation Theory for Linear Operators", Springer-Verlag, New York (1966).

[4] Linear evolution equations of "hyperbolic" type, J. Fac. Sci. Univ. of Tokyo, Sec. l. XVII (1970), 241-258.

Kato, T. [5] Linear evolution equations of "hyperbolic" type II, J. Math. Soc. Japan 25 (1973), 648-666.

[6] On the initial value problem for quasi-linear symmetric hyperbolic systems, Arch. Rat. Mech. 58 (1975), 181-206.

Kato, T. and Fujita, H. [1] On the nonstationary Navier-Stokes system, Rendiconti Sem. Mat. Univ., Padova, 32 (1961), 243-260.

Keener, J. P. and Keller, H. B. [1] Perturbed Bifurcation Theory, Arch. Rat. Mech. Anal. 50 (1973), 159-175.

Keller, J. [1] Bifurcation theory for ordinary differential equations, in "Bifurcation Theory and Non-linear Eigenvalue Problems", ed. J. Keller and S. Antman, Benjamin (1969).

Kelley, A. [1] The stable, center-stable, center, center-unstable, and unstable manifolds, Appendix C of "Transversal Mappings and Flows by R. Abraham and J. Robbin, Benjamin, New York (1967). (See also J. Diff. Eqns. 3 (1967), 546-570.)

[2] On the Liapunov sub-center manifold, Appendix C of "Foundations of Mechanics" by R. Abraham, Benjamin, New York (1967).

Kielhöfer, H. [1] Stability and semi-linear evolution equations in a Hilbert space, Arch. Rat. Mech. Anal. 57 (1974), 150-165.

Kirchgässner,, K. [1] Die Instabilität der strömmung zwischen rotierenden Zylindern gegenüber Taylor-Wirbeln fur beliebige Spaltbreiten, Zeit, Angew. Math. Phys. 12 (1961), 14-30.

[2] "Multiple Eigenvalue Bifurcation for Holomorphic Mappings", Contributions to Nonlinear Functional Analysis, Academic Press (1971).

Kirchgässner, K. and Kielhöfer, H. [1] Stability and bifurcation in fluid dynamics, Rocky Mountain J. of Math. 3 (1973), 275-318.

Kirchgässner, K. and Sorger, P. [1] Stability analysis of branching solutions of the Navier-Stokes equations, Proc. 12th Int. Cong. Appl. Mech., Standord Univ., August, 1968.

[2] Branching analysis for the Taylor problem, Quart. J. Mech. Appl. Math. 22 (1969), 183-210.

Knops, R. J. and Wilkes, E. W. [1] Theory of Elastic Stability, Handbuch der Physik, Vol. V/a/3 (1973), 125-302.

Kogelman, S. and Keller, J. B. [1] Transient behavior of unstable nonlinear systems with application to the Benard and Taylor Problem, SIAM J. on Appl. Math. 20 (1971), 619-637.

Kolomogorov, A. N. [1] Sulla teoria di volterra della lotta per l'Esisttenza, Giorn. Instituto Ital. Attuari 1 (1936), 74-80.

[2] The local structure of turbulence in incompressible viscous fluid for very large Reynolds numbers, C. R. Acad. Sci. USSR 30 (1941), 301.

[3] Dissipation of energy in locally isotropic turbulence, C. R. Acad. Sci. USSR 32 (1941), 16.

Kopell, N. and Howard, L. N. [1] Horizontal bands in the Belousov reaction, Science 180 (1973), 1171-1173.

[2] Plane wave solutions to reaction-diffusion equations, Studies in Appl. Math. 52 (1973), 291-328.

[3] Bifurcations under non-generic conditions, Adv. in Math. 13 (1974), 274-283.

[4] Wave trains, shock structures, and transition layers in reaction-diffusion equations, in "Mathematical Aspects of Chemical and Biochemical Problems and Quantum Chemistry", Proc. SIAM-AMS Symposium 8 (1974), 1-12.

[5] Pattern formation in the Belousov reaction, in "Lectures on Mathematics in the Life Sciences" AMS, Vol. 7 (1974), 201-216.

[6] Bifurcations and Trajectories joining critical points (preprint).

Kraichnan, R. H. [1] The structure of turbulence at very high Reynolds numbers, J. Fluid Mech. 5 (1959), 497-543.

[2] The closure problem of turbulence theory, Proc. Symp. Appl. Math. XIII, Am. Math. Soc. (1962), 199-225.

[3] Isotropic turbulence and inertial range structures, Phys. Fluids 9 (1966), 1728-1752.

[4] On Kolmogorov's inertial-range theories, J. Fluid Mech. 62 (1974), 305-330.

[5] Remarks on turbulence theory, Advances in Math., 16 (1975), 305-331.

Kryloff, N. and Bogoluiboff, N. [1] "Introduction to Non-linear Mechanics", Annals of Math. Studies #11, Princeton (1947).

Kurzweil, J. [1] Exponentially stable integral manifolds, averaging principle and continuous dependence on a parameter, (I, II), Czech. Math. J. 16 (91) (1966), 380-

Ladyzhenskaya, O. A. [1] "The Mathematical Theory of Viscous Incompressible Flow" (2nd Edition), Gordon and Breach, New York (1969).

[2] Example of nonuniqueness in the Hopf class of weak solutions for the Navier-Stokes equations, Math. USSR-Izvestija 3 (1969), 229-236.

[3] Mathematical Analysis of Navier-Stokes Equations for Incompressible Liquids, Annual Review of Fluid Mech. 7 (1975), 249-272.

Landau, L. D. and Lifshitz, E. M. [1] "Fluid Mechanics", Addison-Wesley, Reading, Mass. (1959).

Lanford, O. E. [1] Bifurcation of periodic solutions into invariant tori: the work of Ruelle and Takens, in "Nonlinear Problems in the Physical Sciences and Biology", Springer Lecture Notes #322 (1973).

[2] The Lorenz Attractor (in preparation).

Lang, S. [1] "Real Analysis", Addison-Wesley, Reading, Mass. (1969).

Langer (ed.) [1] "Nonlinear Problems", Univ. of Wisconsin Press (1962).

LaSalle, J. P. and Lefschetz, S. (ed.) [1] "Nonlinear Differential Equations and Nonlinear Mechanics", Academic Press, New York (1963).

[2] "Stability by Lyapunov's Direct Method with Applications", Academic Press, New York (1961).

Lasota, A. and Yorke, J. A. [1] Bounds for periodic solutions of differential equations in Banach spaces, J. Diff. Eq. 10 (1971), 83-91.

Lefever, R. and Nicolis, G. [1] Chemical Instabilities and Sustained Oscillations, J. Theoret. Biol. 30 (1971), 267-284.

Lefschetz, S. [1] "Contributions to the Theory of Nonlinear Oscillations", Princeton Univ. Press (1950).

[2] "Differential Equations: Geometric Theory" (2nd Edition), Interscience, New York (1963).

Leray, Jean [1] Etude de diverses equations integrales non-linearies et de qualques problemes que pose l'hydrodynamique, Jour. Math. Pures, Appl. 12 (1933), 1-82.

[2] Essai sur les mouvements plans d'un liquide vesqueux que limitent des parois, J. de Math. 13 (1934), 331-418.

[3] Sur le mouvement d'un liquide visqueux emplissant l'espace, Acta Math 63 (1934), 193-248.

[4] Problemes non-lineaires, Ensign, Math. 35 (1936), 139-151.

Levin, S. [1] Dispersion and population interactions, American Naturalist (to appear).

Li, T. Y. and Yorke, J. A. [1] Period three implies chaos, Am. Math. Monthly (to appear).

Liapunov, M. A. [1] "Problème générale de la stabilité du mouvement", Annals of Math. Studies 17, Princeton (1949).

Lichtenstein, L. [1] "Grundlagen Der Hydromechanik", Verlag Von Julius Springer, Berlin (1929).

Likova, O. B. [1] On the behavior of solutions of a system of differential equations in the neighborhood of an isolated constant solution, Ukrain. Math. Zh. IX (1957), 281-295 (in Russian).

[2] On the behavior of solutions of a system of differential equations in the neighborhood of closed orbits, Ukrain. Mat. Zh. IX (1957), 419-431 (in Russian).

Lin, C. C. [1] "The Theory of Hydrodynamic Stability", Cambridge University Press (1955).

Lorenz, E. N. [1] Deterministic Nonperiodic Flow, Journal of the Atmospheric Sciences, 20 (1963), 130-141.

Marsden, J. E. [1] Hamiltonian one parameter groups, Arch. Rat. Mech. An 28 (1968), 362-396.

[2] The Hopf bifurcation for nonlinear semigroups, Bull. Am. Math. Soc. 79 (1973), 537-541.

[3] On Product Formulas for Nonlinear Semigroups, J. Funct. An. 13 (1973), 51-72.

[4] "Applications of Global Analysis in Mathematical Physics", Publish or Perish, Boston (1974).

[5] A formula for the solution of the Navier-Stokes equations based on a method of Chorin, Bull. A.M.S. 80 (1974), 154-158.

Marsden, J. and Abraham, R. [1] Hamiltonian mechanics on Lie groups and hydrodynamics, Proc. Pure Math XVI, Amer. Math. Soc. (1970), 237-243.

Marsden, J., Ebin, D., and Fischer, A. [1] Diffeomorphism groups, hydrodynamics, and relativity, Proc. 13th biennial seminar of Canadian Math. Congress, ed. J. R. Vanstone, Montreal (1970), 135-279.

Marsden, J. and McCracken, M. [1] A Product Formula for the Navier-Stokes Equations based on a method of Chorin (in preparation).

Masuda, K. [1] On the Analyticity and the Unique Continuation Theorem for Solutions of the Navier-Stokes Equation, Proc. Japan Acad. 43 (1967), 827-832.

Matkowsky, B. J. [1] A simple nonlinear dynamic stability problem, Bull. A.M.S. 76 (1970), 620-625.

May, R. M. [1] Limit cycles in predator-prey communities, Science 177 (1972), 900-902.

[2] Biological populations with non-overlapping generations: stable points, stable cycles, and chaos, Science (to appear).

McCracken, M. F. [1] Computation of stability for the Hopf bifurcation theorem and the Lorenz equations (preprint).

[2] The Stokes Equation in L_p, Thesis, Berkeley (1975).

McLaughlin, J. B. and Martin, P. C. [1] Transition to turbulence of a statically stress fluid, Phys. Rev. Letters 33 (1974), 1189-1892.

[2] Transition to turbulence in a statically stressed fluid (preprint).

McLeod, J. B. and Sattinger, D. H. [1] Loss of stability and bifurcation at a double eigenvalue, J. Funct. An. 14 (1973), 62-84.

Minorsky, N. [1] "Nonlinear Oscillations", Van Nostrand, Princeton (1962), (Reprinted, R. E. Krieger, Huntington, New York (1974).)

Mitropolsky, Y. A. and Likova, O. B. [1] "Lectures on the Method of Integral Manifolds", Institute of Mathematics of the Ukrainian Academy of Sciences, Kiev (1968).

Morrey, C. B., Jr. [1] "Multiple Integrals in the Calculus of Variations", Springer (1966).

Moulton, F. [1] "An Introduction to Celestial Mechanics", Macmillan, New York (1902).

Naimark, J. [1] On Some Cases of Periodic Motions Depending on Parameters, Dokl. Akad. Nauk SSR 129 (1959), 736-739.

[2] Motions closed to doubly asymptotic motions, Soviet Math. Dokl. 8 (1967), 228-231.

Newhouse, S. E. [1] On Simple Arcs between Structurally Stable Flows, Proc. Liverpool Symposium on Dynamical Systems, Springer Lect. Notes.

[2] Nondensity of Axiom A(a) on S^2, Proceedings of Symposia in Pure Mathematics, Vol. XIV, (1970), 191-202.

Newhouse, S. E. and Palis, S. [1] Cycles and Bifurcation Theory (preprint).

Newhouse, S. E. and Peixoto, M. M. [1] There is a Simple Arc Joining any Two Morse-Smale Flows (preprint). (See also Peixoto [1], p. 303).

Nicholson, A. J. [1] An outline of the dynamics of animal populations, Aust. J. Zool. 2 (1954), 9-65.

[2] The self adjustment of populations to change, in "Cold Spring Harbor Symposia on Quantitative Biology", 22 (1957), 153-173.

Nirenberg, L. [1] "Topics in Nonlinear Analysis", Courant Institute Lecture Notes (1974).

Noyes, R. M., Field, R. J. and Körös, E. [1] Oscillations in chemical systems, J. Am. Chem. Soc. 94 (1972), 1394-1395, 8649-8664, 96 (1974), 2001.

Nussbaum, R. D. [1] A Global Bifurcation Theorem with Applications to Functional Differential Equations, J. Funct. An. 19 (1975), 319-338.

Orszag, S. A. [1] Analytical theories of turbulence, J. Fluid Mech. 41 (1970), 363-386.

Oster, G. and Auslander, D. [1] Deterministic and stochastic effects in population dynamics, Sixth Triennial World Congress Int'l. Fed. Automat. Control (1974).

Oster, G. and Takahashi, Y. [1] Models for age specific interactions in a periodic environment, Ecological Monographs (to appear).

Othmer, H. and Scriven, L. [1] Instability and dynamic pattern in cellular networks, J. Theoret. Biol. 32 (1971), 507-537.

Painlevé, P. [1] Gewöhnliche Differentialgleichungen: Existenz der Lösungen, Encyklopädie der Mathematischen Wissenschaften IIA4a.

[2] Les petit mouvements périodiques des systems, Comptes Rendus Paris, XXIV (1897), 1222.

Peixoto, M. M. (ed.) [1] "Dynamical Systems", Academic Press (1973).

Pimbley, G. [1] Periodic solutions of predator-prey equations simulating an immune response, I. Math. Biosci. 20 (1974), 27-51.

Pliss, V. A. [1] "Nonlocal Problems in the Theory of Oscillations", Academic Press, New York (1966).

Poincare, H. [1] "Les Méthodes Nouvelles de la Mécanique Céleste", Vol. I Paris (1892).

[2] Sur les courbes définie par une équation différentielle, C. R. Acad. Sci. 90 (1880), 673-675.

Ponomarenko, T. B. [1] Occurrence of space-periodic motions in hydrodynamics, J. Appl. Math. Mech. 32 (1968), 40-51, 234-245.

Poore, A. B. [1] A model equation arising in chemical reactor theory, Arch. Rat. Mech. Anal. 52 (1973), 358-388.

[2] On the Theory and Application of the Hopf-Friedrichs Bifurcation Theory, Arch. Rat. Mech. An. (to appear).

[3] On the dynamical behavior of the two temperature feedback nuclear reactor model, SIAM J. Appl. Math. (to appear).

Prigogine, I. and Lefever, R. [1] On symmetry breaking instabilities in dissipative systems, J. Chem. Phys. 46 (1967), 3542, 48 (1968), 1695.

Prigogine, I. and Nicolis, G. [1] Biological Order, Structure and Instabilities, Quarterly Rev. Biophys. 4 (1971), 107-148.

Pyrtli, A. S. [1] The birth of complex invariant manifolds near a singular point of a vector field which depends on parameters, Funct. An. Appl. 6 (1969), 1-6.

Rabinowitz, P. A. [1] Existence and nonuniqueness of rectangular solutions of the Benard problem, Arch. Rat. Mech. An. 29 (1967), 30-57.

[2] Some global results for non-linear eigenvalue problems, J. Funct. Anal. 7 (1971), 487-513.

Rayleigh, Lord [1] On convective currents in a horizontal layer of fluid when the higher temperature is on the under side, Phil. Mag. 32 (1916), 529-546.

Renz, P. [1] Equivalent flows on smooth Banach manifolds, Indiana Univ. Math. J. 20 (1971), 695-698.

Reynolds, O. [1] On the dynamical theory of incompressible viscous fluids and the determination of the criterion, Phil. Trans. Roy. Soc. London A186 (1895), 123-164.

Roseau, M. [1] "Vibrations non Lineáres et Théorie de la Stabilité", Springer Tracts in Nat. Phil. #8 (1966).

Rosenblat, M. and Van Atta, C. (ed.) [1] "Statistical Models and Turbulence", Springer Lecture Notes in Physics #12 (1972).

Rosenzweig, M. [1] Paradox of enrichment: destabilization of exploitation ecosystems in ecological time. Science 171 (1971), 385-387.

Rudin, W. [1] "Functional Analysis", McGraw-Hill, New York (1973).

Ruelle, D. [1] Dissipative systems and differential analysis, Boulder Lectures (1971).

[2] Strange attractors as a mathematical explanation of turbulence, in "Statistical Models and Turbulence", ed. by M. Rosenblatt and C. Van Atta, Springer-Verlag, Berlin-Heidelberg-New York (1972).

[3] Bifurcations in the presence of a symmetry group, Arch. Rat. Mech. An. 51 (1973), 136-152.

Ruelle, D. [4] A measure associated with an Axiom A attractor, Am. J. Math. (to appear).

[5] Some comments on Chemical Oscillations, Trans. New York Acad. Sci. 35 (1973), 66-71.

Ruelle, D. and Takens, F. [1] On the nature of turbulence, Comm. Math. Phys. 20 (1971), 167-192, 23 (1971), 343-344.

Sacker, R. [1] Thesis (unpublished).

Sather, D. [1] Branching of solutions of nonlinear equations, Rocky Mtn. J. 3 (1973), 203-250.

Sattinger, D. H. [1] On global solutions of nonlinear hyperbolic equations, Arch. Rat. Mech. An. 30 (1968), 148-172.

[2] Bifurcation of periodic solutions of the Navier-Stokes equations, Arch. Rat. Mech. An. 41 (1971), 66-80.

[3] The mathematical problem of hydrodynamic stability, J. Math. and Mech. 19 (1971), 797-817.

[4] Stability of bifurcating solutions by Leray-Schauder degree, Arch. Rat. Mech. An. 43 (1971), 154-166.

[5] "Topics in Stability and Bifurcation Theory", Springer Lecture Notes No. 309 (1973).

[6] Transformation groups and bifurcation at multiple eigenvalues, Bull. Am. Math. Soc. 79 (1973), 709-711.

[7] On the stability of waves of nonlinear parabolic systems, Adv. in Math. (to appear).

[8] Six Lectures on the Transition to Instability, Springer Lecture Notes #322 (1973), 261-287.

Schmidt, D. S. and Sweet, D. [1] A unifying theory in determining periodic families for Hamiltonian systems at resonance, J. Diff. Eq. 14 (1973), 597-609.

Segal, I. [1] Nonlinear semigroups, Ann. of Math. 78 (1963), 339-364.

Sel'kov, E. [1] Self-oscillations in glycose, Europe Jour. Biochem. 4 (1968), 79-86.

Sel'kov, E. E., Zhabotinsky, A. M., Shnoll, S. E. (Eds.) [1] "Oscillatory Processes in Biological and Chemical Systems", (Russian) Acad. Sciences, U.S.S.R. (1971).

Serrin, J. [1] "Mathematical Principles of Classical Fluid Dynamics", Encyclopedia of Physics, Vol. 8/1, Springer (1959).

[2] On the stability of viscous fluid motions, Arch. Rat. Mech. An. 3 (1959), 1-13.

[3] The initial value problem for the Navier-Stokes equations in "Non Linear Problems", ed. Langer, Univ. of Wisconsin Press (1962).

Shinbrot, M. [1] "Lectures on Fluid Mechanics", Gordon-Breach (1973).

Shub, M. [1] Topologically Transitive Diffeomorphisms of T^4, Symposium on Differential Equations and Dynamical Systems, Lecture Notes in Mathematics no. 206, 39-40.

Siegel, C. L. and Moser, J. [1] "Lectures on Celestial Mechanics", Springer (1971).

Siegmann W. L. and Rubenfeld, L. A. [1] A nonlinear model for double-diffusive convection, SIAM J. Appl. Math. 29 (1975), 540-557.

Smale, S. [1] Differentiable dynamical systems, Bull. Am. Math. Soc. 73 (1967), 747-817.

[2] On the mathematical foundations of electric circuit theory, J. Diff. Geom. 7 (1972), 193-210.

[3] A mathematical model of two cells via Turing's equation, Am. Math. Soc. Proceedings on Pure and Applied Math (to appear).

[4] Diffeomorphisms with many Periodic Points, Differential and Combinatorial Topology, Princeton Univ. Press, Princeton, N. J., 63-80.

Smale, S. and Williams, R. [1] The qualitative analysis of a difference equation of population growth (preprint).

Sobolewskii, P. E. [1] Equations of parabolic type in a Banach space, Trans. A.M.S. 49 (1965), 1-62.

Sotomayor, J. [1] Generic Bifurcations of Dynamical Systems, p. 561 of Peixoto [1]; see also Publ. IHES 43 (1974).

Spangler, R. A. and Snell, F. M. [1] Sustained Oscillations in a Catalytic Chemical System, Nature 191 (1961), 457-458.

[2] Transfer Function Analysis of an Oscillatory Model Chemical System, J. Theor. Biol. 16 (1967), 381-405.

Stakgold, I. [1] Branching of solutions of nonlinear equations, SIAM Review 13 (1971), 289-332.

Stern, T. E. [1] "Theory of Nonlinear Networks and Systems - An Introduction", Addison-Wesley, Reading, Mass. (1965).

Stoker, J. J. [1] "Nonlinear Vibrations", Interscience, New York (1950).

Takahashi, Y., Rabins, M. and Auslander, D. [1] "Control and Dynamic Systems", Addison Wesley, Reading, Mass. (1970).

Takens, F. [1] Unfolding of certain singularities of vector fields: generalized Hopf bifurcations, J. Diff. Eq. 14 (1973), 476-493.

[2] Singularities of vector fields, I.H.E.S. Publications Mathematiques 43 (1974), 47-100.

Temam, R. [1] On the Euler Equations of Incompressible Perfect Fluids, J. Funct. An., 20 (1975), 32-43.

Temam, R. et. al. [1] Proceedings of the Conference on Theoretical and Numerical Methods in Turbulence Theory (Orsay (1975)), Springer Lecture Notes (to appear).

Thom, R. [1] "Structural Stability and Morphogenesis", Addison Wesley, Reading, Mass. (1974).

Turing, A. M. [1] The chemical basis of morphogenesis, Phil. Trans. Roy. Soc. (B) (1925), 37-72.

Vainberg, M. M. and Trenogin, V. A. [1] "Theory of branching of solutions of non-linear equations", Noordhoff, Leyden (1974).

[2] The Methods of Lyapunov and Schmidt in the Theory of Nonlinear Equations and their Further Development, Russ. Math. Surveys 17 (1962), 1-60.

Varley, G., Gradwell, G. and Hassell, M. [1] "Insect Population Ecology", Univ. of California Press, Berkeley (1973).

Velte, W. [1] Uber ein Stabilitatskiterium der Hydrodynamik, Arch. Rat. Mech. An. 9 (1962), 9-20.

[2] Stabilitats verhalten und Verzweigung stationaret Losungen der Navier Stokesschen Gleichungen, Arch. Rat. Mech. An. 16 (1964), 97-125.

[3] Stabilitat und verzweigung stationaret Losungen der Navier-Stokesschen Gleichungen beim Taylorproblem, Arch. Rat. Mech. An. 22 (1966), 1-14.

Von Neumann, J. [1] Recent theories of turbulence, in "Collected Works", VI, Macmillan, New York (1963), 437-472.

Walter, C. [1] Oscillations in controlled biochemical systems, Biophys. J. 9 (1969), 863-872.

Weiss, L. [1] "Ordinary Differential Equations: Proceedings of NRL Symposium", Wiley (1973).

Weissler, F. [1] Thesis, Berkeley, Calif. (in preparation).

Williams, R. F. [1] One Dimensional Non-Wandering Sets, Topology 6 (1967), 473-487.

Whitham, C. [1] "Linear and Nonlinear Waves", Wiley (1972).

Wolibner, W. [1] Un théorème sur l'existence du mouvement plan d'un fluide parfait homogène, incompressible, pendant un temps infiniment longue, Math. Z. 37 (1933), 698-726.

Yorke, J. A. [1] Periods of periodic solutions and the Lipschitz constant, Proc. Am. Math. Soc. 22 (1969), 509-512.

Yosida, K. [1] "Functional Analysis", Springer-Verlag, Berline-Heidelberg-New York (1971).

Zeeman, E. C. [1] Topology of the Brain, in "Mathematics and Computer Science in Biology and Medicine", London: Medical Research Council (1965).

[2] Differential Equations for the Heartbeat and Nerve Impulse, p. 683 of Peixoto [1].

Zhabotinskii, A. M. and Zaikin, A. N. [1] Auto-wave processes in a distributed chemical system, J. Theor. Biol. 40 (1973), 45.

Ziegler, H. [1] "Principles of Structural Stability", Ginn-Blaisdell (1968).